RESEARCH HANDBOOK ON CLIMATE CHANGE ADAPTATION LAW

RESEARCH HANDBOOKS IN ENVIRONMENTAL LAW

This highly topical series addresses some of the most important questions and areas of research in Environmental Law. Each volume is designed by a leading expert to appraise the current state of thinking and probe the key questions for future research on a particular topic. The series encompasses some of the most pressing issues in the field, ranging from climate change, biodiversity and the marine environment through to the impacts of trade, regulation, and sustainable development.

Each Research Handbook comprises specially-commissioned chapters from leading academics, and sometimes practitioners, as well as those with an emerging reputation and is written with a global readership in mind. Equally useful as reference tools or high-level introductions to specific topics, issues and debates, these Handbooks will be used by academic researchers, post-graduate students, practising lawyers and lawyers in policy circles.

Research Handbook on Climate Change Adaptation Law

Edited by

Jonathan Verschuuren

Professor of International and European Environmental Law, Tilburg Sustainability Center, Tilburg University, the Netherlands

RESEARCH HANDBOOKS IN ENVIRONMENTAL LAW

Edward Elgar

Cheltenham, UK • Northampton, MA, USA

Published by
Edward Elgar Publishing Limited
The Lypiatts
15 Lansdown Road
Cheltenham
Glos GL50 2JA
UK

Edward Elgar Publishing, Inc.
William Pratt House
9 Dewey Court
Northampton
Massachusetts 01060
USA

A catalogue record for this book
is available from the British Library

Library of Congress Control Number: 2013930506

This book is available electronically in the ElgarOnline.com Law Subject Collection, E-ISBN 978 1 78100 008 3

FSC
www.fsc.org
MIX
Paper from
responsible sources
FSC® C013056

ISBN 978 1 78100 007 6

Typeset by Columns Design XML Ltd, Reading
Printed and bound in Great Britain by T.J. International Ltd, Padstow

Contents

Contributors

Robert W Adler holds the James I. Farr Chair in Law at the University of Utah S. J. Quinney College of Law. After completing a BA from Johns Hopkins University and a JD from Georgetown University Law Center (*cum laude*), Professor Adler practised environmental law for 15 years. He has published dozens of articles and reports in law, policy and science journals, including Vanderbilt Law Review, Harvard Environmental Law Review, Utah Law Review and George Washington Law Review. He has published two books with Island Press, one on the history and impact of the Clean Water Act and the other on restoring Colorado River ecosystems. He also published an environmental law casebook with Aspen Publishers (with David Driesen and Kirsten Engel).

Rebekah Byrne is a student in the Master of Environmental Law at Sydney Law School, The University of Sydney and was a student in Rosemary Lyster's 2012 Energy and Climate Law and Comparative Climate Law units of study. She is an environmental consultant in the construction industry.

Michael Faure is a professor of Comparative and International Environmental Law and director of the Maastricht European Institute for Transnational Legal Research (METRO) at Maastricht University. His expertise lies in the field of environmental (criminal) law, tort and insurance and economic analysis of (accident) law. Recent publications on the topic include: 'Insurability of Damage caused by Climate Change: A Commentary' (University of Pennsylvania Law Review, Vol. 155 (2007)) and 'International Liability as an Instrument to Prevent and Compensate for Climate Change' (Stanford Journal of International Law, Vol. 43 (2007)). One of his recent books is *Disaster Law*, edited together with Farber (Edward Elgar 2010).

Mariya Gromilova is a PhD candidate in the field of climate induced displacement at Tilburg Law School. Among the most recent activities is participation in the 2012 ClimMig Conference in Vienna, where she presented the paper 'Right to Development of People Induced to Displacement: Is Migration a Fair Adaptation Strategy from the Human Rights Perspective?'.

Keith H Hirokawa is associate professor of environmental law at the Albany Law School. He currently focuses his research on green building and construction and has extensively published on urban sustainability issues, such as his book *Legal Strategies for Greening Local Government*, edited with Salkin (American Bar Association 2011), as well as on green buildings, such as his article 'At Home with Nature: Early Reflections on Green Building Laws and the Transformation of the Built Environment' (Environment Law, Vol. 39 (2009)).

Nicola Jägers is the professor in human rights law at the Centre for Transboundary Legal Development of Tilburg University in the Netherlands, primarily focusing her research on the role of non state actors in human rights violations. Her publications include such book chapters as 'Protecting the Victims of the Privatization of War?' in Rianne Letschert and Jan van Dijk (eds), *The New Face of Victimhood: Globalization, Transnational Crimes and Victim Rights* (Springer 2011) and 'The WTO: An Obstacle to Enforce Human Rights Obligations?' in Jan Wouters et al. (eds), *Accountability for Human Rights Violations by International Organizations* (Intersentia 2010).

Rosemary Lyster is the professor of climate and environmental law at the University of Sydney and Director of the Australian Centre for Climate and Environmental Law (ACCEL). In the area of environmental law, Rosemary specialises in energy and climate law, water law and GMOs and environmental law. She has published several books in the field of energy and climate law, such as (with MacKenzie and McDermott) *Law, Tropical Forests and Carbon* (Cambridge University Press 2013) and *In the Wilds of Climate Law* (Australian Academic Press 2010), as well as a large number of journal articles. These include 'Smart Grids: Opportunities for Climate Change Mitigation and Adaptation' (Monash University Law Review, Vol. 36, No. 1 (2010)) and 'The New Frontier of Climate Law: Reducing Emissions from Deforestation (and Degradation)' (Environmental and Planning Law Journal, Vol. 26, No. 6 (2009)).

Dewald van Niekerk is the director of the African Centre for Disaster Studies at North West University in South Africa. Although trained in social science, he has been focusing his research on various issues relating to disaster risk reduction, including disaster management policies and legislation. Van Niekerk has been involved in the drafting of the South African policy and legislation on disaster risk reduction since 1999 and has published extensively on management of climate and other environmental disasters. Recent publications include 'Denial, Rhetoric or

Action: Are South African Municipalities Planning for Climate Change?' (Development Southern Africa Journal, Vol. 29, No. 2, 2011) and (with other authors) 'National Planning and Disaster' in Wisner et al., *Routledge Handbook of Natural Hazards and Disaster Risk Reduction and Management* (Routledge 2011).

Aurelia Marina Pohrib graduated *cum laude* from Albany Law School in 2012. During her years at Albany Law School she has pursued studies related to international and environmental law. Her essay 'The Market for Egg Donation: Its Economic Inequalities & Lack of Regulation' won the 2011 New York Bar Foundation's Judge Bernard S Meyer Scholarship, and her article 'Frustrating Noble Intentions: The Clash between ICC's Deterrent Effects and the Prosecutor's Policy of Focused Investigations on Perpetrators Situated in the Highest Echelons of Responsibility' in the International Community Law Review (ICLR 15.2, 2013).

Jonathan Rosenbloom is a professor of environmental law and sustainability at Drake Law School. Besides law (LLM Harvard, JD New York Law School *magna cum laude*), he also studied architecture and is especially interested in meeting social, economic and environmental challenges in the built environment. Recent publications include 'New Day at the Pool: State Pre-emption, Common Pool Resources, and Non-Place Based Municipal Collaborations' (Harvard Environmental Law Review, Vol. 36, (2012)), and 'Government Entrepreneurs: Incentivizing Sustainable Businesses as Part of Local Economic Development Strategies' in P Salkin and K Hirokawa, *Greening Local Government* (ABA Publishing 2012).

Tim Stephens is an international lawyer and human geographer at the University of Sydney's School of Law and Co-Director of the Sydney Centre for International Law. He has published widely on issues of public international law, national and international environmental law and the law of the sea. He is Co-Editor in Chief of the Asia Pacific Journal of Environmental Law. In 2010, Dr Stephens was awarded the IUCN Academy of Environmental Law Junior Scholarship Prize for his contribution to environmental law scholarship. Recent publications in the field of climate law include: (with other authors) *Climate Change and Australia: Warming to the Global Challenge* (Federation Press 2011) and *The International Law of the Sea* (Hart 2010), and the book chapter 'A Slow Burn: The Emergence of Climate Change Law in Australia' in Birch (ed.), *Water, Wind, Art and Debate: How Environmental Concerns Impact on Disciplinary Research* (Sydney University Press 2007).

Arie Trouwborst is a senior researcher at the Tilburg Sustainability Center specialising in nature conservation law. He obtained a PhD with distinction from Utrecht University in 2006, with a thesis on the precautionary principle. He has been awarded several research grants and prizes, including the François Prize 2001 (Netherlands Society of International Law), Environmental Thesis Prize 2002 (Netherlands Society for Environmental Law), and Erasmus Research Prize 2007 (Praemium Erasmianum Foundation). In 2009 Trouwborst obtained a prestigious 'Veni' grant from the Netherlands Organisation for Scientific Research (NWO) Innovational Research Incentives Scheme for a research project on international nature conservation law and climate change adaptation. Relevant publications include 'International Nature Conservation Law and the Adaptation of Biodiversity to Climate Change: A Mismatch?' (Journal of Environmental Law, Vol. 21, No. 3 (2009)), and 'Conserving European Biodiversity in a Changing Climate: The Bern Convention, the EU Birds and Habitats Directives and the Adaptation of Nature to Climate Change' (Review of European Community and International Environmental Law, Vol. 20, No. 1 (2011)).

Jonathan Verschuuren is the professor of international and European environmental law at the Tilburg Law School and the Tilburg Sustainability Center (both at Tilburg University, the Netherlands), and extraordinary professor of law at North West University, at Potchefstroom, South Africa. He has been the vice dean for research of the law school (2008–2010), after which he was a visiting professor at the Australian Centre for Climate and Environmental Law, University of Sydney (2011). He recently edited books titled *The Impact of Legislation* (Brill 2009) and *International Governance and Law* (Edward Elgar 2008). His current research focus is on climate change adaptation, leading to publications such as 'Climate Change: Rethinking Restoration in the European Union's Birds and Habitats Directives' (Ecological Restoration, Vol. 28, No. 4 (2010)) and 'Towards a Legal Framework for Coastal Adaptation: Assessing the First Steps in Europe and Australia' (Transnational Environmental Law, Vol. 1, No. 2 (2012)).

Lindsay F Wiley is a professor at the Washington College of Law. Her current research focuses on access to health care and healthy conditions in the US and globally. She also works on various law and policy issues at the intersection of public health, food systems, and environmental change. Prior to joining the faculty at WCL, Professor Wiley was the Global Health Law Program Director at the O'Neill Institute for National and Global Health Law at Georgetown University. She had also previously worked at the Center for Law and the Public's Health at the Johns

Hopkins Bloomberg School of Public Health, the American Society for Law, Medicine and Ethics, and Gordon, Feinblatt, Rothman LLC in Baltimore, Md. She received her AB and JD from Harvard, where she served on the Harvard Law Review, and her MPH from Johns Hopkins.

Preface and acknowledgements

Climate change is here to stay, at least for the time being. If we stopped all greenhouse gas emissions today (which, of course, is highly unlikely), we would only see marginal improvements in 30 to 40 years. So we have to adapt to changing climate. The Working Group II report of the Intergovernmental Panel on Climate Change (IPCC) concludes that adaptation will be necessary to address impacts resulting from warming of the Earth's atmosphere, which is already unavoidable due to past emissions. Around the world, adaptation policies are still in their infancy. It appears to be much more difficult to devise and implement adaptation policies than it is to devise and implement mitigation policies. It requires an immense, coordinated effort to adapt all the policies and laws in the many fields of governance involved in such a way that society at large will be prepared for the impact of climatic change events and the associated extreme weather incidences that we will experience in the course of the twenty-first century. We cannot simply apply existing laws. A whole new approach is to be pursued in many fields of law.

What are the legal challenges and barriers to climate change adaptation and how can they be overcome? That is the main question that we address in this book. The book discusses all the fields of law that are faced with challenges brought about by the need to adapt to the changing climate and to severe weather conditions, such as tort law, insurance law, disaster law, water law, marine law, planning law, biodiversity law, building and construction law, environmental law, refugee law, agricultural law and energy law. For each field of law, the challenges and barriers to adaptation are discussed, as well as ways to overcome these barriers. The book does not focus on a specific legal system, such as that of the US or the EU. Instead, we take a global view, focusing at the international, the regional and the domestic level, using examples from around the world. While such an approach obviously does not allow discussing the legal implications of adaptation measures for each country, it does offer an opportunity to transcend national legal specifics and thus get a more objective and juxtaposed view of the main issues to be solved as well as potential solutions. Through this approach, we aim to ensure that this book will become the definitive text on climate adaptation law for the coming years, to be used in the many existing and emerging programmes on climate change at universities around the world, as a principal text for policy-makers and governments, as a main source of scholarly reference and as a general reference for those interested in climate governance issues.

This book is the result of cooperative efforts of the Tilburg Sustainability Center (TSC) at Tilburg University, the Netherlands and the Australian Centre for Climate and Environmental Law (ACCEL) at the University of Sydney, Australia. I would like to thank everyone at ACCEL for their support and hospitality, particularly Rosemary Lyster, Susan Shearing, Andrew Edgar, Tim Stephens, Bernard Dunne and Jo Brady.

Jonathan Verschuuren
Tilburg, 1 December 2012

1. Introduction

Jonathan Verschuuren

1 WHAT IS CLIMATE CHANGE ADAPTATION?

Climate change is here to stay, at least for the time being. If we stopped all greenhouse gas emissions today (which, of course, is highly unlikely), we would only see marginal changes for the better in 30 to 40 years. So we have to adapt to the changing climate. Adaptation is defined as, in human systems, 'the process of adjustment to actual or expected climate and its effects, in order to moderate harm or exploit beneficial opportunities' and, in natural systems, 'the process of adjustment to actual climate and its effects'.[1] For the latter, it is added that human intervention in natural systems may facilitate adjustment to expected climate.[2] The Working Group II report of the Intergovernmental Panel on Climate Change (IPCC) concludes that adaptation will be necessary to address impacts resulting from the warming which is already unavoidable due to past emissions. For some impacts, namely those that already show or will show in the very near future, adaptation is *the only available and appropriate response*, according to the IPCC.[3] Such impacts, for instance, are:[4]

- increased water availability in some regions;
- decreased water availability and increasing drought in others;
- increased ecosystem changes (species shifting their natural range) and risk of extinction of species;
- small scale negative impacts for poor farmers and fishermen;

[1] Ch B Field et al. (eds), *Managing the Risks of Extreme Events and Disasters to Advance Climate Change Adaptation: Special Report of the Intergovernmental Panel on Climate Change* (SREX Report) (Cambridge University Press 2012) 556.

[2] Ibid.

[3] M L Parry et al. (eds), *Climate Change 2007: Impacts, Adaptation and Vulnerability: Contribution of Working Group II to the Fourth Assessment Report of the Intergovernmental Panel on Climate Change* (Cambridge University Press 2007) 19.

[4] Ibid., Table SPM-1 at 18.

- increased damage from floods and storms;
- increased burden from malnutrition and infectious diseases and a changed distribution of some disease vectors such as the mosquitoes vectoring malaria and dengue.

The 2012 special report of the IPCC *Managing the Risks of Extreme Events and Disasters to Advance Climate Change Adaptation* adds more recent data and some more impacts:[5]

- extreme events will have greater impacts on sectors with closer links to climate, such as water, agriculture and food security, forestry, health, and tourism;
- in many regions, the main drivers of future increases in economic losses due to some climate extremes will be socioeconomic in nature;
- increases in exposure will result in higher direct economic losses from tropical cyclones;
- disasters associated with climate extremes influence population mobility and relocation, affecting host and origin communities.

The drafters of the 1992 UN Framework Convention on Climate Change (UNFCCC)[6] treated mitigation and adaptation as being equally important. The UNFCCC mentions adaptation as one of the policies and measures to mitigate the adverse effects of climate change (Art. 3(3)). Under the commitments listed in Article 4 of the UNFCCC there are several that deal with adaptation. Adaptation, however, received much less attention than mitigation in the 20 years that followed the signing of the UNFCCC. This not only goes for policy and legal measures taken on the basis of the UNFCCC, and regional and national implementation thereof, but also for academic research.

Perhaps the lack of attention is one of the reasons why adaptation measures are still in their infancy. It appears to be much more difficult to devise and implement adaptation policies than it is to devise and implement mitigation policies. Partly this is due to the fear that attention to adaptation measures will make it more difficult to get mitigation

---

[5] Field et al., above note 1 at 16.

[6] United Nations Framework Convention on Climate Change (New York, 9 May 1992), (1992) 31 ILM 849, entered into force 21 March 1994.

measures adopted and implemented and promote a defeatist mentality.[7] Partly, the problem with adaptation is that you cannot always take one-size-fits-all measures. Some of the consequences of climate change, particularly sea level rise, are similar in all regions of the world. Mostly, however, the consequences of climate change may dramatically differ from one region to another. Some regions suffer from droughts, whereas others face increased floods. This may be so even within one country, such as, for instance, in big countries like the USA, China and Australia. It may even be so that within the same region, periods of extreme droughts are followed by a period of intense rainfall causing floods. The biggest problem, however, is an inherent difficulty of adaptation law and policy. Whereas mitigation measures primarily can be implemented in one policy field (that of environmental law), adaptation measures have to be implemented through a wide range of policies such as water, marine and coastal, fisheries, biodiversity, energy, building and construction, agriculture, telecommunications, and infrastructure, etc. It requires an immense, coordinated effort, to adapt all the policies and laws in these fields in such a way that the society at large will be prepared for the climatic changes and associated extreme weather events that we are going to experience in the course of the twenty-first century, with a peak expected between 2050 and 2080.

What kinds of measures should we think of? Probably one of the most advanced adaptation plans that exists today, the 2010 report by the New York City Panel on Climate Change (NYPCC), gives a good overview of the measures that should be taken in a modern metropolitan area like New York City. They include, for instance:[8]

- zoning, environmental, water and waste regulations to manage precipitation, flooding and stormwater, to ensure sustained water supply and to protect wastewater infrastructure;
- zoning and land use regulations dealing with sea level rise and storm surges;
- laws and regulations aimed at facilitating increased energy demand during heat waves and designating public buildings that can serve as emergency cooling centres in heat waves;

---

7 Katherine M Baldwin, 'NEPA and CEQA: Effective Legal Frameworks for Compelling Consideration of Adaptation to Climate Change' (2009) 82 Southern California Law Review 769, 783.

8 As discussed by E Sussman et al., 'Climate Change Adaptation: Fostering Progress through Law and Regulation' (2010) 18 N.Y.U. Environmental Law Journal 55–155.

- laws and regulations enabling the development of new power generation resources, and encouraging siting of emergency power generators to supply energy when there is a peak in demand (during heat waves) or in case of a weather-induced electric outage;
- laws and regulations requiring upgrading underground energy and telecommunications infrastructure to withstand flooding and sea level rise;
- various environmental impact assessment regulations requiring assessment of the consequences of the proposed activity on adaptation to climate change;
- amending energy, building, and sewer codes to adapt buildings to high wind conditions, flooding, and high temperatures;
- requirements on green or energy-smart landscaping, leading to energy consumption reductions in buildings, stormwater retention, and tree shading;
- requirements on adapting the transportation infrastructure to deal with flooding, saltwater damage, increased power demands and power outages, overheating of subway platforms, increased stress on infrastructure because of higher temperatures;
- air quality requirements dealing with more frequent periods of elevated concentrations of ground-level ozone, as well as elevated ozone concentrations;
- requirements on waste management sites and brownfields to prevent containment leaking in case of flooding and sea level rise;
- improving laws and regulations concerning emergency preparedness to be able to deal with storm surges, inland flooding and heat waves during summer.

This list deals with just the specific situation in New York City. Other cities around the globe will face other challenges. Coastal defence systems may have to be reinforced or even re-created from the start, for which land reclamation or expropriation of current land owners may be in order; negative impacts on coastal habitats will have to be minimized or mitigated. Parts of a city may even be moved to floating platforms. Other cities may not have the capacity and funds to adopt the kinds of measures New York City is adopting and may have to divert to much more drastic measures such as abandoning parts of the city or the entire city altogether. The same goes for rural areas that are below or on sea level or rural communities that have to cope with increasing droughts in addition to other stresses such as extreme poverty, land degradation, or a large number of HIV/AIDS infections. Bangladesh and parts of southern Africa are the cases in point for situations like this. Mass displacement and mass migration, both internally and cross border, are direct consequences which in turn may lead to food shortages, lack of adequate housing and jobs, and violence between migrants and the existing population. Current international refugee law clearly is not at all equipped to deal with these so-called climate refugees and their problems.

Outside the city, a wide range of adaptation measures are necessary. In order to protect biodiversity, policies should be aimed at making protected areas climate proof by making sure that these areas are large, robust, stable, and interconnected enough to adapt to the changing climate. Protected areas should be able to live through occasional flooding, wild fires, and extreme weather events, such as heavy storms. They should have enough variety in habitat types to host new species in search of a more suitable new climate zone. Current biodiversity law, both at the domestic and at international level, is hardly able to provide for these kinds of measures. In addition, in most legislatures, biodiversity law is aimed at specific 'flag species'. However, as a consequence of climate change, species are appearing and disappearing within just a few years, rendering the idea of having a specific flag species for a certain site useless.

Measures in the field of planning law and water law may be necessary to deal with the increased risk of inland flooding. Land may be designated to serve as controlled flooding areas (or flood control reservoirs) to protect more sensitive parts of the land against flooding. Land may also have to be designated for fresh water storage, so as to have buffer capacity in times of drought. In those regions with increased winter precipitation and increased risk of drought during summer, such as will be the case in northwestern Europe, farmers may have to store fresh water on their agricultural lands in winter, to be used in summer. It is obvious that here agricultural policies and water policies are closely interlinked, as water in controlled flooding areas, which are often agricultural lands, can be used for this purpose. Farmers may even engage in recreational activities in summer with water thus stored on their lands.

The agricultural sector will also have to adjust crop variety depending on the changing climate and weather conditions. In those areas of the world where local agriculture is essential for local food production, the government will have to intervene in order to keep or restore food security. The kinds of measures that have to be taken vary greatly throughout the world. Farmers may have to shift to crops that are better suited in a wetter, or dryer, or warmer climate, or crops that are better suited in an environment of increasing salinity. More or less the same goes for the forestry sector. The fisheries sector will have to get used to regular adjustments of fish quotas due to climate change induced decreased fish stocks. The surplus of water on agricultural lands described above may foster the introduction of aquaculture on agricultural lands.

Planning law and building law are among the fields of law that are applied by government authorities seeking to reduce potential harms from wild fires in those areas that are prone to increasing risks of bush fires.

Climate change is not only impacting government policies in the fields mentioned above, but also has a direct impact on private actors and thus on private law. Insurance companies are assessing the consequences of climate change for their line of business, the tourist industry is shifting its focus to new, yet to be developed, areas, and law firms are setting up climate litigation divisions as the number of law suits is likely to increase with the increasing damage by climate change and with increasing adaptation costs. With the increase of court cases, it is clear that litigation law, insurance law, and even property law, are affected by climate law and are, in a way, adapting to climate change as well.

2 CATEGORIES OF ADAPTATION

In literature and policy documents, various forms of adaptation are distinguished. One of these distinctions is autonomous adaptation (also called passive or spontaneous adaptation) versus planned adaptation (also called active, or directed, adaptation). Autonomous adaptation is defined by the IPCC as adaptation that does not constitute a conscious response to climatic stimuli but is triggered by ecological changes in natural systems and by market or welfare changes in human systems.[9] An example of this is earlier migration back to the breeding grounds of migratory birds (or even migratory birds that stop migrating altogether), or, in a human development context, developers in the recreation sector starting hotels and restaurants in areas that have become more suitable for mass recreation. Autonomous, or passive, adaptation most likely is unsuitable to address climate change impacts because 'adaptations to slowly evolving trends embedded in a noisy background of inherent

[9] Parry et al., above note 3 at 869.

variability are likely to be delayed for decades' and may lead to maladaptations because of misperceptions about weather anomalies.[10]

Planned adaptation is the result of a deliberate policy decision, based on an awareness that conditions have changed or are about to change and that action is required to return to, maintain, or achieve a desired state.[11] A variant of the latter form of adaptation is proactive or anticipatory adaptation, which is defined as adaptation that takes place before impacts of climate change are observed. Most of the attention is focused on planned adaptation for two reasons. First, it is obvious that autonomous adaptation has its limits and cannot be relied upon in most cases. Second, from a legal and policy point of view, this form of adaptation does not need much attention because it usually happens within existing legal and policy frameworks. Therefore, this book almost entirely deals with planned adaptation.

Adger defined three 'cornerstones' of planned adaptation: reducing the sensitivity of affected systems (example: crop change); altering the exposure of systems to the effects of climate change (example: increase hazard preparedness); and increasing the resilience of social and ecological systems (example:. green buildings).[12] Often, however, increasing resilience is used as the general goal of planned adaptation, as also seems to be the case in the definition of that term offered by the IPCC: resilience is 'the ability of a system and its component parts to anticipate, absorb, accommodate, or recover from the effects of a hazardous event in a timely and efficient manner, including through ensuring the preservation, restoration, or improvement of its essential basic structures and functions'.[13]

Another form of adaptation that should be mentioned here is maladaptation. The Organisation for Economic Co-operation and Development (OECD) defines maladaptation as 'business-as-usual development which, by overlooking climate change impacts, inadvertently increases exposure and/or vulnerability to climate change. Maladaptation could also include actions undertaken to adapt to climate impacts that do not succeed in

---

[10] Stephen H Schneider and Janica Lane, 'Dangers and Thresholds in Climate Change and the Implications for Justice' in W Neil Adger et al., *Fairness in Adaptation to Climate Change* (MIT Press 2006) 46.

[11] Ibid.

[12] W Neil Adger, Nigel W Arnell and Emma L Tompkins, 'Successful Adaptation to Climate Change across Scales' (2005) 15 Global Environmental Change 77, 79.

[13] Field et al., above note 1 at 563.

reducing vulnerability but increase it instead.'[14] An example is the construction of elevated roads so as to make them resilient to floods, which consequently, however, trigger new development in a zone that is vulnerable to flooding.

The opposite of maladaptation is successful adaptation.[15] The success of adaptation measures is not easy to assess, as adaptation measures may be successful in the short term, but not in the long run, or may be successful in one location but have negative side effects in another location, for instance downstream.[16] Important factors for success are not only the effectiveness of measures, but also their efficiency and equity and legitimacy.[17] The latter, typically, are for a large part determined by the legal framework for adaptation policies and, therefore, are central to this book.

3 LINKS BETWEEN ADAPTATION AND MITIGATION

Adaptation cannot be regarded in isolation from mitigation. Not only are they both important elements of any climate change policy,[18] they are also positively and negatively interlinked. First, many adaptation measures equally serve mitigation goals. Afforestation, reforestation, and preserving and restoring mangroves, with the goal to protect the land against flooding, against landslides following intense rainfall, or against the negative impact of storms, are all measures that can also be part of a mitigation strategy. The same goes for adaptation measures in the field of biodiversity conservation. Creating corridors between protected areas and enlarging protected areas through restoration lead to an increase in vegetation, and thus to additional carbon uptake. This is even true for the construction of green buildings and green roofs: not only are they naturally cooler and thus an effective measure against heat waves, they also lead to less energy consumption and, again, to additional carbon uptake. In fact, any measure to reduce the amount of energy consumption

---

[14] OECD, *Integrating Climate Change Adaptation into Development Co-operation: Policy Guidance* (Paris: OECD Publishing, 2009) 53.

[15] See extensively Susanne Moser and Maxwell Boykoff, *Successful Adaptation to Climate Change: Linking Science and Policy in a Rapidly Changing World* (Routledge 2013).

[16] Adger et al., above note 12 at 80.

[17] Ibid. at 82.

[18] W Neil Adger, Irene Lorenzoni and Karen L O'Brien, 'Adaptation Now' in W Neil Adger, Irene Lorenzoni and Karen L O'Brien (eds), *Adapting to Climate Change: Thresholds, Values, Governance* (Cambridge University Press 2009) 4.

to avoid an energy fall out during heat waves is also a mitigation measure.

The second link between adaptation and mitigation is a negative one. Adaptation measures can be harmful to the climate in the sense that they lead to: a) more greenhouse gas emissions; or b) a reduction of carbon uptake. The examples are obvious. Installing air conditioners to combat the heat leads to more energy consumption and thus, if the energy comes from a coal-fueled energy installation, to higher emissions. Replacing a natural coastal habitat by a large sea wall to combat sea level rise and storm surges leads to a loss of natural carbon uptake. This example, coincidently, shows that an adaptation measure in one policy field may be detrimental to adaptation policy in another field, as the large sea wall that replaces natural coastal habitat probably will lead to a loss of biodiversity adaptation opportunities.

Mitigation measures, on the other hand, can be harmful for adaptation as well. Afforestation in arid and semi-arid regions strongly reduces water yields and thus has a negative impact on local agriculture and biodiversity. Switching to hydropower may reduce irrigation options for farmers and thus deprive them of adaptation opportunities.

It is clear that adaptation and mitigation policies have to be developed together, so that they are mutually beneficial.[19] Therefore, although this book focuses on adaptation law, we cannot and will not ignore mitigation law.

4 DEALING WITH RISKS AND UNCERTAINTY AND OTHER BARRIERS TO ADAPTATION

Although thousands of scientists around the globe have been researching climate change for many years now, we are still faced with many uncertainties. Uncertainties as to the exact nature and intensity of the changes that we can expect, the timeframe within which they are to be expected, and the locations that will be hardest hit. Uncertainties present

---

[19] See extensively chapter 18 of the IPCC's Working Group II report: R J T Klein et al., 'Inter-relations Between Adaptation and Mitigation' in M L Parry et al. (eds), *Climate Change 2007: Impacts, Adaptation and Vulnerability: Contribution of Working Group II to the Fourth Assessment Report of the Intergovernmental Panel on Climate Change* (Cambridge University Press 2007) 745–777.

themselves in two ways: a) uncertainties caused by our lack of knowledge; and b) uncertainties caused by randomness inherent to the phenomenon at hand. In literature on risks, these are usually described as epistemic risks (a) and aleatory risks (b). Both types of risk are relevant in climate adaptation law and policies.

Obviously, risks inherent in the randomness of floods, storms and other events influenced by climate change will never disappear. We can only predict that in certain areas storm intensity will increase, rainfall will increase etc., without being able to tell exactly when and where a storm will hit and how strong that storm will be.

Uncertainties as a consequence of a lack of knowledge do gradually become smaller because of increasing scientific knowledge as to what is happening to atmospheric and climatic processes due to higher levels of greenhouse gases in the Earth's atmosphere. We must, however, not be so optimistic as to think that this type of uncertainty will eventually disappear altogether, at least not in the foreseeable future. Our lack of knowledge is still quite large, especially because, as a consequence of climate change, *everything* is changing. Increasing carbon emissions leads to temperature rise, which leads to a great number of subsequent changes in weather patterns. Many of these changes probably do not occur in a linear way. Instead, it becomes increasingly clear that there are many tipping point effects and non-linear effects. Such effects are much more difficult to predict. Paradoxically, increasing scientific knowledge has led to an increase in the epistemic risks. We now know better what we do not know. Our lack of knowledge also concerns the effect of these changes on human behaviour and on that of the other living organisms on Earth. How are living organisms, including human beings, going to adapt to these changes once they occur, gradually or suddenly? In addition to all of this, we should keep in mind that scientists will keep disagreeing on certain aspects of their findings. This is simply what scientists do: they question findings in order to establish a theory on the basis of which the most accurate predictions can be made. Policy makers have to deal with this 'tug of war among competing theories and qualifications of theories'.[20]

Kleindorfer shows how both types of risks are tightly intertwined with choice. A homeowner or a business might consider options such as insurance or mitigation before the fact in order either to reduce or pay for

---

[20] Paul Kleindorfer, 'Interdependency of Science and Risk Finance in Catastrophe Insurance and Climate Change' (2010) INSEAD working paper 10, http://papers.ssrn.com/sol3/papers.cfm?abstract_id=1538161, accessed 5 October 2012.

losses resulting from an extreme weather event. In order to reach such a decision, the homeowner or business may want to gain knowledge to reduce the epistemic risk. Knowledge is thus a fundamental input to rational choice under uncertainty.[21] The same processes take place at the macro level of state authorities making climate adaptation policies and regulations. Authorities are faced with the huge challenge of regulating for an uncertain future and with dealing with 'low probability, high consequence events'. It requires a proactive and long term approach of policy makers and regulators; an approach in which they can rely less on front-end methodologies,[22] such as environmental impact assessments, regulatory impact assessments, and cost–benefit analyses.[23] Laws and regulations have to leave room for long-term decision-making under uncertainty and still give the decision-makers the tools to reach legitimate decisions. Craig translated this approach based on information and taking long policy horizons into five principles for climate change adaptation law: 1) monitor and study everything all the time;[24] 2) eliminate or reduce non-climate change stresses and otherwise promote resilience; 3) plan for the long term with much increased coordination across media, sectors, interests, and governments; 4) promote principled flexibility in regulatory goals and natural resource management; and 5) accept – really accept – that climate change adaptation will often be painful.[25] What this means in practice in the various fields of law is the core issue in each of the chapters of this book.

---

[21] Ibid. at 12–13.

[22] J B Ruhl, 'Climate Change Adaptation and the Structural Transformation of Environmental Law' (2010) 40(2) Environmental Law 363, 416.

[23] Climate change decision-making truly shook up discussions on cost–benefit analyses because it forced us to think about how to value the benefits for future generations in today's currency. See extensively Richard L Revesz and Matthew R Shahabian, 'Climate Change and Future Generations' (2010) N.Y.U. School of Law, Public Law & Legal Theory Research Paper Series, http://papers.ssrn.com/sol3/papers.cfm?abstract_id=1666423, accessed 5 October 2012.

[24] The 2012 climate change impacts report by the European Environment Agency has a similar recommendation: 'On-going and planned monitoring and research at national and EU level can improve assessments of past and projected impacts of climate change, thereby enhancing the knowledge base for adaptation', European Environment Agency, *Climate Change, Impacts and Vulnerability in Europe 2012. An Indicator-based Report* (EEA 2012) 18.

[25] Robin Kudis Craig, '"Stationarity is Dead" – Long Live Transformation: Five Principles for Climate Change Adaptation Law' (2010) 34 Harvard Environmental Law Review 9.

It is increasingly concluded that current risk analysis and risk management practice will not be effective as a tool to provide effective adaptation as they do not necessarily lead to strategies that facilitate the required changes in existing political rationalities.[26] Risks of business-as-usual should be prioritized over the risks of change, and systemic changes should be targeted instead of stability.[27] This requires a radical departure from the conventional approach to environmental regulation, dominated by the risk regulation paradigm.[28]

Another challenge to adaptation is that it can only be successful if adaptation governance occurs at all levels of government, as well as at the level of private parties, through multiple policy responses. Peel et al. conclude that adaptation policy and regulation need to link effectively across different levels, both horizontally and vertically, from individual or local actor responses to national laws and international treaties.[29] Again, this is a common feature that we find in each of the chapters of this book.

5 THE FOCUS OF THIS BOOK

What are the legal challenges and barriers to climate change adaptation and how can they be overcome? What can be done within existing legal frameworks, and where are new or adapted frameworks needed? These are the main questions that we address in this book. The book discusses all the fields of law that are faced with challenges brought about by the need to adapt to the changing climate and to extreme weather conditions: tort law, property law, insurance law, disaster law, water law, marine law, planning law, biodiversity law, building and construction law, environmental law, migration law, agricultural law, and energy law. It does so in an integrated way, which is the only way given the strong connections between these fields of law. For each field of law, the challenges and barriers to adaptation are discussed, as well as ways to overcome these barriers. The book does not focus on a specific legal system, such as that

---

[26] Nick Pidgeon and Catherine Butler, 'Risk Analysis and Climate Change' (2009) 18(5) Environmental Politics 670, 686.

[27] Veerle Heyvaert, 'Governing Climate Change: Towards a New Paradigm for Risk Regulation' (2011) 74(6) Modern Law Review 817.

[28] Ibid.

[29] Jacqueline Peel, Lee Godden and Rodney J Keenan, 'Climate Change Law in an Era of Multi-Level Governance' (2012) 1(2) Transnational Environmental Law 245, 275.

of the European Union (EU) or the United States,[30] or on a specific level of governance, such as the international level.[31] Instead, we take a transnational, multi-level governance view, focusing both at the international, the regional, and the domestic level, using examples from around the world. While such an approach obviously does not allow discussion of the legal implications of adaptation measures for each country, it does offer an opportunity to transcend national legal specifics and thus get a better view of the main issues to be solved as well as potential solutions.

The set-up of the book is as follows. Chapter 2 first describes how the international community has been dealing with adaptation within the frameworks of the UNFCCC and the Kyoto Protocol. Chapter 3 then takes an ethical approach to adaptation by focusing on climate justice and the relationship between climate adaptation law and development. Human rights and migration law are then dealt with in Chapter 4 when focusing on climate induced displacement. Chapter 5 focuses on compensation. How can victims of climate change be compensated for their losses caused by climate change, both direct losses and losses arising out of the need to take adaptation measures? Liability law, insurance law, compensation funds, tax law, and direct compensation by the government are fields of law that are addressed here. Disaster law is the topic of Chapter 6, dealing with disaster risk reduction as an important means of making society resilient to extreme weather events and other disasters associated with climate change.

From Chapter 7 onwards, the focus will be on specific sectors: public health law (Chapter 7), agricultural and forestry law (Chapter 8), water law (Chapter 9), marine and coastal law (Chapter 10), biodiversity law (Chapter 11), land use and planning law (Chapter 12), law dealing with the built environment (Chapter 13), pollution control law (Chapter 14), and energy law, with a focus on electricity supply (Chapter 15).

---

30 A comprehensive study of climate change adaptation law in the US is offered in Michael B Gerrard and Katrina Fischer Kuh (eds), *The Law of Adaptation to Climate Change: United States and International Perspectives* (ABA 2012).

31 See Rosemary Rayfuse and Shirley V Scott (eds), *International Law in the Era of Climate Change* (Edward Elgar 2012).

6 CONCLUSIONS

The contributions to this book show that, although adaptation receives a growing amount of attention, both in practice and in academia, adaptation law is only just starting to emerge. In most instances, there are some plans or policies aimed at adaptation in various fields, usually those fields that already have to deal with increasing problems, such as storm water management, and flood management. An adaptation of the laws still has to start. It is obvious that existing laws have to be assessed on their ability to facilitate adaptation. This is a huge undertaking because there is hardly any field that is not affected by climate change. All laws and regulations that in any possible way organize society have to be 'climate proofed': laws regarding agriculture, forestry, fisheries, energy and telecommunications infrastructure, water management, air quality, industrial installations, nature conservation, buildings, transport infrastructure, public health, migration, disaster management, coastal defences, etc. It is not really possible to draw more specific overarching conclusions than the ones already mentioned in this introductory chapter: every field faces specific climate change impacts and needs specific adaptations, adaptations that also need to vary according to local circumstances. In various chapters, examples are presented of how existing laws are effectively applied to create resilience or to otherwise prepare for extreme weather events or other climate change impacts. Often, though, existing legislation needs to be adapted so that the competent authorities are obliged to plan for and take adaptation measures. The EU, for example, has just embarked on an ambitious programme to climate proof all existing Directives and Regulations. In 2013, the first climate proofed piece of EU legislation is expected to be adopted (a revised Directive on environmental impact assessment). It will probably take at least ten years before the entire body of EU law has been climate proofed. Similar programmes will have to be set up at all levels of government: international, regional, national/federal, provincial/regional, and local. Since many impacts of climate change will be local impacts, and since these impacts can vary greatly from one location to another, it is important that, at the local level, the authorities take the lead in local adaptation programmes. At that level, planning law is probably the most important instrument in the authorities' adaptation toolkit. Higher levels of government have to ensure that the authorities at the local level have sufficient room for manoeuvre. For adaptation issues at the higher levels, i.e., at the level of transboundary river basins, national or transboundary coastal areas, international marine areas, regional or international migration, and other levels, international institutions will have to take the lead and coordinate international adaptation efforts. At all levels, issues of equity and justice arise and need to be incorporated into the law-making process.

2. Climate change adaptation under the United Nations Framework Convention on Climate Change and related documents

Jonathan Verschuuren

1 INTRODUCTION

It is fair to say that adaptation law originates at the international level. Both the 1992 United Nations Framework Convention on Climate Change (hereafter: UNFCCC),[1] and the 1997 Kyoto Protocol[2] comprise various obligations for the parties to these international agreements to adopt and implement adaptation policies. The documents also impose upon developed countries the duty to, both financially and practically, assist developing countries with their adaptation actions. In the international arena, most attention is going to the latter issue, as will be obvious from the overview of the most important adaptation related actions both under the UNFCCC (section 2) and under the Kyoto Protocol (section 3). In section 4, some remarks will be made as to the impact of the UNFCCC's work on adaptation on regional and domestic law and policy.

2 INTERNATIONAL ADAPTATION LAW UNDER THE UNFCCC

2.1 Adaptation in the UNFCCC

The primary objective of the UNFCCC, as laid down in Article 2, is to achieve stabilization of greenhouse gas concentrations in the atmosphere at a level that would prevent dangerous anthropogenic interference with

[1] United Nations Framework Convention on Climate Change (New York, 9 May 1992), (1992) 31 ILM 849, entered into force 21 March 1994 .

[2] Kyoto Protocol to the United Nations Framework on Climate Change, UN Doc FCCC/CP/1997/7/Add. 1 (Kyoto, 10 December 1997), (1998) 37 ILM 22, entered into force 16 February 2005.

the climate system. Such a level should be achieved within a timeframe sufficient to allow ecosystems to adapt naturally to climate change, to ensure that food production is not threatened and to enable economic development to proceed in a sustainable manner. Mitigation, according to the objective of the UNFCCC, is supposed to be successful so that ecosystems, food production, and the economy are more or less automatically kept as they were before.

Fifteen years later, the Intergovernmental Panel on Climate Change (IPCC) sounded the alarm bell. Adaptation is necessary to address impacts resulting from the warming which is already unavoidable due to past emissions. For impacts that already show or will show in the very near future, adaptation is *the only available and appropriate response*, according to the IPCC in 2007.[3]

Fortunately, we did not have to start from scratch after these alarming words of the IPCC had been published. As a matter of fact, other than the objective of the UNFCCC cited above suggests, the 1992 Convention focuses not just on combating climate change, but also on combating the adverse effects of climate change.[4] This is obvious from the principles of the UNFCCC, laid down in Article 3. Developed country Parties, for instance, should take the lead in combating the adverse effects of climate change.[5] Developing country Parties that are particularly vulnerable to the adverse effects of climate change should be given full consideration.[6] Precautionary measures should be taken to mitigate the adverse effects of climate change.[7]

Under the commitments listed in Article 4 of the UNFCCC there are several that deal with adaptation. Parties to the Convention have to:

- formulate, implement, publish and regularly update national and regional programmes containing measures to facilitate adequate adaptation to climate change;[8]

---

[3] M L Parry et al. (eds), *Climate Change 2007: Impacts, Adaptation and Vulnerability: Contribution of Working Group II to the Fourth Assessment Report of the Intergovernmental Panel on Climate Change* (Cambridge University Press 2007) 19.

[4] M J Mace, 'Adaptation under the UN Framework Convention on Climate Change: the International Legal Framework' in W Neil Adger et al., *Fairness in Adaptation to Climate Change* (MIT Press 2006) 53.

[5] Article 3(1).

[6] Article 3(2).

[7] Article 3(3).

[8] Article 4(1)(b).

- cooperate in preparing for adaptation to the impacts of climate change; develop and elaborate appropriate and integrated plans for coastal zone management, water resources and agriculture, and for the protection and rehabilitation of areas, particularly in Africa, affected by drought and desertification, as well as floods;[9]
- take climate change considerations into account in the relevant social, economic and environmental policies and actions, and employ appropriate methods, for example impact assessments, with a view to minimizing the adverse effects on the economy, on public health and on the quality of the environment, of projects or measures taken to adapt to climate change;[10]
- promote and cooperate in research (scientific, technological, technical, socio-economic and other) intended to further the understanding and to reduce the remaining uncertainties regarding the economic and social consequences of various response strategies,[11] as well as exchange information on this;[12] and
- promote and cooperate in education, training and public awareness related to climate change in general, thus including adaptation issues.[13]

In addition to this impressive list of adaptation duties, further obligations have been imposed on developed country Parties. They also have to help developing countries with their adaptation policies by:

- assisting developing countries that are particularly vulnerable to the adverse effects of climate change in meeting the costs of adaptation to those adverse effects;[14] and by
- taking all practicable steps to promote, facilitate and finance the transfer of, or access to, environmentally sound technologies and know-how to developing country Parties, to enable them to implement the provisions of the UNFCCC;[15] this obligation, thus, also relates to the adaptation provisions mentioned above. Hence, the transfer of technology not only applies to mitigation technologies,

---

9 Article 4(1)(e).
10 Article 4(1)(f).
11 Article 4(1)(g).
12 Article 4(1)(h).
13 Article 4(1)(i).
14 Article 4(4).
15 Article 4(5).

but also to technologies and know-how necessary to implement adaptation measures.

An especially interesting provision on adaptation commitments is Article 4(8), in which all Parties are compelled to focus on the specific needs and concerns of developing country Parties arising from the adverse effects of climate change and the impact of the implementation of response measures. A whole range of countries especially vulnerable to the effects of climate change is mentioned here: small island countries, countries with low-lying coastal areas, countries with arid and semi-arid areas, forested areas and areas liable to forest decay, countries with areas prone to natural disasters, countries with areas liable to drought and desertification, countries with areas of high urban atmospheric pollution, countries with areas with fragile ecosystems, including mountainous ecosystems, countries whose economies are highly dependent on income generated from the production, processing and export, and/or on consumption of fossil fuels and associated energy-intensive products, and landlocked and transit countries.

All subsequent provisions of the UNFCCC, on such issues as research and systematic observation, education, training and public awareness, the financial mechanism, etc., apply to both mitigation and adaptation.[16] The two subsidiary bodies that are established by the UNFCCC,[17] the Subsidiary Body for Scientific and Technological Advice (hereafter: SBSTA), and the Subsidiary Body for Implementation (hereafter: SBI), are both competent to deal with adaptation issues.

Finally, it should be mentioned that the UNFCCC stipulates some general principles that apply to mitigation as well as adaptation: equity, solidarity, precaution, sustainability, and good neighbourliness.[18]

2.2 National Adaptation Programmes of Action

Many countries have developed national adaptation programmes pursuant to Article 4(1)(b) of the UNFCCC. Developing countries, however, appeared to be largely unable to draft such programmes. Since such a programme is considered to be the first step towards financing and implementation of adaptation measures, the Conference of the Parties

---

16 Articles 5, 6 etc.

17 Articles 9 and 10.

18 Article 3, sections 1 to 5 respectively. See further P J J Driessen and H F M W van Rijswick 'Normative Aspects of Climate Adaptation Policies' (2011) 2 Climate Law 559–581.

(COP), in its 2001 session in Marrakesh, adopted guidelines for the preparation of National Adaptation Programmes of Action (hereafter: NAPA) to assist the least developed countries in the process,[19] and established a Least Developed Countries Fund (hereafter: LDCF).[20] This has been quite successful. By 2011, 45 least developed countries had submitted their NAPAs, containing detailed priority policies to respond to their urgent and immediate needs to adapt to climate change.[21] Each NAPA reports on the urgent and immediate needs for which further delay could increase vulnerability to climate change or lead to increased costs at a later stage.

The 2010 Nepal NAPA, for instance, formulates extensive policies to increase communities' resilience in the fields of water management, agriculture, disaster management, forest and ecosystem management, public health, and others.[22] Associated with these policies is an impressive list of over one hundred specific climate adaptation options for agriculture and food security, for the water sector, for the energy sector, for forests and biodiversity, for public health, for urban settlements and infrastructure, and to address climate-induced disasters. Concrete measures range from simple practical measures, such as tree planting around farm lands and water resources in mountainous regions[23] and better rain water collection,[24] to complex legal measures such as a zoning programme to adapt to water induced disasters and the associated activation of an inundation committee,[25] and the enforcement of planning regulations and building codes in urban areas incorporating climate change dimensions.[26]

The process of drafting and implementing NAPAs in the least developed countries always involves issues of equity and justice, as the

---

19 Decision 28/CP. 7, FCCC/CP/2001/13/Add. 4, 7–13.

20 Decision 7/CP. 7 FCCC/CP/2011/13 Add. 1, 43–45.

21 The secretariat of the UNFCCC makes all of these available through their website, http://unfccc.int/cooperation_support/least_developed_countries_portal/submitted_napas/items/4585.php, accessed 5 October 2012.

22 Government of Nepal, National Adaptation Programme of Action (NAPA) to Climate Change, Kathmandu 2010, available at the UNFCCC's website mentioned in the previous footnote.

23 Ibid. at 69.

24 Ibid. at 68.

25 Ibid. at 70.

26 Ibid. at 77.

interests of the poorest and most vulnerable may be neglected by those in charge of the process. A participatory process is therefore advocated.[27]

After submission to the secretariat, the NAPA enters the so-called GEF cycle, which allows for funding of the projects identified in the NAPA under the LDCF and which is administered by the Global Environment Facility (hereafter: GEF). The GEF is a financial organization instituted by governments, international organizations, such as the World Bank and the United Nations Environment Programme (UNEP), non-governmental organizations (NGOs) and the private sector, to fund environmental projects in developing countries.[28]

2.3 The Buenos Aires Programme of Work on Adaptation and Response Measures and the Nairobi Work Programme on Impacts, Vulnerability and Adaptation to Climate Change

Since its adoption in 1992, adaptation has been an issue during several Conferences of the Parties, particularly those held in Buenos Aires in 2004 and in Cancun in 2010. COP10 in Buenos Aires led to the adoption of the Buenos Aires Programme of Work on Adaptation and Response Measures, which is a rather weak attempt to push adaptation, especially in developing countries, forward. Developing countries were requested to make use of the NAPA opportunities mentioned above; developed countries were asked to make available additional funding for adaptation measures under NAPAs; and the GEF was requested to step up its activities in the implementation phase of the NAPAs. In addition to this, the Buenos Aires Programme of Work initiated additional information gathering and capacity-building measures. Finally, the SBSTA was requested to develop a five-year programme of work on scientific, technical and socio-economic aspects of impacts, vulnerability and adaptation to climate change.[29]

This five-year programme was adopted one year later during COP11 in Montreal,[30] and had as its main objective to assist the parties, especially developing countries, to improve their understanding of impacts, vulnerability and adaptation, and to make informed decisions on practical

---

[27] Saleemul Huq and Mizan R Khan, 'Equity in National Adaptation Programs of Action (NAPAs): The Case of Bangladesh' in W Neil Adger et al., *Fairness in Adaptation to Climate Change* (MIT Press 2006) 199.

[28] For more information, see the GEF's website, http://www.thegef.org, accessed 5 October 2012.

[29] Decision 1/CP.10, FCCC/CP/2004/10/Add. 1.

[30] Decision 2/CP.11, FCCC/CP/2005/5/Add. 1, 5–11.

adaptation actions. The SBSTA operationalized this again one year later during COP12 in Nairobi,[31] hence its current name: Nairobi Work Programme on Impacts, Vulnerability and Adaptation to Climate Change (NWP) 'Understanding Vulnerability, Fostering Adaptation'. This programme has been running ever since and acts as a coordination mechanism for a wide variety of capacity-building and information dissemination activities by some 200 organizations, including private companies such as insurance firms and consultancy firms, universities and other research institutes, development banks and other financial organizations, NGOs and charitable organizations.[32]

2.4 The Cancun Adaptation Framework and the Adaptation Committee

It was only in 2010 that the COP placed adaptation high on its agenda. In the Cancun Agreements, adopted in 2010, the conference dealt with adaptation first (before mitigation), stating that enhanced action on adaptation was urgently needed.[33] The conference decided to establish the Cancun Adaptation Framework under which all parties to the UNFCCC will:[34]

- plan, prioritize and implement adaptation actions;
- assess adaptation actions, including economic, social and environmental evaluation of adaptation options;
- strengthen institutional capacities for adaptation;
- build resilience of socioeconomic and ecological systems;
- enhance climate change related disaster risk reduction strategies, taking into consideration early warning systems, risk assessment and management, and sharing and transfer mechanisms such as insurance;
- take measures to enhance understanding, coordination and cooperation with regard to climate change induced displacement, migration and planned relocation;
- research, develop and diffuse technologies, practices and processes for adaptation;

---

[31] FCCC/SBSTA/2006/11, 5–13.

[32] Information on activities and organizations involved can be found at the UNFCCC's website, http://unfccc.int/adaptation/nairobi_work_programme/items/3633.php, accessed 5 October 2012.

[33] Decision 1/CP.16, FCCC/CP/2010/7/Add. 1.

[34] Ibid.

- strengthen information, education and public awareness;
- improve climate related research in order to help decision-makers at the national and regional levels.

The Adaptation Committee was established to promote the implementation of all this[35] and to function as the overall advisory body to the COP on adaptation.[36] During COP17 in Durban, further decisions were taken on the role and composition of the Adaptation Committee.[37] From the instructions of the COP to the Adaptation Committee it seems that the committee will have to function as the principal coordinator in the field of adaptation. It has to engage and develop linkages with all adaptation related work programmes, bodies and institutions under the UNFCCC,[38] and it has to seek input from intergovernmental, international, regional, national and subnational organizations, centres and networks, the private sector and civil society.[39] Given the recent enormous rise in adaptation institutions and policies at international and other levels, such a role seems very timely. The Adaptation Committee has to comprise 16 members, two from each of the five UN regional groups, one from a small island developing country, one from a least developed country, two from Annex I parties and two from non-Annex I parties.[40]

3 INTERNATIONAL ADAPTATION LAW UNDER THE KYOTO PROTOCOL

3.1 Adaptation in the Kyoto Protocol

Five years after the adoption of the UNFCCC, the Kyoto Protocol mainly focused on the reduction of greenhouse gas emissions and not on adaptation. Given the stage of international discussions on climate change at that time, this was to be expected: a substantial first step towards greenhouse gas emission reductions was within reach. This explains why the Kyoto Protocol's provisions on adaptation are a bit meagre.

---

35 Ibid.
36 Decision 2/CP.17, FCCC/CP/2011/9/Add. 1, 19.
37 Ibid. at 19–22.
38 Ibid. at 20.
39 Ibid. at 21.
40 Ibid. at 20.

The Protocol does work out the general provisions of the UNFCCC mentioned above a little further. It states that the programmes to facilitate adequate adaptation to climate change have to concern the energy, transport and industrial sectors as well as agriculture, forestry and waste management, and that adaptation technologies as well as spatial planning are important elements in such adaptation policies.[41] In addition, the Kyoto Protocol regulates that information on such national adaptation policies has to be integrated into the communication that has to be submitted to the secretariat of the UNFCCC.[42] From a practical point of view, probably the most important provision on adaptation in the Kyoto Protocol is a section in the provision on the Clean Development Mechanism (CDM), stating that money generated through CDM projects should be used to assist developing countries to meet the costs of adaptation.[43] It was this provision that, four years later, formed the basis for the establishment of the Adaptation Fund.

3.2 Adaptation Fund, Special Climate Change Fund, Least Developed Countries Fund and Green Climate Fund

The Adaptation Fund was established during COP7 of the UNFCCC in Marrakesh in 2001,[44] but it operates under the Kyoto Protocol. Its aim is to finance concrete adaptation projects in developing country Parties to the Kyoto Protocol, as well as some specific adaptation activities that are not limited to developing countries. The latter is a consequence of another decision taken in Marrakesh which aimed at the following activities:[45]

- starting to implement adaptation activities promptly where sufficient information is available to warrant such activities, in the areas of water resource management, land management, agriculture, health, infrastructure development, fragile ecosystems, including mountainous ecosystems, and integrated coastal zone management;
- improving the monitoring of diseases and vectors affected by climate change, and related forecasting, and improving disease control and prevention;

41 Article 10(b)(i).
42 Article 10(b)(ii).
43 Article 12(8).
44 Decision 10/CP.7, FCCC/CP/2001/13/Add. 1, 52.
45 Decision 5/CP.7, FCCC/CP/2001/13/Add. 1, 32.

- supporting capacity-building for preventive measures, planning, preparedness and management of disasters relating to climate change, including contingency planning, in particular for droughts and floods in areas prone to extreme weather events;
- strengthening and, if necessary, establishing national and regional information networks for rapid response to extreme weather events.

It was not until the COP serving as the meeting of the Parties to the Kyoto Protocol first convened in 2005 (in Montreal), that the Adaptation Fund materialized.[46] In the next few years, various decisions were taken to operationalize the Fund, first, in 2006, by adopting its guiding principles,[47] then, in 2007, more importantly, by instituting the Adaptation Fund Board (hereafter: AFB) as the operating entity of the Fund.[48] Although the COP serving as the meeting of the Parties to the Kyoto Protocol remains in control over the Adaptation Fund, the AFB has considerable power over the actual funding process.[49] The AFB is a legal entity under German law,[50] with involvement from both the GEF and the World Bank, acting as secretariat and trustee respectively.[51] The Adaptation Fund is financed by a fixed share of 2 per cent of the proceeds of all CDM project activities (to be precise, 2 per cent of the Certified Emission Reductions, or CERs, issued for a CDM project), as well as individual donations by countries.

Finally, in 2010, almost ten years after the decision was taken to establish an adaptation fund, the AFB took its first decision to fund an adaptation project, namely an adaptation project to combat coastal erosion due to sea level rise in Senegal. Since then, many other projects have been financed by the Adaptation Fund, mostly on coastal adaptation

---

46 Decision 28/CMP. 1, FCCC/KP/CMP/2005/8/Add. 4, 3.

47 Decision 5/CMP. 2, FCCC/KP/CMP/2006/10/Add. 1, 28.

48 Decision 1/CMP. 3, FCCC/KP/CMP/2007/9/Add. 1, 3. On the functioning of the Adaptation Fund, see in more detail R Czarnecki and K Guilanpour, 'The Adaptation Fund after Poznań' (2009) 1 Carbon and Climate Law Review 79–87.

49 Its rules of procedure were adopted in 2008, Decision 1/CMP. 4, FCCC/KP/CMP/2008/11/Add. 2.

50 Decision 5/CMP. 6, FCCC/KP/CMP/2010/12, Add. 1, 13.

51 MOU's with both institutions have been concluded to formalize this involvement. Decision 1/CMP. 4, FCCC/KP/CMP/2008/11, Add. 2, 11 and 12 respectively.

and flood prevention, but also on adaptation in agriculture.[52] A troublesome review process of the AFB has been underway since 2010, which was concluded at COP18 in Doha in December 2012.[53] One of the issues at hand is the relationship to two other funds (SCCF and LDCF – see below) and the administrative costs involved in maintaining these three funds. It was concluded that the interim institutional arrangements are extended until 2015, and that another review needs to have been conducted by COP20 in December 2014.[54]

Besides the Adaptation Fund, there are – indeed! – two more funds, both of which operate under the UNFCCC. Like the Adaptation Fund, the Special Climate Change Fund (hereafter: SCCF) and the Least Developed Countries Fund (hereafter: LDCF) were established during COP7 of the UNFCCC in Marrakesh in 2001.[55] The SCCF is a general fund, aimed at financing activities, programmes and measures relating to climate change in a number of areas, one of which is adaptation (the others being transfer of technology, energy, transport, industry, agriculture, forestry, waste management, and activities that assist developing countries to diversify their economies). The LDCF, as its name already indicates, is specifically aimed at financing projects in the least developed countries, particularly, as indicated above in section 2.2, projects related to their NAPAs. Both the SCCF and the LDCF are operated by the GEF, their councils have joint meetings, and they both rely mainly on voluntary contributions from Annex I states.

Examples of projects funded under the joined funds are, for instance: a project to promote the implementation of national and transboundary integrated water resource management that is sustainable and equitable given expected climate change in Swaziland; a project integrating climate change risks into water and flood management by vulnerable mountainous communities in the Greater-Caucus region of Azerbaijan; and a project to promote a value chain approach to adaptation in agriculture in Ghana.[56]

---

52 Information on the projects funded under the Adaptation Fund is available from the AFB's website, http://www.adaptation-fund.org, accessed 5 October 2012.

53 As reported at UNFCCC website, http://unfccc.int/cooperation_and_support/financial_mechanism/adaptation_fund/items/3659.php, accessed 21 January 2013.

54 Ibid.

55 Decision 7/CP.7, FCCC/CP/2001/13/Add. 1, 43.

56 GEF/LDCF/SCCF.9/Joint Summary, 18 November 2010.

With the proliferation of funds, there is a clear risk of overlap and inefficiency. The GEF, involved in all of them as well as in their own GEF Trust Fund, which initially also had funding of adaptation measures as one of its main goals,[57] is aware of this risk. In its 2010 Revised Programming Strategy on Adaptation to Climate Change, the GEF Council tried to show how the different adaptation related funds were all complementary.[58] However, the arguments used are not particularly strong. For instance, the fact that the LDCF and the SCCF have a mandate under the UNFCCC and the Adaptation Fund a mandate under the Kyoto Protocol, is but a formal argument. The fact that the funds' revenues come from different sources is not a strong argument either; nor is the argument that the LDCF and the SCCF, other than the Adaptation Fund, are aimed at more issues than just adaptation.

The biggest problem of the funds, however, is the lack of available funds. In the 42 NAPAs that had been submitted by 2009, urgent adaptation measures have been laid down of a total cost of around USD 2 billion.[59] In 2010, however, the LDCF received total pledges of USD 131.5 million, and the SCCF a total of USD 49.8 million from Annex I states. The largest contributors for both funds in 2010 were the United States and Germany, both with donations of around USD 50 million in total (for both funds together). Current funding is totally inadequate to address even the most urgent and immediate adaptation needs of the least developed countries. It is, therefore, not surprising that the least developed countries raise this issue in every COP, and not without success. At the Copenhagen and Cancun meetings in 2009 and 2010, pledges were made by developed countries to provide new and additional resources approaching USD 30 billion for the period 2010 to 2012, with a balanced allocation between adaptation and mitigation, and with prioritized funding for adaptation for the most vulnerable developing countries, such as the least developed countries, small island developing states and Africa.[60] In addition, the developed countries even committed to the goal of jointly mobilizing USD 100 billion per year by 2020 to address the needs of

---

[57] Through its Strategic Priority on Adaptation, a USD 50 million allocation inside the GEF Trust Fund. As of 2010, the SPA was merged with the SCCF/LDCF; see http://www.thegef.org/gef/adaptation, accessed 5 October 2012.

[58] GEF/LDCF.SCCF.9/4/Rev. 1, October 2010, 7–9.

[59] See the LDC Expert Group report *The Least Developed Countries: Support needed to fully implement national adaptation programmes of action (NAPAs)* (UNFCCC 2009) 17.

[60] FCCC/CP/2010/7/Add. 1, 16.

developing countries.[61] These funds should be channelled through, again, a new fund, the Green Climate Fund (hereafter: GCF), established in Cancun in 2010.[62]

The setting up of the GCF, however, saw a slow start, with the first meeting of the Transitional Committee, entrusted with the task of drafting the operating documents of the GCF, postponed to April 2011 because of lack of agreement over the composition of this committee.[63] During COP17, convened in Durban in December 2011, the GCF was officially designated as an operating entity of the financial mechanism under Article 11 of the UNFCCC.[64] The GCF has a Board, which oversees the operation of the fund and a whole range of other more specific tasks[65] and which consists of an equal number of members from developing and developed countries.[66] The GCF's main goal is to 'promote the paradigm shift towards low-emission and climate-resilient development pathways by providing support to developing countries to limit or reduce their greenhouse gas emissions and to adapt to the impacts of climate change'.[67] It will do so by 'channeling new, additional, adequate and predictable financial resources to developing countries' and by catalysing 'climate finance, both public and private, and at the international and national levels'.[68] In the governing instrument, specific mention is made of the role of the GCF on financing adaptation measures:[69]

> In allocating resources for adaptation, the Board will take into account the urgent and immediate needs of developing countries that are particularly vulnerable to the adverse effects of climate change, including LDCs [Least Developed Countries, JV], SIDS [Small Island Developing States, JV] and African States, using minimum allocation floors for these countries as appropriate. The Board will aim for appropriate geographical balance.

---

61 FCCC/CP/2010/7/Add. 1, 17.

62 Ibid.

63 The Transitional Committee delivered its final report to the COP in Durban, FCCC/CP/2011/6.

64 Decision 3/CP.17, FCCC/CP/2011/9/Add.1, 55.

65 Ibid. at 60.

66 Ibid. at 59.

67 Ibid. at 58.

68 Ibid. at 58.

69 Ibid. at 64.

Final arrangements on the location of the GCF were concluded at COP18 in Doha, in December 2012. Here Songdo, Incheon, Republic of Korea, was selected to host the GCF.[70]

4 THE UNFCCC's IMPACT ON REGIONAL AND DOMESTIC ADAPTATION LAW AND POLICY

The call for cooperation in addressing climate change adaptation in Article 4 of the UNFCCC has not been unheard, although regional adaptation efforts have been slow and limited, especially when compared to mitigation efforts. By 2012, many regional organizations drafted policies and programmes detailing their adaptation efforts. One of the first examples is the ClimDev Africa initiative ('Climate for Development in Africa') prepared by the African Union Commission, the UN Economic Commission for Africa and the African Development Bank, and endorsed by African leaders under coordination of the African Union in 2006. The initiative's main aim is to enhance information on Africa's climate, and to integrate climate risk assessment into planning and other decision-making processes throughout the continent, so as to address Africa's limited adaptation capacities due to widespread poverty. In 2010, the African Development Bank established a Special Fund to finance projects under the programme.[71] It was not until December 2011, however, that the inaugural ClimDev Africa conference was held (in Durban, coinciding with UNFCCC COP17).[72]

The EU adopted its first policy document on adaptation only in 2009.[73] The White Paper on adaptation is a rather brief document that mainly proposes to integrate adaptation into existing policies, such as agriculture, fisheries, water and biodiversity. It does not, however, do more than briefly review these policies and conclude that most of them are already well equipped to integrate adaptation measures. Academic literature on

---

70 As reported at UNFCCC website, http://unfccc.int/cooperation_and_support/financial_mechanism/green_climate_fund/items/5869.php, accessed 21 January 2013.

71 See the African Development Bank's website on this fund, http://www.afdb.org/en/topics-and-sectors/initiatives-partnerships/climate-for-development-in-africa-climdev-africa-initiative/, accessed 5 October 2012.

72 See the African Union's website, http://www.au.int/pages/cop17/events/ccda, accessed 5 October 2012.

73 European Commission, White Paper – Adapting to Climate Change: Towards a European Framework for Action, COM 2009/0147 final.

several of these fields, however, shows that this conclusion is too optimistic.[74] It seems that changes to existing EU law are unavoidable, and it also seems that legally binding instruments aimed at the Member States cannot be excluded any longer, since not all EU Member States have adaptation policies in place.[75] Both conclusions now appear to have also been reached within the European Commission, as in October 2012, the European Commission published a proposal for a revised Directive on environmental impact assessment within which several provisions focus on climate change adaptation.[76] In the coming years, when up for general revision, EU legislation will be 'climate proofed'. In addition, in April 2013 the European Commission published its Adaptation Strategy, which, however, did not include a proposal for an Adaptation Directive, requiring EU Member States to develop national adaptation policies.[77] Instead, non-binding guidelines will be issued.

Meanwhile, the COP of the UNFCCC, in Durban in 2011, adopted a lengthy decision on national adaptation plans (NAPs).[78] The decision, among other things, detailed the elements that should be considered while developing NAPs, such as the design and development of plans, policies and programmes to address the weaknesses and gaps in adaptation enabling environments, the assessment of medium- and long-term adaptation needs, the integration of adaptation into national and subnational development and sectoral planning, participatory stakeholder consultations, communication, awareness raising, and education.[79]

---

[74] For example in the field of water management: Andrea M Keessen and Helena F M W van Rijswick, *Adaptation in European Water Law and Policy* (Centre for Environmental Law and Policy, Utrecht University 2012), and on EU biodiversity law: Jonathan Verschuuren, 'Climate Change: Rethinking Restoration in the European Union's Birds and Habitats Directives' (2010) 28(4) Ecological Restoration 431–439.

[75] An overview of the existing domestic adaptation policies is available at the European Commission's website on adaptation, http://ec.europa.eu/clima/policies/adaptation/documentation_en.htm, accessed 5 October 2012.

[76] Proposal of 26 October 2012 for a Directive amending Directive 2011/92/EU, COM(2012) 628 final. See for more detail, Chapter 14.

[77] See http://ec.europa.eu/clima/policies/adaptation/what_is_eu_doing/index_en.htm, accessed 24 May 2013.

[78] Decision 5/CP.17, FCCC/CP/2011/9/Add. 1.

[79] Ibid. at 85.

5 CONCLUSION

Although adaptation to climate change is, together with mitigation, the main aim of the 1992 UNFCCC, adaptation remained a neglected topic within the international arena until well beyond 2007, when the IPCC indicated that, for some impacts, adaptation is the only available and appropriate response. The international community did do some early work on adaptation, such as the adoption of guidelines in 2001 to help developing countries to draft NAPAs and the establishment of the Adaptation Fund in the same year. The lack of attention to adaptation was generated by fear among decision-makers that attention to adaptation would deflect attention from mitigation. It is only since the adoption of the Cancun Adaptation Framework and the establishment of the Adaptation Committee in 2010, that the topic has been placed high on the agenda. Under the legal framework offered by the UNFCCC and Kyoto Protocol, international discussions on adaptation mainly revolve around the distribution of funds to developing countries to finance adaptation measures in these countries. Many such funds now exist, such as the Adaptation Fund, the Least Developed Countries Fund, the Special Climate Change Fund and the newly established Green Climate Fund. All of these funds finance adaptation projects, primarily in developing countries.

3. Adaptation and climate justice

Rosemary Lyster

1 INTRODUCTION

This chapter will contribute a unique perspective on theoretical approaches to climate change justice in the context of climate change adaptation,[1] relying on Amartya Sen's latest book *The Idea of Justice*,[2] which admittedly espouses a general theory of global justice. Sen's conceptualization of global justice, and his insistence that the idea of justice should extend to the actual lives of those affected by injustice, is highly relevant to the issue of climate change adaptation.

Within this overarching framework, the author will acknowledge the many attempts of political philosophers and environmental lawyers to address specific questions of corrective, distributive and procedural justice.[3] Yet the author will argue that Sen's insistence on impartial public reasoning for the achievement of global justice is intrinsically important to notions of climate justice. This builds on the admission by developed countries, at the July 2012 Bonn negotiations, that one of the reasons they find it difficult to commit to the higher end of their pledges is the lack of domestic support for action on climate change. The author will propose that an essential aspect of climate change justice is a 'willingness to pay' on the part of negotiating parties and that this can be derived via Sen's notion of a deliberative global justice. Of course Sen's idea of justice is deeply embedded in his Capability Approach which is highly

---

[1] See also Rosemary Lyster, 'Towards a global justice vision for Climate Law in a time of "unreason"' (2012) 4(1) Journal of Human Rights and the Environment (Rio+20: A Critique) forthcoming.

[2] Amartya Sen, *The Idea of Justice* (Penguin Books 2009).

[3] A view of procedural justice, i.e., recognition, rights of participation in decision-making about climate change treaties, and the distribution of power is discussed in Marco Grasso, 'An ethical approach to climate adaptation finance' (2010) 20 Global Environmental Change 74, 77.

relevant to vulnerability and resilience to climate change and attempts to adapt to it.[4]

The adoption of Sen's Capability Approach necessarily invokes Martha Nussbaum's distinctive Capabilities Approach encapsulated in *Creating Capabilities: The Human Development Approach*.[5] However, Sen's approach is preferred over Nussbaum's because it avoids the risk of listing some Capabilities[6] and missing highly significant others, such as the inherent rights of ecosystems and their role in supporting human capabilities.[7] Also, Sen's rejection of a Rawlsian faith in 'just institutions' for achieving justice must surely call into question Nussbaum's faith in legal institutions such as the Constitution, judicial review and legislative debate to satisfy the procedural aspects of a Capabilities Approach.[8]

In establishing this argument, the author will engage with: the latest consensus on climate science; recent climate change multilateral negotiations; the expected impacts of climate change on cities and food security in developed and developing economies; the projected costs of adapting to these human and ecological impacts; and the emerging convergence between climate change, disaster reduction strategies and sustainable development at the international level. The concerns for sustainable development in the face of climate change bring Sen's Capability Approach to the fore. The truth is that there is no way for any jurisdiction to avoid the costs of climate change. Even if some economies attempt to shy away from their obligations to bear the costs of mitigation, none will escape paying the adaptation and disaster response costs, particularly for developing countries which are most vulnerable. From a climate justice perspective, the question is who should pay and why.

---

4 Note that a Senian approach to the allocation of adaptation funds on account of a 'lack of human security' has been developed elsewhere; see Marco Grasso, above note 3.

5 Martha C Nussbaum, *Creating Capabilities: The Human Development Approach* (The Belknap Press of Harvard University Press 2011).

6 These include: Life; Bodily health; Bodily integrity; Senses, imagination, and thought; Emotions; Practical reason; Affiliation; Other species (being able to live with concern for and in relation to animals, plants, and the world of nature); Play; Control over one's environment; ibid. at 34.

7 See, for example, Breena Holland, 'Justice and the Environment in Nussbaum's "Capabilities Approach"' (2008) 61(2) Political Research Quarterly 319.

8 Sen, above note 2 at 67.

2 THE CURRENT CONSENSUS ON CLIMATE SCIENCE

Any attempt to construct a climate justice vision for adaptation through impartial public reasoning must proceed from a sound understanding of climate science and the expected impacts of climate change on humanity and the planet. This is because science still seems to have the authority to make definitive and universal statements about what is and what is not dangerous for people and societies, and ultimately the world.[9]

Between 2009 and 2011, the climate science presented by the Intergovernmental Panel on Climate Change (IPCC) in its Fourth Assessment Report (hereafter: AR4)[10] was confirmed and indeed updated by many of the world's leading scientists and science institutions in the world.[11]

The most recent iteration of the climate science is that by the Australian Climate Commission entitled *The Critical Decade: Climate science, risks and responses*.[12] Crucially, the Report states that recent observations of changes in the climate system strengthen the conclusions of the IPCC's AR4 that contemporary climate change is real, and that, compared with geological time scales, it is occurring at a rapid rate. The

---

[9] Ibid. at 74.

[10] M L Parry et al. (eds), *Climate Change 2007: Impacts, Adaptation and Vulnerability. Contribution of Working Group II to the Fourth Assessment Report of the Intergovernmental Panel on Climate Change* (Cambridge University Press 2007).

[11] See James Hansen, *Storms of My Grandchildren: The Truth About The Coming Climate Catastrophe and our Last Chance to Save Humanity* (Bloomsbury 2009); US Environmental Protection Agency, *Endangerment Finding for Greenhouse Gases* (2009); US National Oceanic and Atmospheric Administration's report *2009 The State of the Climate*; Australian National Academy of Science, *The Science of Climate Change: Questions and Answers* (2010); US National Academies of Sciences, *Advancing the Science of Climate Change* (2010); The Royal Society, *Climate change: a summary of the science* (2010). Meanwhile, two of the most prominent rebuttals of climate science sceptics include the Earth Institute Columbia University's *Climate Change: Addressing the Major Skeptic Arguments*, commissioned by Deutsche Bank, and *Climate Scientists Respond to Written Testimony of Monckton before U.S. Select Committee on Energy Independence and Global Warming,* http://www.skepticalscience.com/Monckton-response.pdf, accessed 24 January 2013. Both of these were published in September 2010.

[12] Available at http://climatecommission.gov.au/topics/the-critical-decade/, accessed 20 September 2012.

rate of climate change is already discernible, creating significant risks today and more serious risks in the coming decades.[13]

2.1 Basic Climate Science Facts

The basic facts are the following:

- the average air temperature at the Earth's surface continued to rise at a rate of 0.17°C per decade over the past three decades;
- the temperature of the upper 700m of the ocean continues to increase, with most of the excess heat generated by the growing energy imbalance at the Earth's surface stored in this compartment of the system;
- the alkalinity of the ocean is decreasing steadily as a result of acidification by human-induced CO_2 emissions;
- net loss of ice from the Greenland and the West Antarctic ice sheets is confirmed by recent observations; the extent of Arctic sea ice cover continues on a long-term downward trend. Most land-based glaciers and ice caps are in retreat;
- sea level has risen at a higher rate over the past two decades, consistent with ocean warming and an increasing contribution from the large polar ice sheets;
- the biosphere is responding in a consistent way to a warming Earth, with observed changes in gene pools, species ranges, timing of biological patterns and ecosystem dynamics.[14]

The Report states that '[t]here is a very large body of internally consistent observations, experiments, analyses, and physical theory that points to the increasing atmospheric concentration of greenhouse gases, with carbon dioxide (CO_2) the most important, as the ultimate cause for the observed warming'.[15] The Report also notes that claims that changes in incoming solar radiation can be the cause of the current warming trend lack any credible evidence. Furthermore, the globally coherent warming trend observed since the middle of the twentieth century is not explained by either multi-decadal or century-scale patterns of natural variability, such as the Medieval Warm Period, or shorter term patterns of variability,

---

13 Ibid. at 6.
14 Ibid.
15 Ibid. at 13.

such as ENSO (El Niño-Southern Oscillation) or the North Atlantic Oscillation.[16]

2.2 Biodiversity

Although the Report does not refer specifically to biodiversity, according to the IPCC[17] there is high confidence that climate change will result in extinction of many species and reduce the diversity of ecosystems, partly because the expected rapid rate of climate change outpaces the resilience of many such systems. Their resilience is already substantially compromised as a result of human development. Climate change will exacerbate this through blocked migration routes, fragmented habitats, reduced populations, introduction of alien species and stresses related to pollution. Biodiversity losses are likely to increase as further warming is likely to cause additional adverse impacts to many ecosystems. Some examples of mass mortalities with warming above 2°C and 3°C include: coral reefs globally, polar bears and walruses – up to 30 per cent of known species being committed to extinction.

In recognition of this, the Conference of the Parties to the Convention on the Conservation of Biological Diversity has recently called on all Parties to 'identify, monitor and address the impacts of climate change and ocean acidification on biodiversity and ecosystem services, and assess the future risks for biodiversity and the provision of ecosystem services using the latest available vulnerability and impact assessment frameworks and guidelines'.[18] Writers, like Eric Posner and David Weisbach,[19] who claim that climate treaties like the Kyoto Protocol are focused on distributing wealth to the poor, fail to adequately acknowledge that their objective is in fact the survival of the planet, all of its human and non-human species and associated ecosystems.

---

16 Ibid.

17 See section 19.3.4 'Ecosystems and biodiversity' in S H Schneider et al., 'Assessing key vulnerabilities and the risk from climate change' in Parry et al., above note 10 at 779–810.

18 Decision X/33. Biodiversity and climate change, UNEP/CBD/COP/DEC/X/33, 29 October 2010.

19 Eric Posner and David Weisbach, *Climate Change Justice* (Princeton University Press 2010).

2.3 Extreme Events and Abrupt, Non-linear and Irreversible Changes in the Climate System

Modest changes in average values of climatic parameters – for example, temperature and rainfall – can lead to disproportionately large changes in the frequency and intensity of extreme events. Also, a number of potential abrupt changes in large sub-systems or processes in the climate system, known as 'tipping elements', have been identified primarily through palaeo-climatic research. If triggered, many of these would lead to catastrophic impacts on human societies. Abrupt shifts in atmospheric circulation can occur very quickly and have large impacts on regional climates as evidenced by the recent cold, snowy winters in northern Europe, and their possible link to climate change.[20]

2.4 Where are the Uncertainties?

Although the fundamental features of climate change are very well known, significant uncertainties surround scientists' understanding of the behaviour of important parts of the climate system. These include: the ways in which the large polar ice sheets on Greenland and Antarctica are responding and will respond in future; the acceleration of the hydrological cycle as the climate warms; and the influence of climate change on spatial patterns of precipitation across the Earth's surface and on the temporal patterns of precipitation, such as droughts and intense rainfall events. These uncertainties, however, in no way diminish climate scientists' confidence in the observation that the Earth is warming and that this is primarily human-induced.[21]

As the Earth Institute at Columbia University observes, uncertainty is inevitable when trying to understand any system for which perfect knowledge is unattainable.[22]

3 WHAT HAVE THE LATEST CLIMATE NEGOTIATIONS DELIVERED?

The recent scientific consensus on climate science must now be read together with the IPCC's Special Report Managing the Risks of Extreme Events and Disasters to Advance Climate Change Adaptation (SREX

---

20 Ibid. at 48.
21 Ibid. at 41.
22 Above note 11 at 7.

Report),[23] released in March 2012. This Report states that there is sufficient evidence to suggest that climate extremes such as heat waves, record high temperatures and, in many regions, heavy precipitation have changed due to climate change in the past half century. Climate-related disasters can result from climate extremes, or even a series of non-extreme events, which occur in combination with social vulnerabilities and exposure to risks. The impact of these events can be reduced, and the resilience of people exposed to extreme events increased, if policies to avoid, prepare for, respond to and recover from the risks of disaster are adopted. However, when thresholds or tipping points associated with social and/or natural systems are exceeded, limits to resilience will be reached, posing severe challenges for adaptation.

The SREX Report reflects the concern that the current multilateral negotiations for a post-2012 agreement are likely to be insufficient for avoiding or even reducing certain features of climate change. So what have the most recent negotiations delivered? The Conference of the Party (COP) negotiations in December 2010 concluded in the Cancun Agreements,[24] which essentially confirmed the parameters of the December 2009 Copenhagen Accord.[25] The Ad Hoc Working Group on Long-term Cooperative Action under the Convention (AWG-LCA) text,[26] emerging from COP 17, essentially built on agreements reached at COP 16. COP 16 and COP 17 both recognized the need to hold the increase in global average temperatures to below 2°C above pre-industrial levels, as well as the need to consider strengthening the long-term goal in relation to a global average temperature rise of 1.5°C on the basis of the best available scientific knowledge.[27] A decision on the future of the Kyoto Protocol was not made at COP 16 but agreement on this was reached at COP 17. A decision was taken under the auspices of the Ad Hoc Working Group on Further Commitments for Annex I Parties under the Kyoto Protocol (AWG-KP)[28] that a second commitment period of the Protocol

---

[23] Ch B Field et al. (eds), *Managing the Risks of Extreme Events and Disasters to Advance Climate Change Adaptation: Special report of the Intergovernmental Panel on Climate Change* (Cambridge University Press 2012).

[24] Decision 1/CP.16, FCCC/CP/2010/7/Add.1; Decision 1/CMP.6, FCCC/KP/CMP/2010/12/Add.1; Decision 2/CMP.6, FCCC/KP/2010/12/Add.1.

[25] FCCC/CP/2009/11/Add.1, 30 March 2010.

[26] Decision 2/CP.17, FCCC/CP/2011/9/Add.1.

[27] Ibid. Art. 4. See also Establishment of an Ad Hoc Working Group on the Durban Platform for Enhanced Action, Decision 1/CP.17, FCCC/CP/2011/9/Add.1.

[28] Decision 1/CMP.7, FCCC/KP/CMP/2011/10/Add.1.

will begin on 1 January 2013 and end on 31 December 2020, as was decided by the AWG-KP in December 2012.[29] It was decided at COP 17 to establish a new body, the Ad Hoc Working Group on the Durban Platform for Enhanced Action.[30] This Working Group started its work as a matter of urgency in the first half of 2012 and must complete its work no later than 2015: 'to adopt a Protocol, or other legal instrument or an agreed outcome with legal force under the UNFCCC'. The Protocol must be adopted at COP 21 in 2015 and come into effect from 2020.

At COP 16 and COP 17 it was acknowledged that developed country Parties must take the lead on emissions reductions because of their historical responsibilities. The COPs took note of the voluntary Copenhagen quantified economy-wide emissions reduction targets of Annex I countries. It is well known that developing countries gave no commitments to emissions reductions under the Kyoto Protocol, consistently with the principle of common but differentiated responsibility. While the principle of common but differentiated responsibility is still alive and well, there has been a distinct shift with regard to the negotiations for a post-2012 world. While COP 17 recognizes that social and economic development and poverty eradication are the first and overriding priorities of developing economies,[31] and that the share of global emissions originating in developing countries will grow to meet their social and developmental needs, a low-emissions development strategy is nevertheless regarded as central to sustainable development.[32] COP 17 noted the voluntary commitments made by developing countries under the Copenhagen Accord, and confirmed at COP 16, to prepare and implement nationally appropriate mitigation actions (NAMAs).[33] COP 17 agreed to continue consideration of a general framework for cooperative sectoral approaches and sector-specific action with a view to making a decision at COP 18 in 2012.[34] Also COP 16 and COP 17 referred to the possibility of a new market-based mechanism to assist developed countries to meet their commitments under the UN Framework Convention on Climate Change (UNFCCC).[35] This acknowledges that the Clean Development Mechanism (CDM) market has experienced difficulties over the years

---

29 The so-called Doha amendment, see http://unfccc.int/kyoto_protocol/doha_amendment/items/7362.php, accessed 24 January 2013.

30 Decision 1/CP.17, FCCC/CP/2011/9/Add.1.

31 Ibid. Art. 87.

32 Ibid. Part B, Preamble.

33 Ibid. Part B, Preamble.

34 Ibid. s. 74.

35 Ibid. Art. 83.

with registering projects, issuing credits, and matters of governance. These continue in spite of efforts of the CDM Executive Board to reform the mechanism.[36]

Understandably, then, there is considerable uncertainty about whether the global community will attain sufficient emissions reduction before the worst impacts of climate change eventuate.

4 UNDERSTANDING VULNERABILITY, ADAPTIVE CAPACITY AND RESILIENCE

What has been achieved through mitigation in the context of current multilateral negotiations is likely to be insufficient for avoiding or even reducing certain features of climate change. Although as Chapter 2 indicates, adaptation is now at the forefront of multilateral negotiations on climate change, there are a number of reasons why adaptation was not given as much attention as mitigation in the early stages of negotiations. A focus on adaptation was regarded as 'defeatist' and a distraction from building a case for stringent mitigation responses. Also, in 1992 when the UNFCCC was established, there was a view that identifying adaptation options would be tantamount to accepting that climate change was occurring at a time when the science was more tenuous that it is now. This was compounded by a belief that ultimately adaptation would come from technological solutions rather than as a result of mitigation and multilateral negotiations. From a developed country perspective there was a concern that accepting the need for adaptation amounted to an implicit assumption of responsibility, with the associated duty to compensate. Finally, many developing countries were reluctant to discuss adaptation lest it derail developed country commitments to mitigation.[37]

However, interest in adaptation has surged given that 'the world has recently witness[ed] a series of weather events so extreme that they are at

---

[36] Note that COP 17 has issued a decision on 'Further guidance to the CDM' which calls for enhanced action on governance, baseline and monitoring methodologies and additionality, the efficiency of the project cycle, and support for underrepresented host countries, Decision 8/CP.7, FCCC/KP/CMP/2011/10/Add.2. A 'materiality standard' under the CDM has also been decided on, Decision 9/CP.7, FCCC/KP/CMP/2011/10/Add.2. The modalities and procedures for carbon capture and storage as a CDM were also established at COP 17, Decision 10/CP.7, FCCC/KP/CMP/2011/10/Add.2.

[37] See E Lisa F Schipper, 'Conceptual History of Adaptation in the UNFCCC Process' in E Lisa F Schipper and Ian Burton (eds), *Adaptation to Climate Change* (Earthscan 2009) 362.

the limits of modern human experience'.[38] These extremes are undercutting development efforts where many affected communities are already poor and vulnerable.[39]

Before analysing the intimate relationship between climate justice and climate change adaptation, there are a number of important definitions that need to be addressed, including: vulnerability, capacity, adaptive capacity, coping and resilience. Vulnerability is defined in the SREX Report as the propensity or predisposition to be adversely affected. It must be viewed in the context of a wide range of factors, including diverse historical, social, economic, political, cultural, institutional, natural resource and environmental conditions. These all affect a group's, or individual's, capacity to anticipate, cope with, resist and recover from the adverse effects of physical events.[40] While the physical causes of vulnerability, such as climate change, are an explicit aspect of vulnerability, notions of sensitivity and adaptive capacity are integral to the social context of climate change.[41]

How are vulnerability and capacity related? Capacity refers to the combination of all the resources available to humanity to achieve established goals. These include the conditions and characteristics that give society access to and use of social, economic, psychological, cultural and natural resources, as well as to the institutions of governance necessary to reduce vulnerability and deal with the consequences of disaster. A relative, rather than an absolute, lack of capacity may be regarded as one dimension of overall vulnerability, in that it increases vulnerability.[42]

Adaptive capacity, on the other hand, is used in a specific context to explain the capacity to adapt to the causes and impacts of climate change. The notions of resilience and coping are fundamental to understanding vulnerability to, and adaptive capacity to cope with, climate change.[43] Coping means the use of available skills, resources and opportunities to address, manage and overcome adverse conditions, in the short to medium term, with the aim of achieving basic functioning.

Meanwhile, resilience is the ability of a system and its component parts to anticipate, absorb, accommodate or recover from the effects of a

---

38 See *Decision Making in a Changing Climate: Adaptation Challenges and Choices* (Word Resources 2010–2011) 13.

39 Ibid. at 29.

40 Above note 23 at 32.

41 Ibid. at 33.

42 Ibid.

43 Ibid.

potentially hazardous event in a timely and efficient manner. This includes ensuring the preservation, restoration or improvement of a system's essential basic structures and functions. While older conceptions of resilience refer to 'bouncing back', this is an increasingly insufficient goal for climate disaster risk management. This is because of the dynamic and often uncertain consequences of climate change (as well as development trends such as urbanization) for hazard and vulnerability profiles. Recent conceptions of resilience of social-ecological systems focus more on process than outcomes, including the ability to self-organize, learn and adapt over time.[44]

There is also a link between adaptation and the coping range. Within its coping range, a community can survive and even thrive with significant natural hazards, especially where the historical distribution of hazard intensity is well known and relatively stable. A community's coping range is partly determined by prior adaptation and, when adaptation efforts have matched its coping range with the range of hazards it typically encounters, a community has the best chance to survive and thrive. As climate change alters the occurrence of extreme events and the vulnerability of humans is shaped by societal trends, adaptation is required to adjust the coping range so as to maintain societal functioning within an expected or acceptable range of risk.[45]

Invoking the Capability Approach to climate justice is especially justified when one considers the fact that the capacity to manage risks and adapt to change is unevenly distributed within and across nations, regions, communities and households. When disasters strike, the poor quickly exhaust limited resources, further undermining household sustainability. In the long run, this reduces capital and increases hazard exposure or vulnerability, while the poverty and vulnerability trap renders recovery to pre-disaster levels of well-being increasingly difficult. Children, the elderly and women are more vulnerable to extreme climate and weather events. Individual and household vulnerability also arises due to other instances of inequality such as race, caste, religious affiliation and physical disability, which may intersect with gender and age effects.[46]

---

[44] Ibid. at 34.
[45] Ibid. at 52.
[46] Ibid. at 456.

5 DESTROYING CAPABILITIES, VULNERABILITY AND RESILIENCE

The SREX Report states that its definition of capacity extends the definition of capabilities referred to in Amartya Sen's 'capabilities approach to development'. Yet, as already indicated, the author finds Sen's Capability Approach highly relevant to the present analysis of climate change justice in the context of climate adaptation.

Sen's Capability Approach, found in an extensive collection of writings, has developed over many decades during which time his approach has been gradually refined. The other main protagonist and theorist in this area is Martha Nussbaum, although as the ensuing discussion will show her definitions and emphases differ in certain respects from Sen's. The life work of Sen and Nussbaum cannot be adequately represented here. However, for present purposes, Sen's work *The Idea of Justice* is relied on extensively, as is Nussbuam's *Creating Capabilities*.

Sen's and Nussbaum's essential question is: 'What is this person able to do and to be?' The Capability, or Capabilities (as preferred by Nussbaum), Approach answers this question.[47] Capability has two essential tenets: freedom and functionings ('beings and doings that people value and have reason to value').[48] For Sen, freedom has two aspects – process aspects (human beings are seen as active agents, directing their own lives and acting as agents that further larger social goals and objectives), and opportunity aspects (ability to choose and achieve valued functionings in their own lives).[49] The Approach differs from a tradition in economics that measures the real value of a set of options by the best use that can be made of them. Rather, options are regarded as freedoms, and freedom has intrinsic value.[50]

As Nussbaum has recently explained, the Approach takes each person as an end and asks about the opportunities available to each person. It is not concerned with total or average well-being and people's powers of self-definition are respected.[51] It is resolutely pluralist about value in the sense that the central capability achievements are different in quality, not just quantity. Finally, it is concerned especially with capability failures

---

[47] Nussbaum, above note 5 at 20.

[48] See Sabine Alkire 'The Capability Approach to Quality of Life' (2008) 5, http://www.stiglitz-sen-fitoussi.fr/documents/capability_approach.pdf, accessed 20 September 2012.

[49] Ibid.

[50] Nussbaum, above note 5 at 25.

[51] Ibid. at 18.

that are the result of discrimination or marginalization and which result in entrenched social injustice and inequality. Government and public policy are vested with an urgent task to improve the quality of life for all people, as defined by their capabilities.

Some of the essential differences between Sen's and Nussbaum's approaches may be loosely summarized as follows: Nussbaum is more explicit about including non-human animals in the Capabilities Approach; Nussbaum provides a list of ten basic Capabilities while Sen focuses rather on 'quality of life' issues;[52] Nussbaum's selection of capabilities is closely bound up with dignity and regards Constitutional Law and public policy as agents in a nation's quest for social justice (or for deriving goals for the community of nations);[53] and Nussbaum's focus on legal institutions leads her to conceive of the procedural value of the Capability/ies Approach very differently from Sen. This last difference stems largely from Nussbaum's belief in a large overlap between human rights and Capabilities, compared with Sen's view that a Capability Approach can include matters of procedure (where one is able to engage in a certain sort of process) whereas rights are always a matter of substantive opportunity (what one is actually able to have).[54]

Suffice it to say that when the science on climate change and consideration of vulnerability and the ability to adapt are taken into account it is very likely that climate change will affect in very real ways what 'a person is able to be and to do'.

6 WHAT ARE SOME OF THE CHALLENGES TO CAPABILITIES AND ADAPTATION?

While the 2011 Critical Decade report speaks to the science of climate change, and the SREX Report looks at questions of adaptation and disaster management, there are two issues which are currently dominating the concerns of multilateral international agencies. These include the impacts of climate change on cities, especially coastal and river cities, and food security. The implications of this for both developed and developing economies are obvious and should, like climate science, be factored into impartial public reasoning on climate change justice.

---

52 Ibid. at 19.
53 Ibid. at 29.
54 Ibid. at 67.

6.1 Cities

UN Habitat's *Cities and Climate Change: Global Report on Human Settlements 2011*,[55] released in March 2011, sets out some of the possible effects of climate change on the world's cities including the following:

- by 2050, climate change will displace as many as 200 million people;
- by the 2080s sea level rise and its associated impacts will affect five times as many coastal residents as in 1990;
- in coastal North African cities, 6 to 25 million people could be exposed to flooding from sea level rise if temperatures rise between 1 and 2°C;
- by 2070, the top ten cities exposed to flooding will be located in developing countries (particularly in China, India and Thailand);
- by 2070, up to 150 million people could be living in a 100-year flood plain with the financial impact rising from USD 3 trillion in 1999 to USD 38 trillion by this time;
- by the 2020s, 12 to 81 million residents in Latin America could experience increased water stress and by the 2050s this could reach to 79 to 178 million.[56]

Meanwhile, the World Bank's 2010 report *Climate Risks and Adaptation in Asian Coastal Megacities*[57] states that 13 of the world's 20 largest cities are located on the coast, and more than a third of the world's population live within 100 miles of a shoreline.[58] By 2070, nine of the top ten most populous and exposed cities are expected to be in Asian developing countries. Adaptation poses an enormous challenge for flood-prone cities, such as Ho Chi Minh City, Kolkata, Dhaka and Manila, facing potential sea level rise and increased frequency and intensity of extreme weather events, with the urban poor most at risk from exposure to hazards. Despite this, few developing country cities have attempted to incorporate climate change systematically into their decision-making processes.[59]

---

[55] UN-HABITAT, *Cities and Climate Change. Global Report on Human Settlements 2011* (Earthscan 2011).

[56] Ibid. at 73.

[57] The World Bank, *Climate Risks and Adaptation in Asian Coastal Megacities: A Synthesis Report* (The World Bank 2010).

[58] Ibid. at xi.

[59] Ibid. at xi.

The report selects three cities – Manila, Ho Chi Minh City and Bangkok – for close study as all are coastal megacities with populations (official and unofficial) ranging from 8 to 15 million people.[60] For Bangkok, the actual costs of a 1-in-30-year flood – including costs resulting from both climate change and land subsidence – will be close to USD 4.6 billion in 2050, with climate change (in a high emission scenario) adding USD 1.5 billion, or approximately 2 per cent of gross regional domestic product (GRDP). In Manila, a similar 1-in-30-year flood can lead to costs of flooding ranging from USD 0.9 billion – given current flood control infrastructure and climate conditions – to USD 1.5 billion with similar infrastructure but a high emission climate scenario. Thus, the additional climate change costs would be approximately USD 0.65 billion or 6 per cent of GRDP. Although the Ho Chi Minh City study adopts a different methodology to analyse costs and so cannot be directly compared to the costs of Manila and Bangkok, the flood costs to Ho Chi Minh City, in present value terms, range from USD 6.5 to 50 billion.[61]

6.2 Food Security

To meet the expected demand for food in 2050 without significant price increases, and in light of the growing impacts of climate change, amongst other external impacts on food production, the world needs to increase food production by 70 to 100 per cent.[62] This needs to be viewed against the background of the 2008 food-price hike leading to food riots and political change in several countries, as well as the excessive heat and drought in Russia leading to wildfires[63] and a grain embargo and the unprecedented floods in Pakistan. Food demand will inevitably increase

---

[60] Ibid. at xii.

[61] Ibid. xiv–xv.

[62] See J Pretty et al., 'The top 100 questions of importance to the future of global agriculture' (2010) 8(4) International Journal of Agricultural Sustainability 219, 220.

[63] See also Jerry Williams et al., *Findings and Implications from a Coarse/Scale Global Assessment of Recent Selected Mega-Fires*, 5th International Wildland Fire Conference, Sun City, South Africa, 9–13 May 2011, http://foris.fao.org/static/pdf/fm/5thIWFConference2011.pdf, accessed 20 September 2012. The authors report on 'mega-fires', which are extraordinary conflagrations unprecedented in the modern era for their deep and long-lasting social, economic, and environmental impacts. Global warming is identified as one of the cumulative effects changing wildfire protection in many countries.

as the world population reaches 9 billion by 2050.[64] Since food production is critically dependent on local temperatures, and precipitation climate change will only add to these pressures, the capacity for farmers to adapt their practices is questionable.[65] By 2050, climate change is likely to increase the number of malnourished children in all developing countries by 8.5 per cent in the optimistic scenario and 10.3 per cent in the pessimistic scenario.[66]

One of the most recent reports on this issue is Food Security and Climate Change: A Report by The High Level Panel of Experts on Food Security and Nutrition[67] published in June 2012. The report acknowledges that food security vulnerability caused by climate change begins at the individual farm level with biophysical impacts on plants and animals, and the systems in which they are managed. Livelihoods are directly altered by these impacts in rural areas and indirectly altered in urban areas. The effects of climate are transmitted elsewhere via international markets, which can affect local food security, for better or worse, by altering domestic prices and influencing livelihoods. Climate change involving changes in temperature and rainfall regimes may have a considerable impact on agricultural productivity. It will result in multiple stresses for animals and plants in many agricultural and aquatic systems in the coming decades.

The poor and other vulnerable groups are likely to be at high risk to climate change-induced food insecurity. The poor includes small-holders and landless people in rural areas, most likely women and children, and marginalized ethnic and indigenous peoples, although urban poverty is growing with the rapid urbanization of the poor. Geographically, the vast majority of the poor are located in two regions – sub-Saharan Africa and South Asia – where climate change is likely to impact most profoundly.[68] Vulnerability to food insecurity is determined by both biophysical and socio-economic factors making the poor more exposed to the effects of climate change.[69]

---

64 Ibid. at 1.

65 Ibid. at 2.

66 Ibid. at 7.

67 See http://www.fao.org/fileadmin/user_upload/hlpe/hlpe_documents/HLPE_Reports/HLPE-Report-3-Food_security_and_climate_change-June_2012.pdf, accessed 20 September 2012; see also IFPRI, *2011 Global Hunger Index: The Challenge of Hunger: Taming Price Spikes and Excessive Food Price Volatility* (IFPRI 2011).

68 FAO report, above note 67 at 12.

69 Ibid.

Dryland agriculture in arid and semi-arid regions is particularly vulnerable to the risks of climate change and variability, particularly drought. Yet this is where over 40 per cent of the world's population and more than 650 million of the poorest and most food insecure people live. At the same time, significant agricultural production takes place in low-lying coastal areas in some regions of the world, where current population densities are high. Here, saline intrusion, sea level rise and increased flooding caused by climate change are a major threat. Given some of the uncertainties about the local and regional impacts of climate change, adaptation should be directed at building a more resilient food system.[70]

Consequently, the report makes the following recommendations: integrate food security and climate change concerns; increase resilience of food systems to climate change; develop low-emissions agricultural strategies that do not compromise food security; collect information locally, share knowledge globally and refocus research to address a more complex set of objectives; and facilitate the participation of all stakeholders in decision-making and implementation.[71]

7 WHO WILL PAY THE COSTS OF ADAPTATION AND WHAT ARE THEY LIKELY TO BE?

In 2007, the UNFCCC estimated that by 2030 the global costs of adaptation would be USD 49 to 171 billion per annum, with USD 22 to 105 billion attributed to adaptation in developed countries and USD 27 to 66 billion to developing countries. However, a recent study has found that the UNFCCC has underestimated investment needs by a factor of between 2 and 3 for the included sectors. If ecosystems are included, a further USD 65 to 300 billion per year must be added. Mining and manufacturing, energy, the retail and financial sectors, and tourism are not included in estimates.[72] Judging by these estimates, the financial commitments made under the Cancun Agreements are woefully inadequate.

Essentially, as discussed in Chapter 2, the COP has taken note of the collective commitment by developed countries to provide USD 30 billion

---

[70] Ibid. at 12.

[71] Ibid. at 14–22.

[72] Martin Parry et al., *Assessing the costs of adaptation to climate change: A review of the UNFCCC and other recent estimates* (Imperial College and International Institute for Environment and Development 2009).

in fast-start funding for the period 2010 to 2012.[73] In the interests of transparency, developed countries are invited to report in May 2011, 2012 and 2013 to the UNFCCC secretariat on the resources they have provided.[74] With regard to long-term finance, the COP has decided in accordance with the UNFCCC that scaled-up, new, predictable and adequate funding must be provided to developing country Parties, taking into account the urgent and immediate needs of developing countries that are particularly vulnerable to the adverse effects of climate change.[75] Developed countries will commit USD 100 billion per year by 2020 to address these needs,[76] with the funds coming from public and private, bilateral and multilateral, and alternative sources.[77] The COP has decided that a significant share of new multilateral funding for adaptation should flow through the Green Climate Fund, discussed in Chapter 2.[78]

8 CLIMATE CHANGE, DISASTER RISK REDUCTION AND SUSTAINABLE DEVELOPMENT CONVERGE

The justification for relying upon a Capability Approach and a global justice vision for climate change adaption emerges not only from the SREX Report and concerns about cities and food security but also from the convergence at the international level between climate change, disaster risk reduction and sustainable development.

The *2011 Global Assessment Report on Disaster Risk Reduction*[79] indicates that disasters continue to destroy the lives and livelihoods of millions of people. The January 2010 earthquake in Haiti and July 2010 floods in Pakistan highlighted the close link between disaster risk and poverty. Meanwhile, the 2011 floods in Australia, the earthquakes in New Zealand and Japan, and the tsunami and nuclear disaster in Japan are stark reminders that developed countries are also highly exposed. Hundreds of smaller climate variability disasters also caused extensive

---

73 Above note 24, Art. 95.
74 Ibid. Art. 96.
75 Ibid. Art. 97.
76 Ibid. Art. 98.
77 Ibid. Art. 99.
78 Ibid. Art. 100.
79 United Nations International Strategy for Disaster Reduction Secretariat, *2011 Global Assessment Report on Disaster Risk Reduction* (United Nations 2011). See also *Natural Hazard Unnatural Disasters: The Economics of Effective Prevention* (The World Bank and the United Nations 2010). See also the SREX Report mentioned above note 23.

damage in Benin, Brazil, Colombia, the Philippines and other countries in 2011.

The report states that mortality risk for all weather-related hazards continues to be concentrated in countries with low gross domestic product (GDP) or weak governance, although globally the risk is declining. Since 1970, populations 'at risk' to exposure from tropical cyclones have tripled while the challenges inherent in reducing flood risks were evident in the August 2010 Pakistan floods, which resulted in 1,700 fatalities.

Economic loss associated with floods and cyclones increased in all regions with the Pakistan flood alone accounting for USD 9.7 billion of damage to infrastructure, farms, homes and other direct and indirect losses. Of extensive disaster losses, defined as low severity losses associated with high-frequency events, 97 per cent are weather related. While mortality is usually low, these disasters are responsible for major damage to local infrastructure, housing, and the livelihoods of low-income households and communities. The damage associated with highly localized flooding, landslides, fires and storms in low- and middle-income countries impacts on economic growth. For example, since the 1990s, the number of houses destroyed relative to population growth in 21 countries has increased six-fold, far faster than the economic losses caused by major hazards. The downstream impacts of extensive disasters on health, education, structural poverty and displacement go unaccounted for in most countries, so masking the real costs of disasters.[80] Disasters also lead to large-scale internal displacement, with the Pakistan floods leaving 6 million people in need of shelter.[81]

The report finds that, globally, drought risk is a hidden disaster risk with impacts on agriculture affecting other economic sectors and yet it is poorly understood and badly managed. Only a few countries document drought losses or have a national policy to address the risks even though recent droughts saw agricultural yields reduced by 20 to 40 per cent in the Caribbean, losses of USD 2.34 billion in Australia and 75 per cent of farmers suffering total crop failure in the Syrian Arab Republic in one season.[82] The report notes that along with other factors, such as poverty and vulnerability, decreasing rainfall, climate variability and climate change are key factors of drought risk.[83]

---

80 *2011 Global Assessment Report*, above note 79 at 4.
81 Ibid. at 5.
82 Ibid.
83 Ibid. at 6.

The financial exposure that governments face from extensive disasters is staggering. Calculations for disaster loss in Colombia show that to address losses in publicly owned assets as well as uninsured private assets of low-income groups, ranges from USD 100,000 some 100 times per year for smaller events, to USD 1 billion once every 30 years for big disasters. In Mexico, excluding impacts from droughts, the government is likely to incur weather-related disaster losses of more than USD 1 million at least 50 times a year and more than USD 1 billion at least once every six years. The government in Nepal will need to make good losses amounting to USD 1 million ten times each year for smaller disasters, to USD 100 million every second year for large scale events.[84] It is no surprise that, as the report notes, in reality few governments meet their responsibilities for covering substantial losses and these are usually transferred to and borne by low-income households.[85]

In May 2011, the Global Platform for Disaster Risk Reduction reported at its bi-annual conference[86] that disaster-related losses are increasing across all regions, threatening the economies of low- and middle-income households as well as outpacing wealth gains across many of the world's more affluent nations. Disasters happen because development patterns are not adjusting to more frequent and extreme catastrophes and emerging risks such as climate change, rapid urbanization and insufficient disaster management and planning. At the conference, major insurers[87] called on the industry to assist government with: analysing the root causes of non-resilient activity; encouraging the private sector to share its expertise in construction, communications, financing, transport and contingency planning; sharing knowledge about risk prediction, forecasting and early warning; conducting risk assessments; and helping establish standards and procedures to enhance resilience. Insurers' interest in this area is obvious given that between 2010 and 2011, the payout for global catastrophes was USD 37 billion.[88]

Not surprisingly, in February 2011, the United Nations General Assembly adopted a Resolution on an International Strategy for Disaster Reduction.[89] This followed a report presented to the General Assembly in

---

84 Ibid. at 8.

85 Ibid. at 8.

86 See http://www.unisdr.org/archive/19873, accessed 20 September 2012.

87 These include WillisRe, Credit Suisse, Titan America, Cisco Internet Business Solutions Groups and MunichRe Foundation.

88 E Johnston 'Global disasters take toll on local insurers', *Sydney Morning Herald*, 24 March 2011.

89 A/RES/65/157, 25 February 2011.

December 2010 entitled Sustainable development: International Strategy for Disaster Risk Reduction.[90] The General Assembly expressed 'its deep concern at the number and scale of natural disasters and their increasing impact in recent years, which have resulted in massive loss of life and long-term negative social, economic and environmental consequences for vulnerable societies throughout the world, … in particular in developing countries'. Likewise it expressed 'its deep concern … at the increasing challenges facing the disaster response and preparedness capacity of Member States and the United Nations system as a result of the combined impacts of … the global economic and financial crisis, climate change and the food crisis'.

The General Assembly recognized: 'the clear relationship between sustainable development, poverty eradication, climate change, disaster risk reduction, disaster response and disaster recovery'. It recognized also: 'the urgent need to … build resilience to natural disasters', while bearing in mind 'the importance of addressing disaster risks related to changing social, economic, environmental conditions and land use, and the impact of hazards associated with geological events, weather, water, climate variability and climate change, in sector development planning and programmes as well as in post-disaster situations'. The General Assembly identified urban areas as particularly at risk given the concentration of population and economic assets while stressing 'that the impacts of natural disasters are severely hampering efforts in achieving the internationally agreed development goals, including the Millennium Development Goals'.

The Resolution 'invites Governments and relevant international organizations to integrate adaptation to climate change with relevant disaster risk reduction measures'.[91] The Conference of the Parties to the UNFCCC and the Parties to the Kyoto Protocol are encouraged 'to continue to address the adverse effects of climate change, especially in developing countries that are particularly vulnerable, in accordance with the provisions of the Convention, and also encourages the Intergovernmental Panel on Climate Change to continue to assess the adverse

---

[90] Distr.Gen. A/65/436/Add.3 7 December 2010. See also *Sustainable development: protection of global climate for present and future generations of humankind*, A/65/436/Add.4, 2 December 2010 and *Implementation of the International Strategy for Disaster Reduction: Report of the Secretary-General*, A/65/388, 22 September 2010.

[91] A/RES/65/157, 25 February 2011 Art. 6.

effects of climate change on the socio-economic and natural disaster reduction systems of developing countries'.[92]

More recently, the Rio+20 United Nations Conference on Environment and Development concluded in June 2012 with *The Future We Want*[93] as the principal outcome of the Conference. Shortly before that, in May 2012, the Conference of the Parties to the UNFCCC met in Bonn to continue their negotiations prior to the Eighteenth Conference of the Parties (COP 18) scheduled to take place in December 2012. It should come as no surprise that *The Future We Want* makes frequent reference to global climate change as a key disruptor of sustainable development. The international community identifies eradicating poverty as 'the greatest global challenge facing the world today and an indispensable requirement for sustainable development' which must be addressed 'as a matter of urgency'.[94] However it acknowledges that 'climate change is a cross-cutting and persistent crisis' and expresses its concern 'that the scale and gravity of the negative impacts of climate change affect all countries and undermine the ability of all countries, in particular, developing countries, to achieve sustainable development and the Millennium Development Goals and threaten the viability and survival of nations'.[95] The international community 'underscore that combating climate change requires urgent and ambitious action'[96] and note with grave concern the significant gap between the aggregate effect of mitigation pledges by Parties in terms of global annual emissions of greenhouse gases by 2020 and aggregate emission pathways consistent with having a likely chance of holding the increase in global average temperature below 2°C, or 1.5°C above pre-industrial levels.[97]

9 EXISTING THEORIES OF CLIMATE CHANGE JUSTICE

Before continuing to develop a 'willingness to pay' thesis, based on impartial public reasoning, current theories of climate change justice, which tend to ask who should pay and why, must be discussed. These

---

[92] Ibid. Art 25.
[93] A/CONF.216/L.1, http://www.uncsd2012.org/thefuturewewant.html, accessed 20 September 2012.
[94] Ibid. Art. 2.
[95] Ibid. Art. 25.
[96] Ibid.
[97] Ibid. Art. 121.

generally consider notions of corrective and distributive justice, while the issue of procedural justice is underdeveloped.

A reading of the current climate justice theories[98] must acknowledge that they have evolved organically over a period of time. This is easily explained given that the emergence of these theories in the 1990s coincided with a time when the climate science presented by the IPCC was far more hesitant than it was in 2012. No one could have known how emissions in the BRICS countries (Brazil, Russia, India, China and South Africa) would grow or what would emerge from the multilateral negotiations on climate change. In a sense, then, climate justice theories have had both to propose how these negotiations should deal with mitigation and adaptation issues as well as constantly to evolve to accommodate them. The author's 'willingness to pay' thesis is no different. Current climate justice theories might be, albeit telegraphically for present purposes, conveniently grouped as: contribution to the problem (polluter

---

[98] There are an extensive number of articles on this topic written both by philosophers and environmental lawyers. Those written by political philosophers include: Simon Caney, 'Cosmopolitan Justice, Responsibility and Global Climate Change' (2005) 18(4) Leiden Journal of International Law 747; Simon Caney, 'Cosmopolitan Justice, Rights and Global Climate Change' (2006) 19 Canadian Journal of Law and Jurisprudence 255; W Neil Adger et al. (eds), *Fairness in Adaptation to Climate Change* (MIT Press 2006); Sverker C Jager and Goran Duus-Otterstrom, 'Dual Climate Change Responsibility: on Moral Divergences between Mitigation and Adaptation' (2008) 17(4) Environmental Politics 576; David Miller, 'Global Justice and Climate Change: How Should Responsibilities Be Distributed?' presented at Tsinghua University, Beijing, 24–25 March 2008; Edward A Page, 'Distributing the Burdens of Climate Change' (2008) 17(4) Environmental Politics 556; Goran Duus-Otterstom and Sverker C Jager 'Identifying Burdens of Coping with Climate Change: a Typology of the Duties of Climate Justice' (2012) 22(3) Global Environmental Change 746; Simon Caney, 'Climate Change and the Duties of the Advantaged' (2010) 13(1) Critical Review of International Social and Political Philosophy 203; Marco Grasso, 'An ethical approach to Climate Adaptation Finance' (2010) 20 Global Environmental Change 74; Paul G Harris and Jonathan Symons, 'Justice in Adaptation to Climate Change: Cosmopolitan Implications for International Institutions' (2010) 19(4) Environmental Politics 617. Those written by environmental lawyers include Daniel A Farber, 'Adapting to Climate Change: Who Should Pay?' (2007) 23(1) Journal of Land Use 1; Rebecca Tsosie, 'Indigenous People and Environmental Justice: The Impact of Climate Change' (2007) 78 University of Colorado Law Review 1625; Amy Sinden, 'Allocating the Costs of the Climate Crisis: Efficiency Versus Justice' (2010) 85(2) Washington Law Review 293; Daniel Bodansky, 'Introduction: Climate Change and Human Rights: Unpacking the Issues' (2010) 38(3) Georgia Journal of International and Comparative Law 511.

pays principle); ability to pay principle; a hybrid principle; and a beneficiary pays principle. Distinctions are sometimes made between a theory of justice regarding obligations to mitigate one's own emissions and an obligation to pay for adaptation.[99]

9.1 Contribution to the Problem – a Corrective Approach

The contribution to the problem approach to climate justice is a very familiar approach given the incorporation of the polluter pays principle into modern environmental law to acknowledge its incorporation by way of Principle 16 in the Rio Declaration.[100] The idea is that countries should contribute costs, in proportion to their share of global cumulative greenhouse emissions, to manage climate change. As the common but differentiated principle insists, developed countries have an ethical responsibility to reduce their emissions given their ongoing and cumulative emissions.[101] Yet the climate justice literature demonstrates considerable machinations about this approach, particularly about whether liability should be *conditional* on *knowledge* of the harm. Different writers propose different dates on which this knowledge might have arisen, including 1900,[102] the mid-1980s,[103] 1985, ever since the 1990s[104] or pre-1990,[105] or 1992 with the establishment of the UNFCCC. A key concern here is that it seems morally objectionable to burden descendants with the responsibility for compensation when citizens and policy makers, responsible for historic emissions, are now dead. On this basis, it is suggested that a country's historical emissions should either be discounted or disregarded entirely.[106] Or, should liability rather be *strict*, i.e., irrespective of knowledge of the harm?

Concerns about the corrective implications of this approach are encapsulated in Eric Posner and David Weisbach's book *Climate Change Justice*. There the authors argue that a corrective justice vision of climate change justice, as expressed by the common but differentiated principle,

---

[99] Jager and Duus-Otterstrom, above note 98.

[100] Distr.Gen. A/CONF.151/26 (Vol. I) 12 August 1992.

[101] See Page, above note 98 at 557.

[102] Ibid. at 560, citing E Neumayer, 'In defence of historical responsibility for greenhouse gas emissions' (2000) 33(2) Ecological Economics 185, 189.

[103] Miller, above note 98 at 136.

[104] Caney (2010), above note 98 at 208.

[105] Ibid.

[106] Page, above note 98 at 560.

is: backward looking and focuses on historical emissions;[107] punishes individuals; cannot be applied to nation-states; and that nation states will only do what makes them financially better off 'by their lights'. Such an approach would 'force many people who have not acted wrongfully to provide a remedy to many people who have not been victimized'.

The authors base their objections to the principle of common but differentiated responsibility for climate change principally on the Kyoto Protocol. However, as discussed above, under the Cancun and Durban Agreements developing countries have all made commitments to reduce emissions within their respective capabilities.

The final issue with regard to corrective justice which must be interrogated is whether the US has any culpability giving rise to moral responsibility to commit to legally binding emissions reductions of the scale required of it as the second biggest emitter globally. Posner and Weisbach assert that intentional, reckless or negligent action is usually thought to be required for a corrective justice claim so it is worthwhile asking whether the US can be blamed for contributing to climate change.[108] They conclude that the claim that US policy has been negligent is far-fetched.[109] How so, one wonders? It is a legal fact that the reason the 1997 Kyoto Protocol could only enter into force in 2005 is that, with the US refusing to ratify the Protocol, the provisions of Article 25 were only satisfied upon Russia's ratification.[110] The US's withdrawal also resulted, in large part, in Australia's refusal to ratify and in Bush and Howard deliberately attempting over many years to undermine multilateral negotiations on climate change.[111]

Finally, in December 2007, the United States House of Representative's Committee on Oversight and Government Reform, based on 27,000 pages of documents from the White House Council on Environmental Quality and the Commerce Department, stated: '[t]he evidence before the Committee leads to one inescapable conclusion: the Bush Administration has engaged in a systematic effort to manipulate climate change science

---

107 Posner and Weisbach, above note 19 at 102.

108 Ibid. at 110.

109 Ibid. at 113.

110 Article 25 provides that '[t]his Protocol shall enter into force on the ninetieth day after the date on which not less than 55 Parties to the Convention, incorporating Parties included in Annex I which accounted in total for at least 55 per cent of the total carbon dioxide emissions for 1990 of the Parties included in Annex I, have deposited their instruments of ratification, acceptance, approval or accession'.

111 See Lyster, above note 1.

and mislead policymakers about the dangers of global warming ... The Bush Administration has acted as if the oil industry's communications plan were its mission statement.' Little wonder then that James Hansen states that by 2004 he became aware that NASA press releases on global warming were sent to the White House where they 'were edited to appear less serious or discarded entirely'.[112]

The Bush years, at the very least, resulted in years of inaction on climate change in the US and significantly undermined multilateral negotiations and progress on climate change at a critical time. In the author's view the US's culpability and causation, required by Posner and Weisbach,[113] are well established.

Posner and Weisbach also reject a distributive approach to climate justice based on the common but differentiated principle. They regard the principle as resulting in a massive, and unjustified, distribution of wealth from rich to poor countries. This is of deep concern for the fact that it, either mistakenly or deliberately, reduces the tenets of the UNFCCC, *ad absurdum*, largely to economic ones where rich countries should reject the idea that they are obliged to reduce the likelihood of harm solely because they are rich.[114] They go on, curiously, to assert that reducing carbon emissions is not a way to help today's poor and if costs are incurred to reduce emissions, these resources are not available to help today's poor. Further, poor nations may be better off if the resources that would be dedicated toward abatement are used in other ways, and in any case the future poor will be less poor.[115] They claim that 'development aid is likely to be more effective than greenhouse gas restrictions as a method of helping poor people in poor nations'.[116]

The author is concerned that Posner and Weisbach's work, which by their own admission contains a limited view of morality,[117] permits current generations to give less scrutiny to their natural moral inclinations to require their governments to take action on climate change, and permits them to 'pass the buck' to the next generation. The concern is

---

[112] Hansen, above note 11 at 125. To read more about the politics of climate science in the White House read James Hansen's account 'A Visit to the White House' in James Hansen, *Storms of My Grand-Children: The Truth About the Coming Climate Catastrophe and Our Last Chance to Save Humanity* (Bloomsbury 2010) 28.

[113] Posner and Weisbach, above note 19 at 104.

[114] Ibid. at 79.

[115] Ibid. at 27.

[116] Ibid. at 78.

[117] Ibid. at 6.

that their arguments may 'subvert our understanding of what is at stake'[118] with the coming climate catastrophe.[119]

9.2 Ability to Pay Principle – a Distributive Approach

The 'ability to pay principle' also leans on the common but differentiated principle to claim that developed countries have greater wealth and capacity to cover the mitigation and adaptation costs of climate change. This approach is indifferent as to contribution to the problem and focuses only on who can rectify the harm,[120] and indeed there may be remedially responsible states without causal responsibility and states without remedial responsibility that are nevertheless causally responsible.[121] It has been suggested that the ability to pay should not focus on resources but rather on the 'excess capacity' of developed countries to pay up to the point where their citizens' well-being is not compromised.[122] Of course the question of ability to pay might be regarded rather differently in 2012 following the recent global financial crisis and the shift of geo-economic power from North to South and West to East, as recognized by the World Economic Forum in January 2011.[123]

9.3 A Hybrid Approach – a Corrective and Distributive Approach

A hybrid approach to justice[124] suggests that climate justice might combine elements of the 'contribution to the problem' and 'the ability to pay' principles. A number of problems with the polluter pay principle are noted, the principal of which is that to apply the principle it is necessary to specify the harm and establish the causal link between the emissions and the harm. From a practical perspective, it is often claimed this cannot be done if the nature of the harm is uncertain or unpredictable, or the link between climate change and harm is uncertain.[125] Nevertheless, the polluter pays principle is not abandoned but modified to include a

---

[118] Ibid. at 45.

[119] See the title of James Hansen's book, above note 11.

[120] Caney (2010), above note 98 at 213.

[121] Jager and Duus-Otterstrom, above note 98 at 581.

[122] Page, above note 98 at 561.

[123] See Lisa Schlein, 'World Economic Forum to Focus on Global Power Shifts', *Voice of America*, 19 January 2011, http://www.voanews.com/english/news/economy-and-business/World-Economic-Forum-to-Focus-on-Global-Power-Shifts-114218864.html, accessed 8 November 2012.

[124] See Caney (2005) and (2010), above note 98.

[125] Caney (2010), above note 98 at 207.

'Poverty Sensitive Polluter Pay Principle', i.e., persons should bear the burden of climate change they have caused but should not be required to compromise on a decent standard of living.[126] Also, the 'ability to pay' principle is adopted as a principle of justice to deal with 'the Remainder', defined as harmful climate changes that arise from: '(a) the emissions of earlier generations; (b) non-human-induced climate change; and (c) the (legitimate) emissions of the disadvantaged'.[127] However, the principle is modified by 'the History Sensitive Ability to Pay Principle', so that those whose wealth came about in unjust ways should have a greater responsibility to pay than those whose wealth did not come about in unjust ways.[128] This approach is 'cosmopolitan' in that it seeks to pierce the 'statist' veil and place an obligation to pay on anyone that possesses wealth, including individuals and corporations. It also places obligations on the least advantaged as well as the most advantaged, requiring the least advantaged to pursue a low carbon development path if they can do so without great cost to themselves.[129] According to the hybrid approach the global poor will not have to bear the cost of adaptation.[130] The approach also requires that the financial benefits agreed to in multilateral negotiations reach individuals and are not simply distributed to states. The 'cosmopolitan' implications of this approach also resonate with Sen's cosmopolitan view of global, rather than state-based, justice.

9.4 Beneficiary Pays Principle

Yet another approach is to say that countries that have benefited the most from emitting greenhouse gases historically bear the greatest responsibility of climate justice. There is no attempt here to hold countries liable for causing climate change, either strictly or conditionally, but they are held strictly liable to pay for the negative externalities from which they have benefited. This principle is different from 'ability to pay', as it is concerned with how a country's wealth arose; and is distinct from 'contribution to the problem', as it is concerned with the effects, rather than the causes, of climate change-inducing activities.[131]

---

126 Ibid. at 218.
127 Ibid. at 213.
128 Ibid.
129 Ibid. at 219–220.
130 Ibid. at 221–222.
131 See Page, above note 98 at 562.

9.5 Greenhouse Development Rights

This approach has two key tenets: the specification of an 'emergency climate programme' (ECP); and the incorporation of poverty reduction and development goals within the global climate architecture. The ECP is consistent with current multilateral negotiations in that it proposes to limit global warming to below 2°C above pre-industrial levels as well as extensive investment to help vulnerable populations cope with unavoidable climate damages. This means that: a global greenhouse gas emissions budget will be fixed to meet this goal and a share in the global budget will be assigned to each country on the basis of a weighting of 40 per cent for historical responsibility and 60 per cent based on an ability to pay; a detailed national emissions budget reflecting these principles. The poverty reduction and development goals allow countries to emit to achieve a 'dignified level of sustainable development' set at USD 9,000 a year (in terms of 2005 USD) for all individuals. The overall financial burden of each country to contribute towards funding climate policy goals is calculated based on a Responsibility and Capacity Indicator.[132] It would seem that this approach to climate justice most closely aligns with a Capability Approach given its concern with dignity, poverty reduction and development goals.

9.6 Are Current Justice Theories Deceptive?

It is worth reflecting at this point on a point made earlier, which is the inevitably historical evolution of climate justice theories. The current justice theories that might seem to converge to suggest that the developed world ought to take the lion's share of the climate burden might be deceptive, precisely because they are historical. The emergence of relatively poor yet significant polluters already seems to threaten this view. Consequently, the global community 'must be prepared to analyse the new set of burden-sharing questions that will then arise, including how different types of burdens can apply to a *different and wider set of agents*'.[133]

---

132 Ibid. at 571.

133 Duus-Otterstrom and Jager, above note 98 at 752.

10 INVOKING A 'WILLINGNESS TO PAY' APPROACH

As mentioned earlier, at the first meeting of the Ad Hoc Working Group on the Durban Platform for Enhanced Action,[134] held on 20 July 2012, some developed countries identified the need for domestic political support as being crucial if they are to move to the higher end of their pledge ranges.[135] The need to examine the question of equity with respect to historical and current and future emissions both within nations as well as between nations was also highlighted.[136] The failure of the international community to respond adequately to an issue which it itself regards as 'one of the greatest challenges of our time' must reflect what Sen would regard as 'unreason'. It now seems clear that a focus on the actual lives of those most likely to suffer from climate change, both now and in the future, has many far-reaching implications for the nature and reach of the idea of justice.[137] A process of practical reasoning about climate change can evolve into appropriate responses in the interest of global justice, and is regarded by the author as an aspect of procedural climate change justice. The local preconceptions and prejudices, which have so dominated the climate change debate, surely cannot survive reasoned confrontation with information not restricted by the same parochialism.[138] As Sen has argued 'unreason is mostly not the practice of doing without reasoning altogether, but of relying on very primitive and very defective reasoning'.[139] What matters most in the context of climate change is the examination of what reasoning would demand in the pursuit of justice,[140] and recognition that positional illusions can impose serious barriers to broadening the informational basis of evaluations. This is one of the reasons why perspectives about the threat of climate change from elsewhere, including from far away, have to be systematically invoked.[141]

Sen's dissatisfaction with Rawls' celebrated approach of 'justice as fairness' is that it is concerned exclusively with setting up 'just institutions', which inescapably limits the involvement of participants in the pursuit of justice to members of the nation-state. Yet what happens in any

---

134 Distr.Gen. FCCC/ADP/2012/2012/INF.1 20 July 2012.
135 Ibid. at Art. 27.
136 Ibid. at Art. 30.
137 Ibid. at viii.
138 Sen, above note 2 at vii.
139 Ibid. at xix.
140 Ibid.
141 Ibid. at 169.

country cannot but have effects and sometimes enormous consequences for the rest of the world. Each country may have parochial beliefs that call for more global examination and scrutiny, or globally sensitive questioning.[142] Furthermore, Rawls' presumption that once the social contract has been arrived at, people would abandon any narrow pursuit of self-interest does not translate well into guidance about 'justice in the world we live [in] rather than in the imagined world with which Rawls is primarily concerned'.[143] As Sen argues, if democracy is seen in terms of the possibility and reach of public reasoning, the task of advancing, rather than perfecting, both global democracy and global justice can be seen as eminently understandable ideas that can plausibly inspire and influence practical actions across borders.[144] Further, 'bringing reasoning to the world becomes the enterprise of morality rather than metaphysics and the work as well as the hope of humanity'.[145] This view of democracy can also have an impact on the pursuit of it at the global level – not just within the nation-state – because public reasoning can be used for advancing both global democracy and global justice.[146]

Surely, the failure of nation-states to commit to significant emissions reductions is partly due to the neglect of the interests and perspectives of those who are not parties to the social contract of a polity, but who bear some of the consequences of the decisions taken in the particular polity.[147] As Sen recognizes, liberation from positional sequestering, for example on climate change, may not always be easy but it is a challenge that has to be met for the sake of ethical, political and legal thinking.[148] Moreover, there is a long history of attempts to go beyond the positional confinement of our moral concerns to the proximate 'neighbourhood',[149] and a deep fragility in the intellectual basis of thinking of people in terms of fixed communities of neighbours.[150] An understanding of justice, particularly in the contemporary world, must acknowledge that the neighbourhood is constructed by our relations with distant people. In

---

142 Ibid. at 71.
143 Ibid. at 79.
144 Ibid. at xiii.
145 Ibid. at xvii.
146 Ibid. at xiii.
147 Ibid. at 140.
148 Ibid. at 155.
149 Ibid. at 170.
150 Ibid. at 171.

truth, with climate change '[t]here are few non-neighbours left in the world today'.[151]

Sen is not concerned that by engaging in reasoned argument with oneself and others, competing priorities may survive despite the confrontation of reason. Reasoned scrutiny cannot remove the risk of plurality or ensure that the answer will be correct, but it insists that public discussion be as objective as it can reasonably be.[152] After all, impartiality is an integral part of justice.[153]

If impartial reasoning and public discussion are essential for a global justice approach to climate change, they have been seriously undermined by a vociferous campaign being waged by climate sceptics, funded largely by the fossil fuel lobby.[154]

10.1 Climate Scepticism, Moral Corruption and a 'Willingness to Pay'

Climate scientists are habitually subjected to vicious and unrelenting email campaigns.[155] They have been threatened with criminal prosecution and investigation by their own universities for research fraud.[156] The attacks on climate science as witnessed by 'Climategate' and attempts to discredit the IPCC's Fourth Assessment Report are now well known. It almost goes without saying that 'well-financed and widely publicized attacks on scientific research and the integrity of the scientists undertaking the research ultimately threaten the viability of all scientific research as public respect for the scientific community fades and public support for science dwindles'.[157] Dwindling public respect for climate

---

[151] Ibid. at 173.

[152] Ibid. at 40.

[153] Ibid. at 42.

[154] Ibid., although the author has adapted more general comments about science to climate science.

[155] See Clive Hamilton 'Bullying, lies and rise of right-wing denial' (22 June 2010), *The Drum*, http://www.abc.net.au/unleashed/32912.html, accessed 20 September 2012.

[156] See Rosemary Lyster 'Rescuing climate science from politics', http://sydney.edu.au/news/law/436.html?newsstoryid=5245, accessed 20 September 2012.

[157] Thomas O McGarity, 'Defending Clean Science from Dirty Attacks by Special Interests' in Wendy Wagner and Rena Steinzor (eds), *Rescuing Science from Politics: Regulation and the Distortion of Scientific Research* (Cambridge University Press 2006) at 25.

science contributes significantly to the state of 'unreason' dominating much of the climate change debate.

Yet it is even more serious. It results in the moral corruption of the citizenry, which in itself undermines Sen's hope for a practical reasoning approach to justice, since 'moral reasoning is just as open to misuse as scientific reasoning'.[158] Stephen Gardiner's thorough treatment of this in *A Perfect Moral Storm*[159] is apposite. Here, Gardiner makes the case that, with climate change, global, intergenerational and theoretical storms converge and encourage a distinct problem for ethical action on climate change – the problem of moral corruption. Given the pronounced temporal dispersion of the causes and effects of climate change, current generations may be attracted to weak or deceptive arguments which permit them to give less scrutiny to their inclination to 'pass the buck' to the next generation. Such arguments corrupt the very terms of the debate, moral and otherwise. 'In other words', says Gardiner, 'the perfect moral storm may work to subvert our understanding of what is at stake'.[160] Yet 'serious moral agents strive to protect themselves against rationalization, self-deception and moral manipulation'.[161]

Consistently with Sen, Gardiner cautions that the potentially catastrophic consequences for the victims of climate change require a special vigilance about our own reasoning.[162] This is difficult, given that moral corruption in its more sophisticated forms seeks to co-opt moral and epistemic values that ought otherwise to be respected.[163] It strikes at people's ability to understand what is going wrong in moral terms and, for other (usually selfish) ends, it subverts moral discourse.[164]

Gardiner provides an insightful and succinct account of the five-pronged 'all-out assault' on the moral demands citizens face. These include: disputing the application of the moral claim (excessive burden, competing special relationships); highlighting the unintended bad consequences of compliance; reducing the magnitude of the moral demand (budget constraints, demanding mutual benefit); undermining the implementation of the duty (discretionary aid, indirect efforts); and breeding resentment on the part of the duty-bearer (recasting oneself as

158 Sen, above note 2 at 337.

159 Stephen M Gardiner, *A Perfect Moral Storm: The Ethical Tragedy of Climate Change* (Oxford University Press 2011).

160 Ibid. at 45.

161 Ibid. at 301.

162 Ibid. at 302.

163 Ibid.

164 Ibid.

the victim).[165] These five tactics are easily recognizable to those who have followed the climate change debate over the years.

One of the most compelling and meticulously referenced accounts of the origins of climate science scepticism, which has profoundly contributed to 'unreason' on climate justice, is that provided by Naomi Oreskes and Eric M Conway in *Merchants of Doubt: How a Handful of Scientists Obscured the Truth on Issues from Tobacco Smoke to Global Warming.*[166] For those who wish to understand the seemingly opaque link between denial about the harmful effects of tobacco and climate change it is essential reading. The authors, who are science historians, trace over four decades the tactics of a small group of prominent US scientists, with deep political and corporate connections, who they claim fought the facts and 'merchandised doubt' on tobacco, acid rain, the hole in the ozone layer, and finally global warming. Cutting their teeth on tobacco their tactics allegedly included: hiring public relations firms; creating the 'Tobacco Industry Research Committee'; cherry picking data and focusing on unexplained or anomalous data to adopt an industry position that there was 'no proof' that tobacco was bad because the science was so unsettled; transmogrifying scientific consensus into a raging scientific debate and convincing mass media that responsible journalists had an obligation to present 'both sides'; and by mid-1980s spending USD 100 million on biomedical research.[167] This led ultimately to a finding by a US Federal Court judge that the tobacco industry, aided by these scientists, had committed the tort of conspiracy to commit fraud – a massive and ongoing fraud to deceive the American public about the health effects of smoking. In *USA v Philip Morris USA Inc.*,[168] the judge stated:

> It is about an industry, and in particular these Defendants, that survives, and profits, from selling a highly addictive product which causes diseases that lead to a staggering number of deaths per year, an immeasurable amount of human suffering and economic loss, and a profound burden on our national

165 Ibid. at 337.

166 Naomi Oreskes and Eric M Conway, *Merchants of Doubt* (Bloomsbury Press 2010).

167 Ibid. at Chapter 1 'Doubt is our Product'.

168 The judgment is available at http://www.library.ucsf.edu/sites/all/files/ucsf_assets/FinalOpinion_full_version.pdf, accessed 20 September 2012. In 2009, a US Court of Appeal upheld the decision, and in 2010, the US Supreme Court declined to hear appeals in the case.

health care system. Defendants have known many of these facts for at least 50 years or more.[169]

Oreskes and Conway's work also tracks the reports and reviews on carbon dioxide and climate change, including sea level rise, polar amplification and temperature change that were being compiled for the US government as early as 1965.[170] The authors nominate 1988 as the year when climate science changed forever with the establishment of the IPCC and the announcement by climate modeller James Hansen, mentioned above, that anthropogenic warming had begun.[171] They go on to document the tactics embraced by certain scientists to create doubt about the consensus on climate science, as they had done with tobacco.[172] In a February 1991 letter to the vice president of the American Petroleum Institute, one of the scientists claimed that '[i]t is generally considered by the scientific community that the Marshall report[173] was responsible for the Administration's opposition to carbon taxes and restrictions on fossil fuel consumption'.[174] As Oreskes and Conway state these 'merchants of doubt' took their claims to the halls of public opinion, rather than the halls of science. As such, these authors argue, they were 'stepping outside the institutional protocols that for four hundred years have tested the veracity of scientific claims'.[175]

The conduct being evinced by the fossil fuel lobby has now led to 'civil conspiracy' litigation against ExxonMobil and 24 other fossil fuel companies, in which USD 400 million in damages is claimed against the defendants for misleading the public about climate science. In *Native Village of Kivalina and City of Kivalina v ExxonMobile & Others* the applicants sought damages[176]

---

169 Oreskes and Conway, above note 166 at 3.

170 Ibid. at 170–173.

171 Ibid. at 183.

172 Ibid. at 190–197. To read more about the politics of climate science in the White House read James Hansen's account 'A Visit to the White House'; Hansen, above note 9 at 28.

173 The report was entitled *Global Warming What Does the Science Tell Us?*

174 Oreskes and Conway, above note 166 at 190.

175 Ibid. at 270.

176 The special injuries suffered by the village of Kivalina are the destruction of its property as storms routinely batter Kivalina. The loss of sea ice leaves Kivalina more vulnerable to waves, storms and erosion leaving Kivalina with the choice either to relocate or face extermination. The applicants claim that the defendants contribute to the damage as a result of their greenhouse gas emissions and that they have engaged in civil conspiracy to undermine climate science. The

against the defendants[177] on the basis of common law public nuisance, while also alleging that the defendants had engaged in the tort of civil conspiracy.[178] In particular, it is alleged that the defendants continued to participate in an agreement with each other to mislead the public with respect to the science of global warming and to delay public awareness of the issue.[179] The claim is that the defendant formed and used front groups, fake citizens' organizations, and bogus scientific bodies to cast doubt on peer review climate science. According to the claim, the most active company in this regard is ExxonMobil, which, between 1998 and 2005, channelled USD 16 million to 42 organizations to promote disinformation on global warming.[180] One of the most prominent of these organizations has been The Advancement of Sound Science Coalition (TASCC) formed originally in 1993 by the public relations company working for Philip Morris allegedly to discredit the science on the health hazards of second-hand tobacco smoke.[181] The applicants claim that:

> Rather than meet its social and legal responsibility, ExxonMobil engaged in a multi-faceted attack on global warming which included exploiting science, denying the consensus on global warming, fostering false science, running misleading advertising denying the existence of global warming or its causes, and funding organizations who attacked global warming on these bases and/or the factor causing global warming.[182]

This matter is currently on appeal before the US Supreme Court.

10.2 The Role of the Media in Undermining a 'Willingness to Pay'

Central to Sen's advancement of public reasoning in the pursuit of justice is a free and independent press. The reasons for this, articulated by Sen, are: the direct contribution of free speech to the quality of our lives; the

---

case was dismissed at first instance by the California District Court on the ground of lack of standing and political doctrine and is currently on appeal.

[177] Co-defendants include BP, Chevron, Shell and a number of energy utilities.

[178] To view the complaint see http://www.climatelaw.org/cases/country/us/kivalina/Kivalina%20Complaint.pdf, accessed 20 September 2012.

[179] Ibid. para. 269 at 65.

[180] Ibid. para. 231 at 56.

[181] Ibid. para. 192 at 48.

[182] Ibid. para. 234 at 57.

informational role in disseminating knowledge for critical public scrutiny; the protective function in giving voice to the marginalized so as to contribute to human security; the formation of values; and facilitating public reasoning in general.[183] While Sen reiterates the normative values underpinning a free press, the media has at times been complicit in the project to misinform the public on climate science and to give a platform for climate sceptics. As Oreskes and Conway demonstrate, the 'merchants of doubt' took their denialist claims to the halls of public opinion where the media became complicit. On a generous view of the media's role, journalists may act out of a concern to provide 'balance' while treating climate change as a scientific controversy. Certainly they are constantly pressured to grant equal time and newsprint or online space to both sides of the 'controversy'.[184] Yet one needs to ask whether, if they really took the time to educate themselves about the consensus on climate science, they would regard it as a matter of 'debate'.

In Australia, the responsibility of the media for undermining the scientific consensus on climate change needs further investigation and analysis. One particular contribution stands out. Entitled 'Climategate gives lord of the sceptics plenty of ammunition',[185] it began thus:

> The visit to Australia this week of Lord Christopher Monckton – the world's most effective global warming sceptic – couldn't have been better timed. Hot on the heels of the 'Climategate' email leak, which called into question the 'tricks' used to sex up the case for the war against global warming, have come back-to-back revelations tarnishing the reputation of the Intergovernmental Panel on Climate Change.

It went on thus:

> Public opinion has changed as the credibility of the IPCC ebbs, the crippling cost of climate change measures becomes apparent and the array of rent-seekers and phonies grows. Monckton is a man whose time has come because he owes nothing to anybody and he has the capacity to interpret the science to a public looking for answers.

These views have never been publicly retracted or modified by the journalist even after the extensive investigations into Climategate and the review of the

---

[183] Sen, above note 2 at 335–337.

[184] Oreskes and Conway, above note 166 at 214. The role of the media in presenting the 'controversy' is discussed in Chapter 6 entitled 'The Denial of Global Warming' 169.

[185] Miranda Devine, 'Climategate gives lord of the sceptics plenty of ammunition', *Sydney Morning Herald*, 28 January 2010.

IPCC. Moreover, Dr Lesley Cannold, a researcher and medical ethicist, reported that Lord Monckton appeared in the media 455 times during his Australian visit while the visit by James Hansen warranted only 61 mentions.[186] This type of media coverage, along with the power of the blogosphere, led James Hansen to comment in March 2010 that 'in the last six months the gap between what is known by scientists and what is understood by the public has widened'.[187] To date, the media has done nothing to 'enrich reasoned engagement through enhancing informational availability and the feasibility of interactive discussions'.[188]

11 LOOKING FORWARD TO THE NEXT CONFERENCE OF THE PARTIES TO THE UNFCCC

This chapter has suggested that to close the gap between the parallel realities that seem to coexist in the international negotiations on a future climate change treaty, the imperatives of the national interest need to be reframed. The voices of vested interests and 'unreason' need to be quietened while the perspectives of those from elsewhere, including from far away, are systematically invoked. In this globalized and interconnected world, the impacts on developing countries now and in the future, including on individuals' 'freedom and functionings', cannot be ignored. However, the focus should not fall solely on developing countries, as developed economies will also be seriously affected, not least by threats to their infrastructure and food security and the spectre of forced climate change migration. All of the latest information on climate change that is currently available to citizens around the world must be considered so that world leaders are given the imprimatur, based on a 'willingness to pay', to enter into meaningful negotiations at subsequent meetings of the Conference of the Parties to the UNFCCC. The media needs to adopt a radically different ethical approach to reporting accurately to the reasoning public.

Perhaps, there is hope that '[e]ven our shared frustrations and shared thoughts on global helplessness can unite rather than divide'.[189]

---

[186] Lesley Cannold, 'Why we've stopped trusting science', *The Sun Herald*, 16 May 2010.

[187] See his presentations and interviews during his visit to the United States Studies Centre at the University of Sydney, http://ussc.edu.au/people/james-hansen, accessed 20 September 2012.

[188] Sen, above note 2 at xiii.

[189] Sen, above note 2 at 173.

4. Climate change induced displacement and international law

Mariya Gromilova and Nicola Jägers

1 INTRODUCTION

Already in 1990, the Intergovernmental Panel on Climate Change (IPCC) predicted that one of the greatest impacts of climate change will be on human migration.[1] Although it is complicated to estimate the precise number of those likely to be displaced as a result of global warming, according to the UN this number ranges between 50 million and 350 million by 2050.[2]

The current situations in Bangladesh and in a number of small island States in the Pacific region are among the most striking examples of what we can expect from climate change in the near future. Bangladesh, due to its geographical and spatial location is already among the most environmentally vulnerable regions, with around 20 percent of the land one meter or less above the sea-level. According to the 2007 UN Fourth Assessment Report of the IPCC, a one meter rise of sea-level will cost Bangladesh up to 17 percent of its land, and will displace at least 35 million people by 2050.[3] The risk for small islands is even more dramatic, as according to the prognoses, such islands as Tuvalu and Maldives will be wiped of the face of the earth by the mid-century.[4]

---

[1] IPCC, Climate Change: The IPCC 1990 and 1992 Assessments (IPCC First Assessment Report Overview and Policymaker Summaries, and 1992 IPCC Supplement) 55.

[2] United Nations, Report of the Secretary-General on Climate Change and its Possible Security Implications, 11 September 2009, UN document A/64/350, 15. For more on the scientific prognoses of the number of people that are likely to be displaced due to climate change see *infra*.

[3] M L Parry et al. (eds), *Climate Change 2007: Impacts, Adaptation and Vulnerability: Contribution of Working Group II to the Fourth Assessment Report of the Intergovernmental Panel on Climate Change* (Cambridge University Press 2007) 488.

[4] Ilan Kelman, 'Island evacuation' (2008) 31 FMR 20.

Furthermore, climate induced displacement is not just a matter of future concern. The evidence illustrates that climate change is already causing population movements. For example, in 1995 half of Bhola Island in Bangladesh became permanently flooded forcing half a million people to relocate.[5] In 1999, two of Kiribati's islands, Tebua Tarawa and Abanuea, disappeared because of sea-level rise.[6] Other unprecedented evidence is the disappearance of Lohachara Island in December 2006, which left 10,000 people homeless.[7]

People who are already relocated, or those who will be, are going to suffer a significant infringement of basic human rights, and need to be protected. At the same time they may present a threat to the neighboring States, as a massive flow of migrants is fraught with the risks of conflicts and economic instability. Therefore, it is crucial that certain pro-active measures are taken to prevent these risks. The search for effective ways to protect people displaced by climate change and to allocate responsibility for their protection attracts increasing attention from scholars and the international community. Yet the way to tackle the problem remains unclear. Numerous obstacles hinder the inclusion of people displaced by climate change into existing legal frameworks. At the same time the urgency of the problem demands that the international community, policy-makers and scholars, come up with a solution concerning the status and protection of people displaced by climate change.

Therefore, the current chapter aims to provide an overview of the contemporary state of certain relevant sections of international law with regard to the issue of climate induced displacement, to outline the most problematic points and to show the limitations and opportunities of the existing approaches to the problem of climate induced displacement.

The first section of the current chapter will focus on the causal relationship between climate change and displacement of people. It will discuss how climate change especially affects the most vulnerable regions in the world, forcing people in these regions to relocate. This section will reveal some of the main obstacles of legal recognition of people displaced by climate change in international law.

Subsequently, the chapter addresses the issue of protection of people displaced by climate change and the issue of responsibility for such

---

5 Emily Wax, 'In Flood-Prone Bangladesh, a Future That Floats' (*Washington Post*, 27 September 2007).

6 Sci/Tech, 'Islands disappear under rising seas' (*BBC News*, 14 June 1999), http://news.bbc.co.uk/2/hi/science/nature/368892.stm, accessed 3 October 2012.

7 'Disappearing World: Global Warming Claims Tropical Island' (*The Independent*, 24 December 2006).

protection under international law. Thereto, the investigation of existing international legal frameworks which can potentially be applied for the protection of these people will be presented. We will consider the limitations and opportunities of international refugee law, international human rights law and international environmental law.

Ultimately, we will draw conclusions on the current state of international law and its capacity to deal with the issue of climate induced displacement, as well as on the most controversial issues which still remain unclear. Additionally, we will outline some possible solutions to the problem and will consider the feasibility of these suggestions.

2 LINKING CLIMATE CHANGE AND MIGRATION

As has been shown, the problem of climate induced displacement is recognized, and currently is on the forefront of many international discussions. It is acknowledged that people who will have to relocate need assistance and protection. However, the way to tackle the problem remains unclear. There are a number of obstacles in conceptualizing the issue: recognizing people displaced by climate change under international law, granting them a timely protection and determining those who are responsible for this protection.

2.1 Causal Relationship

An underlying issue that hampers tackling the issue of climate induced displacement and that runs all through the search for a potentially relevant legal framework, lies in the fact that the relationship between climate change and migration is considered to be highly controversial. Some scholars argue that there is no direct causal link between climate change and relocation. These critics consistently refer to the fact that, especially in cases with slow-onset disasters, it is impossible to claim that climate change was a primary factor in the person's decision to move, as the reasons for relocation are almost always multiple.[8] Environmental factors are closely linked to economic, political and social ones. Therefore, environmental degradation is undoubtedly a driver of displacement, but it is unlikely that it is the unique cause, as other conditions, such as

[8] Graeme Hugo, 'Environmental Concerns and International Migration' (1996) 30(1) International Migration Review 110–113.

unemployment, are always taken into account in people's decisions to migrate.[9]

In case of rapid-onset disasters, as, for example, a tsunami, or industrial accidents, the link is nevertheless identifiable. Still, the critics argue that with respect to these types of climate related impacts, the relocation will have an internal and temporal character, which is not a new occurrence. People displaced internally are able to receive protection and assistance within the existing international legal regime[10] and, therefore, some scholars claim that there is no need to recognize them distinctly.[11]

An associated problem, which only complicates the establishment of the link between climate change and migration, is the lack of statistics on the ongoing climate induced movement. This, of course, hampers the process of drawing further conclusions. The science of climate change is complex enough, and it still cannot accurately predict the way climate change will affect our ecosystem. Making prognoses is even more problematic when it comes to interaction between meteorological and social factors. Methodologically, it is complicated to establish how exactly people will respond to climate change, and to estimate the precise numbers of those who will relocate. Among the problems are: the lack of a definition for people displaced by climate change, which will be discussed *infra*; plausible assumptions about human behavior and ignorance of changes in human behavior and possibility of adaptation; and the above-mentioned issue of multi-causality.[12]

Nevertheless, these complications do not stop scientists trying. Current predictions on climate induced migration range between 25 million and 1 billion people by 2050.[13] According to Myers, the total number of people at risk of sea-level rise in Bangladesh is around 26 million, in Egypt 12

---

9 Benoit Mayer, 'The International Legal Challenges of Climate Induced Migration: Proposal for an International Legal Framework' (2011) 22(3) Colorado Journal of International Environmental Law and Policy 10.

10 Guiding Principles on Internal Displacement 1998, UN Doc. E/CN.4/1998/53/Add.2. Critics, however, argue that the protection offered by the Guiding Principles falls short. See *infra*.

11 David Keane, 'The Environmental Causes and Consequences of Migration: A Search for the Meaning of "Environmental Refugees"' (2004) 16 Georgetown International Environmental Law Review 211.

12 Frank Biermann and Ingrid Boas, 'Preparing for a Warmer World: Towards a Global Governance System to Protect Climate Refugees' (2007) 33 Global Governance Working Paper 9.

13 R Bird et al., *Human Tide: The Real Migration Crisis* (Christian Aid 2007) 5, 6.

million, in China 73 million, in India 20 million, and in several other parts of the world 31 million, making an aggregate total of 162 million. In addition, at least 50 million people could relocate due to droughts and other climate change impacts. In total, Myers predicts 212 million people displaced by climate change by 2050.[14] The Stern Review on the Economics of climate change, even though criticizing the way Myers results have been tested, agrees that there will be 200 million displaced by 2050.[15] According to the UN Secretary-General the potential number lies between 50 and 350 million.[16] The '1 billion' figure is the result of wider interpretation of climate induced displacement, as it also counts people displaced by development projects, conflicts and human rights abuses.[17] Ultimately, it is possible to generalize that the total number of those who relocate will be around 200 million.

Besides the alarming prognoses for the future, there is, as mentioned earlier, some evidence that climate change already causes population movement. According to the International Organization for Migration (IOM) the 'gradual and sudden environmental changes are already resulting in substantial population movements' and in 2008, '20 million persons were displaced by extreme weather events, compared to 4.6 million internally displaced by conflicts and violence over the same period'.[18]

The previously mentioned examples of Bangladesh and several small island States, show that climate induced displacement is not just a matter of future concern, but a currently existing threat which is only advancing. Thus, while the discussions on the complexity of the relationship between climate change and migration, and the paramount importance of the social drivers to environmental changes are bound to go on for decades to come, climate change is clearly a factor that influences migration.[19] Therefore, it seems more important to focus on the particular

---

[14] Norman Myers, 'Environmental refugees: A growing phenomenon of the 21st century' (2002) 357 Philosophical Transactions: Biological Science 609–611.

[15] Nicholas Stern, *The Economics of Climate Change: The Stern Review* (Cambridge University Press 2006) 77.

[16] UN Secretary-General Report, Climate change and its possible security implications (UN doc A/64/350, 11 September 2009) para. 54.

[17] Biermann and Boas, above note 12 at 10.

[18] IOM, 'Migration, Climate Change and Environmental Degradation: A Complex Nexus', http://www.iom.int/jahia/Jahia/complex-nexus, accessed 3 October 2012.

[19] Jon Barnett and Michael Webber, 'Accommodating Migration to Promote Adaptation to Climate Change' (2010) 5270 Policy Research Working Paper 9.

impacts of climate change and to look at the way climate change currently affects and will continue affecting people.

2.2 Impacts of Climate Change on Migration in Different Regions of the World

It is very likely that at a certain point the consequences of climate change will be visible worldwide and that each continent will suffer its grave impacts.[20] Nevertheless, currently and in the nearest future, the influence of climate change on migration will be most notable in already vulnerable regions of the world. Some of these effects can already be observed. The alarming rate of environmental change will sharpen them, and more than likely bring new ones. The majority of scientists agree that global temperatures will continue to rise throughout the following decades. The IPCC forecasts a temperature rise of 2.5 to 5°F (1 to 3°C) over the next century.[21] Seemingly insignificant, such an increase will have dramatic consequences for some regions. Among the most direct effects of the temperature's change are sea-level rise, changes of rainfall patterns, negative implications on human health due to the risks of heat related illness, and many more of the associated effects, such as floods, droughts, increasing numbers of storms, economic losses, land losses, and the changing shapes of landscapes. Notwithstanding the fact that in general the effects of climate change will be felt worldwide, it is possible to allocate several areas which are particularly vulnerable to climate change effects. Among them are the small island States, coastal zones and regions of Africa and Asia.[22] It is these hotspots that will be especially affected by climate induced relocation.

Small island States are greatly threatened by sea-level rise, as some of them lie less than two meters above sea-level. If the sea-level rises to the respective point, such islands as Tuvalu and Kiribati will be wiped off the

---

[20] IPCC, 2007: Summary for Policymakers in M L Parry et al. (eds), *Climate Change 2007: Impacts, Adaptation and Vulnerability: Contribution of Working Group II to the Fourth Assessment Report of the Intergovernmental Panel on Climate Change* (Cambridge University Press 2007) 18.

[21] Ibid.

[22] Bonnie Docherty and Tyler Giannini, 'Confronting a Rising Tide: A Proposal for a Convention on Climate Change Refugees' (2009) 33(2) Harvard Environmental Law Review 355.

face of the earth.[23] Other States, including the Maldives, the Marshall Islands and a number of Caribbean islands, are also threatened. Besides the risk that the sea-level rise poses, numerous associated climate change effects, such as storms, droughts, and more frequent heat waves are expected. For example, the predicted 10 percent decrease in rainfall by 2050 could reduce the availability of fresh water on Kiribati by 20 percent.[24] Any of the above-listed climate change effects could cause inhabitants to flee these islands.

Due to the geographical location and the population density, South and East Asia are among the other spots greatly exposed to large-scale forced migration. Many of the Asian urban cities, such as Shanghai and Calcutta are dangerously threatened by tropical cyclones and storms.[25] There is evidence that the intensity and frequency of many extreme weather events in the region, such as heat waves, tropical cyclones, prolonged dry spells, intense rainfall, tornadoes, snow avalanches, thunderstorms and severe dust storms only increases.[26] Low-lying and shoreline areas, such as Bangladesh, are at the top of the risk group. The coastal zone of Bangladesh, which covers about 30 percent of the country, is home to millions of people. According to the 2007 Fourth Assessment Report of the IPCC, a one meter rise of sea-level will cost Bangladesh up to 17 percent of its land, and will displace at least 35 million people by 2050.[27] Another climate change related threat that Asia faces is the melting of glaciers. This will increase the risk of flooding during the wet season, and reduce dry-season water supplies to one-sixth of the world's population, predominantly in the Indian sub-continent and parts of China.[28]

Africa is another target of climate induced migration, because, while Asia is submerging, low and mid-latitudes are exposed to other extremes. North Africa and the Sahel, the horn of Africa and South Africa are extremely vulnerable to drought. Fourteen African countries experience water scarcity at present, and there are prognoses this number will

---

[23] German Advisory Council on Climate Change, *The Future Oceans: Warming Up, Rising High, Turning Sour* (Berlin: German Advisory Council on Global Change 2006) 46, 50.

[24] IPCC, Analyzing Regional Aspects of Climate Change and Water Resources (Technical Paper IV, Sec 5, 2008) 110.

[25] Munich Re Group, 'Megacities – Megarisks: Trends and Challenges for Insurance and Risk Management' (Münchener Rückversicherungs-Gesellschaft 2004) 41, 76.

[26] IPCC, above note 20 at 469–506.

[27] Ibid. at 484.

[28] Stern, above note 15 at 56.

increase to 25 countries by 2030.[29] Changing patterns of rainfall place the food security in sub-Saharan Africa under a serious danger. According to the IPCC reduced rainfall could lower crop yields by as much as 20 percent by 2020, which will lead to increasing malnutrition.[30]

Above-stated risks and changes eventually will make certain parts of the world unapt for living, by causing food and water supplies to become more unreliable, increasing the frequency and severity of floods and storms and through erosion and inundation of coastal areas. These will force people to search for new places that can provide them with the means of living.

Given the urgency of the problem and existing prognoses of the top science bodies and the UN agencies, the debates on whether climate change is a direct driver for population movement, or the contra arguments that such displacement will be indirect and multi-faceted, do not seem to be appropriate. Directly or indirectly, climate change will create massive displacement and put global and regional stability at risk. What deserves particular attention is that displacement of people is associated with significant deprivation of basic human rights and the inability to acquire sufficient assistance and financial support from host States, as well as the risks of conflicts over the reduced availability of resources. Therefore, these people require attention and protection, and need to be placed on the legal and political agenda. However, the disagreement on causality creates insuperable obstacles to recognizing displaced people under international law and reaching agreement on which elements should be incorporated in the definition of this group of people. Until there is agreement with regard to the issue of causality, the obstacles in approaching people displaced by climate change will remain. In the light of current analysis the lack of a definition presents a major obstacle. Therefore, before moving to the question whether existing legal frameworks are able to deal with the issue of climate induced displacement it is crucial to clarify what group of people is implied in the further analysis.

2.3 People Induced to Displacement by Climate Change

Although terms and concepts such as climate change induced displacement, environmental or climate change refugees, environmental migration and environmental migrants have been widely used in the ongoing

---

29 Tearfund, *Fleeing the Heat* (Tearfund 2006) 12.

30 IPCC, above note 20 at 10.

debates on climate change, there is no universally excepted definition of people displaced by climate change.

There have been a lot of attempts to define the category of people displaced by climate change. El-Hinnawi was the first to use the term environmental refugee in his work for the UN Environmental Programme (UNEP) during the 1980s. According to his definition, environmental refugees are:

> [p]eople who have been forced to leave their traditional habitat, temporarily or permanently, because of a marked environmental disruption (natural and/or triggered by man) that jeopardized their existence and/or seriously affected the quality of their life.[31]

El-Hinnawi includes three major types of environmental refugees in his definition. First of all, the definition includes people who are temporarily dislocated due to disasters, whether natural or anthropogenic. The major characteristic of this group is the ability to return to their habitats once the environmental disruption is over and it is safe to return. The second group consists of people who have been permanently displaced due to drastic environmental changes. Finally, the third type are the people who migrate due to the gradual deterioration of environmental conditions.[32]

Myers suggests an even broader definition and describes environmental refugees as 'people who can no longer gain a secure livelihood in their homelands because of drought, soil erosion, desertification, deforestation and other environmental problems, together with associated problems of population pressures and profound poverty'.[33]

Even though both definitions cover a wide range of factors, as will be shown in the following section, the term 'refugee' seems to be politically and legally inadequate, because it does not reflect the nature of climate induced movement. In fact, the United Nations High Commissioner for Refugees (UNHCR) made an attempt to avoid the term 'refugee' and introduced the definition of environmentally displaced people, describing them as those:

> who are displaced from or who feel obliged to leave their usual place of residence, because their lives, livelihoods and welfare have been placed at

---

[31] Essam El-Hinnawi, *The Environmental Impacts of Production and Use of Energy* (UNEP 1985) 4.

[32] Ibid.

[33] Norman Myers, *Environmental Refugees: An Emergent Security Issue* (OSCE 2005) 6–7.

serious risk as a result of adverse environmental, ecological or climatic processes and events.[34]

In this chapter we build upon the definition developed in the context of the EACH-FOR project, which is a two-year research project devoted to environmental change and forced migration scenarios launched by the European Commission.[35] According to its definition the term Environmentally Displaced Persons (EDPs) covers three categories: environmental migrants (people who chose to move voluntarily from their usual place of residence primarily due to environmental concerns or reasons); environmental displacees (people who are forced to leave their usual place of residence because their lives, livelihoods and welfare have been placed at serious risk as a result of adverse environmental processes and events (natural and/or triggered by people)); and development displacees (people who are intentionally relocated or resettled due to a planned land use change).[36] This approach provides the broad interpretation of factors that constitute environmental displacement, as it covers the wide range of situations in which the displacement might occur, including external and internal, forced and voluntary, permanent and temporary relocation, relocation as a response to rapid-onset and slow-onset environmental disasters as well as relocation instigated by the development projects.[37] Since, in this chapter, we only focus on environmental change related to climate change, we will use the term Climatically Displaced Persons (CDPs).

---

[34] Brian Gorlick, 'Environmentally-Displaced Persons: a UNHCR Perspective, Senior Policy Advisor', Presentation Environmentally Displaced Persons: a UNHCR perspective (UNHCR 2007), http://www.equatorinitiative.org/images/stories/events/2009events/brian_gorlick_environmentally_displaced_persons_unhcr_perspective.pdf, accessed 3 October 2012.

[35] EACH-FOR project is the project which focused on the environmental change and forced migration scenarios, supported by the European Commission and carried out by a consortium of researchers between January 2007 and March 2009. More information available on the official webpage, http://www.each-for.eu, accessed 3 October 2012.

[36] EACH-FOR, *Environmental Change and Forced Migration Scenarios* (Synthesis Report 2009) 8–9, http://www.each-for.eu, accessed 3 October 2012.

[37] First of all it avoids the use of the term 'refugee'; secondly, in comparison to the approach of the UNHCR it includes development displacement.

3 THE EXISTING LEGAL FRAMEWORKS FOR THE PROTECTION OF PEOPLE DISPLACED BY CLIMATE CHANGE

As has been mentioned in the previous section, the category of people induced to displacement by climate change is relatively new, and is not directly recognized under international law. However, as the number of these people is rapidly growing, most of these people are from the developing countries and are already vulnerable, and their human rights are under a great threat, there is an urgent need to have a mechanism for their protection and to establish who is responsible to assist and protect them. This section aims to review existing international legal frameworks, to see whether they offer protection to people displaced due to climatic reasons, whether they provide certain clarity with regard to responsibility towards these people, and whether these frameworks can play a pre-emptive role in dealing with the issue of climate induced displacement.

3.1 International Refugee Law

The most relevant legal framework at the international level, for the protection of displaced populations is refugee law, especially the 1951 United Nations Convention relating to the Status of Refugees (Refugee Convention) and its Optional Protocol.[38] Currently, 147 States have ratified either one or both of these instruments, and thus are bound by the provided definition of refugee. However, as will be observed in this section, when it comes to CDPs, the applicability of the Refugee Convention turns out to be highly problematic. In order to analyze what are the main obstacles for people displaced by climate change to fit into the refugee law regime, the definition of refugees under the Refugee Convention and Protocol needs to be recalled.

Article 1 of the Refugee Convention states that the refugee status applies to any person who:

> owing to well-founded fear of being persecuted for reasons of race, religion, nationality, membership of a particular social group or political opinion, is outside the country of his nationality and is unable or, owing to such fear, is unwilling to avail himself of the protection of that country; or who, not having a nationality and being outside the country of his former habitual

---

[38] UN Convention Relating to the Status of Refugees of 28 July 1951, 189 UNTS 150.

> residence as a result of such events, is unable or, owing to such fear, is unwilling to return to it.[39]

All these requirements have to be met in order to gain refugee status.

To help governments and courts determine who can qualify as a refugee, in 1979 the UNHCR produced the UNHCR Handbook on Procedures and Criteria for Determining Refugee Status (UNHCR Handbook).[40] The interpretation of the refugee definition in the UNHCR Handbook, which is undoubtedly the most authoritative interpretation of the 1951 Refugee Convention and the 1967 Refugee Protocol, excludes victims of natural disasters and rules them out from acquiring refugee status.[41] Although a lot of scholars, lawyers, and governments use the terms: 'environmental refugee' and 'climate change refugee', and try to apply refugee law for their protection,[42] there are a number of significant obstacles in qualifying people displaced by climate change as refugees under international law.

Before turning to these obstacles it is important to note that several regional instruments contain a broader definition of a refugee. Among notable examples are the 1969 Organization of African Unity Convention Governing the Specific Aspects of Refugee Problems in Africa (OAU Convention), the 1984 Cartagena Declaration on Refugees (Cartagena Declaration) of Latin America and the 1994 Arab Convention on Regulating Status of Refugees in Arab Countries (Arab Convention).

All these regional instruments follow the refugee definition found in the 1951 Refugee Convention, but they also provide for additional circumstances. The OAU Convention includes any person compelled to leave his/her country because of 'external aggression, occupation, foreign domination or events seriously disturbing public order in either part or the whole of his country of origin or nationality'.[43]

Similarly, the Cartagena Declaration extends the definition of refugee to persons who have fled their country 'because their lives, safety or

---

39 Ibid.

40 Handbook on Procedures and Criteria for Determining Refugee Status under the 1951 Convention and the 1967 Protocol relating to the Status of Refugees (HCR/IP/4/Eng/REV.1, Geneva 1992).

41 Ibid. at para. 39.

42 Kara K Moberg, 'Extending Refugee Definitions to Cover Environmentally Displaced Persons Displaces Necessary Protection' (2009) 94 Iowa Law Review 1114.

43 Convention Governing the Specific Aspects of Refugee Problems in Africa (OAU Convention) Organization of African Unity 1969, 1001 UNTS 45, Art. 1(2).

freedom have been threatened by generalized violence, foreign aggression, internal conflicts, massive violation of human rights or other circumstances which have seriously disturbed public order'.[44]

The Arab Convention even mentions natural disasters among the reasons for acquiring refugee status, and defines a refugee as 'any person who unwillingly takes refuge in a country other than his country of origin or his habitual place of residence because of the sustained aggression against, occupation and foreign domination of such country or because of the occurrence of natural disasters or grave events resulting in major disruption of public order in the whole country or any part thereof'.[45]

It should be noted, however, that so far no States have ratified the Arab Convention and therefore it has no legal force.[46] The Cartagena Declaration is a non-binding instrument. Nevertheless, most of the Latin American States apply the definition as a matter of practice and some have incorporated the definition into their own national legislation.[47] The OAU Convention is a binding legal instrument and is a part of regional law.

In sum, several regional instruments provide for a broader definition by including persons that flee events that seriously disturb public order. Arguably this may include CDPs. This will be further discussed *infra*. But as these regional instruments also build on the definition as provided in the 1951 Refugee Convention, the next section will focus on that definition. Each component of the definition will be critically accessed in order to identify problems with the application of the refugee definition to people displaced by climate change. Moreover, the above-mentioned regional instruments (the OAU Convention and the Cartagena Declaration) will be discussed as well, in order to analyze whether a broader refugee definition can have some added value for those who have to relocate due to climate change effects.

---

[44] Cartagena Declaration on Refugees, adopted at the Colloquium on the International Protection of Refugees in Central America, Mexico and Panama on 22 November 1984, Section III (3).

[45] Arab Convention on Regulating Status of Refugees in the Arab Countries, adopted by the League of Arab Countries in 1994, Art. 1.

[46] Due to the fact that the Arab Convention has not been ratified, we will not deal with it below.

[47] Eduardo Arboleda, 'Refugee Definition in Africa and Latin America: The Lessons of Pragmatism' (1991) 3(2) International Journal of Refugee Law 189.

3.1.1 Requirement of exile

The refugee definition under the Refugee Convention requires the person to be outside his or her country of origin. This immediately excludes those who have not moved yet, or have been internally displaced. According to prognoses, much of the movement provoked by climate change, especially at the early stages, will be internal. Therefore, international refugee law will not be able to provide its protection to a significant number of people induced to displacement. Internally Displaced People (IDP) are subject to a totally different legal mechanism and will be analyzed within the following section, which addresses the applicability of human rights law to the issue of climate induced displacement.

Another concern that follows from the requirement of exile is that it implies that the person is entitled to protection only when the relocation has already taken place. Yet, as has been emphasized, there is a strong necessity for a pre-emptive approach to the protection of people at risk of climate induced displacement. Therefore there is an obvious drawback to refugee law with regard to the issue of climate induced displacement.

The same is true with regard to the regional instruments, as in this respect they duplicate the Refugee Convention.

3.1.2 Persecution and its grounds

A major obstacle to placing climate induced displacement within the framework of international refugee law, is the difficulty in characterizing natural disasters and other weather related events as persecution. Rising sea levels, salination, and increasingly frequent storms, earthquakes and floods may be harmful, but they do not constitute persecution in accordance with the meaning which has been ascribed in international and domestic law.[48]

The UNHCR Handbook affirms that '[t]here is no universally accepted definition of "persecution", and various attempts to formulate such a definition have met with little success'.[49] Article 33 of the 1951 Refugee Convention states that threat to life or freedom on account of race, religion, nationality, political opinion or membership of a particular

---

48 Jane McAdam and Ben Saul, 'An Insecure Climate for Human Security? Climate Induced Displacement and International Law' (2009) Sydney Centre for International Law Working Paper 4, 9, http://sydney.edu.au/law/scil/documents/2009/SCILWP4_Final.pdf, accessed 5 October 2012.

49 Handbook on Procedures and Criteria for Determining Refugee Status under the 1951 Convention and the 1967 Protocol relating to the Status of Refugees (HCR/IP/4/Eng/REV.1, Geneva, January 1992) para. 51.

social group is always persecution. Furthermore, other serious violations of human rights, based on the same grounds, would also constitute persecution.[50] Whether other prejudicial actions or threats can be considered as persecution will depend on the circumstances of each individual case.[51]

According to the UNHCR 'persecution is normally related to actions by the authorities of a country'.[52] In other words, if one is searching for recognition as a refugee because of environmental impairment, the 'persecutor' should be identified. An applicant must show that the cause of harm lies in the actions of government and that the government is unwilling or unable to prevent from continuing the persecution. However, in the case of climate change it is nearly impossible to establish this link. Moreover, the majority of CDPs live in developing countries, which are not among those countries that have contributed to global warming and should be held responsible for it.[53]

On the other hand, as the UNHCR Handbook does not provide the definition of persecution, there is no direct impediment for considering environmental harm as persecution. There are a number of examples where we can acknowledge environmental harm as persecution. For instance, the desertification of the African Sahel, where it is claimed the governments of the Sahel region 'could have enacted policies and programs to cut population growth, to improve agricultural techniques, or to heighten food production'.[54] Another relevant example is the Chernobyl nuclear disaster that happened in 1986 at the Chernobyl Nuclear Power Plant in Ukraine. The impacts of the Chernobyl disaster were aggravated by the Soviet government's delayed response and apparent disregard for safety and environmental considerations in the country's quest for nuclear power.[55] In these examples we can see the cause-effect relationship between environmental consequences and governmental actions. Nevertheless, this kind of correlation will mostly not be found in many cases of climate induced displacement. Even if there is a case where we can establish this link, usually, it will still not be

---

[50] Convention relating to the Status of Refugees 1951, Art. 33.

[51] Above note 49 at para. 52.

[52] Ibid. at para. 65.

[53] Kara K Moberg, 'Extending Refugee Definitions to Cover Environmentally Displaced Persons Displaces Necessary Protection' (2009) 94 Iowa Law Review 1121, 1122.

[54] Angela Williams, 'Turning the Tide: Recognizing Climate Change Refugees in International Law' (2008) 30(4) Law & Policy 508.

[55] Ibid.

possible to apply the Refugee Convention, because of the other requirements that have to be met.

According to the Refugee Convention, in order to obtain refugee status, alongside being persecuted one also has to show that this persecution is based on certain grounds. As it is specified in the refugee definition, reasons for persecution are limited to 'race, religion, nationality, membership of a particular social group or political opinion'.[56]

Frequently cited refugee situations, such as massive movements in the Middle East (for instance, the results of the Iraq wars and the flight of Palestinian refugees), or movements in Asia, such as the migration of Afghan refugees to Pakistan as a result of the Soviet invasion in 1979, represent the exact scenarios for which the Refugee Convention was designed.[57] In these cases, the identified types of persecution are based upon one of the grounds laid down in Article 1A(2) of the Refugee Convention. This greatly differs from the case of people induced to relocate by climate change. The nature of climate change impacts is to a large extent indiscriminative. This means that it is not tied to particular characteristics such as a person's background or beliefs. Even though some areas are more affected by climate change than others, this is on account of their geographical location and not because of the nationality or race of their inhabitants.[58] The same is true with regard to such categories of people as women and children. Though climate change is expected to injure women and children the most, it is not because of their gender or age, but rather due to their general vulnerability.

The definition does not leave much room for interpreting the reasons for persecution, but instead it includes an exhaustive list of race, religion, nationality, membership of a particular social group or political opinion, clearly setting the boundaries of the legal application of the Refugee Convention.[59]

As has been mentioned above, regional instruments contain a broader refugee definition. Under the OAU Convention the refugee definition includes those fleeing 'events seriously disturbing public order in either

---

[56] Convention relating to the Status of Refugees 1951, Art. 1A(2).

[57] Williams, above note 54 at 508.

[58] Jane McAdam, 'Climate Change displacement and International Law: Complementary Protection Standards' (2011) UNHCR Legal and Protection Policy Research Series 13, http://www.unhcr.org/4dff16e99.html, accessed 5 October 2012.

[59] Williams, above note 54 at 508.

part or the whole of his country of origin or nationality'.[60] The Cartagena Declaration contains a similar provision. Lopez suggests that even though both instruments were not designed for protection of climate migrants, these people can argue that climate change is an event seriously disturbing public order.[61] Yet, this claim raises certain doubts. While sudden-onset disasters (for example, tsunamis or floods) might arguably be considered as events seriously disturbing public order, with regard to the slow-onset disasters, such as droughts and sea-level rise, this hardly is the case. Only if a country affected by a severe drought declares a national emergency or formally identifies the disaster as one disrupting public order, it might be possible to claim that people induced to relocate should receive temporary asylum or refugee status in another country.[62] Otherwise, slow-onset disasters are hard to qualify as events seriously disturbing public order. Although, in Africa it has been common to provide a temporary protection to people who cross an international border to flee a natural disaster (this occurred for example when Congolese fled to Rwanda following the eruption of Mount Nyiragongo in 2002), African governments have never characterized this as an obligation arising from the OAU Convention. Furthermore, the wording of both regional instruments seems to require evidence of an actual threat.[63] The OAU Convention refers to the person that 'was compelled to leave'; the Cartagena Declaration extends to people 'who have fled their country' for certain reasons.[64] This means that these instruments cannot play a pre-emptive role in the protection of people at risk of climate induced displacement.

3.1.3 Right to return

Another obstacle to placing climate displacees within the legal framework of refugee law both at the international and regional level, is linked to the fact that the refugee concept implies a right to return once the

---

[60] Convention Governing the Specific Aspects of Refugee Problems in Africa (OAU Convention) Organization of African Unity 1969, 1001 UNTS 45, Art. 1(2).

[61] Aurelie Lopez, 'The Protection of Environmentally-Displaced Persons in International Law' (1993) 18 Yale Journal of International Law 389.

[62] Michelle T Leighton, *Climate Change and Migration: Key Issues for Legal Protection of Migrants and Displaced Persons* (German Marshall Fund of the United States 2010) 4.

[63] McAdam, above note 58 at 15.

[64] OAU Convention, Art. 1(2), Cartagena Declaration, Section III(3).

persecution that triggered the original flight has ceased.[65] It seems obvious that in extreme cases of the sea-level rise, when the whole State may disappear, there will simply be no possibility for displaced people to go back.

3.1.4 Other obstacles

Aside from the definitional obstacles identified above, complications on the policy level exist. Namely, the UNHCR does not encourage an extensive interpretation of the existing definition. The UNHCR is concerned that expanding the current definition of refugee 'would possibly lead to an erosion of the currently valid international refugee protection regime'.[66] According to the UN agency, a modification of the refugee definition may have as a consequence a renegotiation of the 1951 Refugee Convention, which, in the current political environment, may lead to a lowering of protection standards for refugees under the present definition.[67]

A connected argument of the UNHCR is that the case with environmental refugees involves different types of moral and legal responsibilities in comparison with 'normal' refugees. 'Whereas political and war refugees are victims of their home state or of a regionalized conflict, with no direct responsibility for their plight with the countries that eventually offer refuge, the moral responsibility for climate change is different.'[68] However, while stating this, the UNHCR does not provide further instructions on the ways to tackle the issue.

Furthermore, most of the developed countries are reluctant to recognize the problem. It has been frequently emphasized that the vast majority of people moving as a result of the effects of climate change will come from the countries that are least responsible for climate change, financially weak, have the least capacity to manage its after-effects, and are not able to implement adaptation programs. Therefore, many developed countries in Europe and North America fear that if the term refugee expands to cover people displaced by climate change, it would compel them to offer these people the same protection as political refugees. Clearly, no country is willing to do so, on account of overwhelming costs, lack of territories, and concerns about social order. At

---

65 Oli Brown, *Migration and Climate Change* (IOM 2008) 14.

66 Parliamentary Assembly, Environmentally Induced Migration and Displacement: a 21st Century Challenge, (Report Committee on Migration, Refugees and Population 2008) para. 55.

67 Ibid.

68 Ibid. at para. 56.

the same time, at the international level there are no existing institutions with a direct mandate to address the problem. Currently, the UNHCR is charged with providing for refugees, and, is already overstretched and unable to cope with their current 'stock' of refugees.[69]

It can be concluded, that, the term 'refugee', as provided under the Refugee Convention, does not extend to cover people induced to relocate by climate change. The main reason for that is that the Refugee Convention's definition is not open to interpretations of the grounds for persecution. Thus, it excludes people displaced by climate change from its scope. Furthermore, there is no willingness on the side of the UNHCR and States to extend the protection given to regular refugees to people displaced by climate change, and there is no special institution with a mandate to assist them. Among other crucial limitations are the already mentioned exclusion of people displaced internally and the failure to play a pre-emptive role in dealing with the issue of climate induced displacement.

At the regional level existing instruments do not seem to be a feasible solution either due to the above-mentioned limitations. Even if agreement is reached that the OAU Convention and the Cartagena Declaration extend to people induced to displacement by climate change, their protection only applies after the displacement took place and therefore these instruments cannot play a pre-emptive role.

3.1.5 National refugee legislation

Interestingly, whereas international and regional refugee law can be considered as failing in providing protection to people induced to displacement by climate change, national laws in certain countries have been more successful in dealing with the problem. Currently two States, Finland and Sweden, have adopted asylum law legislation granting subsidiary protection for 'environmental migrants'.

The Swedish Aliens Act offers subsidiary protections to a person who is 'outside the country of the alien's nationality, because he or she … is unable to return to the country of origin because of an environmental disaster'.[70] However, according to the Division for Migration and Asylum Policy at the Swedish Ministry of Justice the legislation is limited to sudden disasters and does not extend to cases of continuous environmental degradation. Moreover, the capacity of the Swedish refugee law to

---

69 Ibid.

70 SFS2005:716, Ch 4, sec 2§3.

grant protection to environmental refugees is hard to estimate.[71] The relevant provision has never been used with regard to people displaced by climate change and therefore it might prove difficult to protect an excessive number of displaced people, which the scientific prognoses predict.

In Finland according to the Aliens Act, asylum may be granted to the person if 'he or she cannot return to his or her country of origin or country of former habitual residence as a result of an environmental catastrophe'.[72] According to the Finnish Immigration Service, the legislation can be extended to cases where the alien's home environment has become too dangerous for human habitation either because of human actions or as a result of natural disaster. Even though this regulation has not been in frequent use, the Finnish example shows that individual States have a potential to issue immigration and asylum policy to provide legal protection for environmental migrants.[73]

Nevertheless, the Scandinavian subsidiary protection of people displaced by environmental disturbance is the exception to the rule,[74] and it does not extend to the international level.

3.2 Human Rights Law

In the previous section we addressed the question whether international refugee law provides protection to those who are displaced or at risk of displacement due to climate change. International refugee law can be seen as a *lex specialis*, as a subset of international human rights law. In principle, human rights protection applies to all irrespective of whether one is displaced or not. Human rights law obligates States to safeguard the human rights, such as life and property, of those within a State's jurisdiction against threats of disaster and foreseeable harm.

Hence, in this section we turn to discuss whether general human rights law[75] provides an adequate legal framework to address the problems

---

71 Benjamin Glahn, 'Climate Refugees? Addressing the International Legal Gaps' (*International Bar News* 2009), available at http://www.ibanet.org/Publications/, accessed 5 October 2012.

72 Finnish Aliens Act 301/2004, Sec 88a(1).

73 Glahn, above note 71.

74 Benoit Mayer 'The International Legal Challenges of Climate Induced Migration: Proposal for an International Legal Framework' (2011) 22(3) Colorado Journal of International Environmental Law and Policy 29.

75 This section is limited to international human rights law. International humanitarian law (IHL) may also be a relevant section of international law given

arising from climate induced displacement. Before turning to the human rights obligations of States relevant to the problems faced by persons dislocated internally or beyond borders, it is useful to distinguish the effects of rapid- and slow-onset climate change from the perspective of human rights law. The conditions facing those relocating due to slow-onset climate change such as droughts are significantly different to those forced to flee following a rapid disaster such as a tsunami. Overall, the latter category finds better protection under international law than those dislocated due to slow-onset climate change, as the displacees might meet the refugee definition or might be given a temporary protected status in certain countries.[76] Those forced to relocate due to slow-onset climate change, however, do not qualify for these narrow exceptions and need to fall back on human rights law in general.

3.2.1 Human rights law and climate induced displacement

Climate change in general will clearly entail a multitude of negative effects on the realization of internationally recognized human rights.[77] Rights such as the right to life, the right to an adequate standard of living, the right to health, the right to food, the right to housing, the right to information and a host of other internationally recognized human rights will be affected by climate change. Moreover, those individuals and groups that are already in a precarious position, such as women, children and indigenous peoples living primarily in poorer regions and countries, are most likely to experience these negative effects.[78] As mentioned *supra*, the IPCC in 1990 first noted that the greatest single impact of climate change will be on human migration. Climate induced relocation may affect a broad range of human rights but certain human rights such

---

the presupposed links between environmental factors and violent conflict. Due to space constraints IHL is not discussed in this chapter.

[76] Nevertheless, the current TPS designations remain highly discretionary and too narrow to apply to most cases of climate induced migration. See GMF, *Climate Change and Migration* (The German Marshall Fund of the United States 2010) 6, 8.

[77] See for a mapping exercise the 2008 report of the International Council on Human Rights Policy, Climate Change and Human Rights: A Rough Guide. See also the 2009 report of the Office of the High Commissioner for Human Rights on the relationship between climate change and human rights, UN Doc. A/HRC/10/61, 15 January 2009.

[78] UN Doc. A/HRC/10/61, paras 42–54.

as the right to housing, the right to property and land rights will specifically be affected.[79]

The fact however that the effects of climate change will clearly affect a broad range of human rights does not automatically imply that climate change as such can be labeled as a human rights violation.[80] As put forward by the Office of the High Commissioner for Human Rights (OHCHR), from a strict legal perspective, labeling the effects of climate change as human rights violations for which States can be held accountable is problematic for several reasons.[81]

Firstly, there is the above-mentioned problem of causality. In practice it will prove difficult for an individual that cannot fully enjoy his or her human rights to prove a causal link with climate change especially in the case of slow-onset climate change. The human right challenges posed by such climate change are subtle and it is problematic to pinpoint a directly responsible State. However, as pointed out by Knox, it is possible to locate the largest emitters of greenhouse gases and therefore, at least in principle it would be possible to establish that some States are far more culpable than others.[82] Nevertheless, allocation of responsibility remains a difficult issue. An additional problem regarding climate induced displacement is, as discussed previously, the fact that global warming is often only just one cause of relocation. Thirdly, procedures to claim compensation and reparation under international human rights law are open to those that have suffered such a violation, in other words after the

---

[79] See, in general: Displacement Solutions, Climate Change Displaced Persons and Housing, Land and Property Rights (Displacement Solutions 2009), available at http://displacementsolutions.org/files/documents/DS_Climate_change_strategies.pdf, accessed 5 October 2012.

[80] Nevertheless, in 2005 the Inuit filed a claim with the Inter American Commission on Human Rights directed against the US for their high emissions of greenhouse gases and the ensuing violations of the human rights of the Inuit. The Inter American Commission decided not to deal with the Inuit petition, stating that the information provided did not make it possible to determine whether the alleged facts would tend to characterize a human rights violation. For more on this case see Randall S Abate, 'Climate Change, the United States, and the Impacts of Arctic Melting: A Case Study in the Need for Enforceable International Environmental Human Rights' (2007) 43A(1) Stanford Journal of International Law 3.

[81] UN Doc. A/HRC/10/61, para. 70.

[82] John H Knox 'Linking Human Rights and Climate Change at the United Nations' (2009) 33 Harvard Environmental Law Review 477, 489.

fact, whereas in the case of climate change the effects often concern projections of future harm.[83]

For these reasons it will prove very difficult to hold a State responsible under international human rights law for the effects of climate change, such as displacement. However, as also noted in the report by the OHCHR 'irrespective of whether or not climate change effects can be construed as human rights violations, human rights obligations provide important protection to the individuals whose rights are affected by climate change or by measures taken to respond to climate change'.[84] Under international human rights law, States have the obligation to protect individuals against harm that affects the enjoyment of their human rights even if they are not directly responsible for those threats. The obligation of States to protect individuals against threats to the enjoyment of human rights even though the State is not causing the threat is firmly grounded in international human rights law. This implies that States have to take certain measures to protect citizens against foreseeable risks connected to climate change that may amount to human rights violations.[85] For example, States have the obligation to provide emergency relief policies and to guarantee access to information and participation in decision-making regarding environmental risks. A failure on the part of a State to implement such measures will amount to a breach of its human rights obligations.[86] This is an important basis for applying international human rights law to the effects of climate change, leaving aside the problematic debate on causality.

3.2.2 Protection for people displaced within a State

The normative framework for those dislocated within a State is better developed than for those that flee their country due to climate change. The 1998 Guiding Principles on Internal Displacement protect displaced

---

[83] UN Doc. A/HRC/10/61/, para. 70.

[84] UN Doc. A/HRC/10/61, para. 71.

[85] UN Doc. A/HRC/10/61, para. 74.

[86] The supervisory human rights bodies have addressed such issues on numerous occasions. For a review of relevant jurisprudence see K Anton and D Shelton, *Environmental Protection and Human Rights* (Cambridge University Press 2011). To mention but one example, in the case of *Budayeva and Others v Russia*, the European Court of Human Rights held that a failure to implement land/planning and emergency relief policies while the authorities were aware of an increasing risk of a large/scale mudslide, amounted to a violation of the right to life as provided in Article 2 of the ECHR. Moreover, the State failed in providing sufficient information on the risk to the community (European Court of Human Rights, No. 15339–02).

persons against arbitrary or forced displacement and rights related to housing and property restitution.[87] The definition of a displaced group in the Guiding Principles is of a descriptive nature and non-exhaustive. According to the Guidelines, internally displaced persons are:

> persons or groups of persons who have been forced or obliged to flee or to leave their homes or places of habitual residence, in particular as a result of or in order to avoid the effects of armed conflict, situations of generalized violence, violations of human rights or natural or human made disasters, and who have not crossed an internationally recognized border.

In the case of forced relocation, the displaced will qualify as a vulnerable group with a right to increased assistance and protection. Nevertheless, the effects of climate change might cause relocation that does not fit in the broad definition of IDPs, for example when people move pre-emptively. In other words, the Guiding Principles fall short of offering protection to persons forced to relocate following slow-onset climate change unless the reasons for their relocation (drought etc.) are considered as disasters. Moreover, the Guiding Principles are also considered weak due to their non-binding nature.

If the Guiding Principles do not apply or fall short of offering adequate protection, general human rights law protects the internally displaced. International human rights law explicitly recognizes certain groups, such as women, children or indigenous peoples, as vulnerable groups. However, this is not the case with those displaced by climate change. Nevertheless, States have the obligation to respect, protect and fulfill the human rights of *all* people within their jurisdiction.[88] This means that persons affected by climate induced displacement within the borders of a country are entitled to the full range of human rights guarantees that the State is obligated to provide in line with the human rights treaties it has succeeded to.[89] People that suffer human rights violations due to climate induced dislocation within national borders will thus have to turn to that given State. However, these countries usually are developing countries

---

87 UN Doc. E/CN.4/1998/53/Add.2, Annex, Principle 1, para. 1 and Principle 6, para. 1.

88 See, for example, Article 2 International Covenant on Civil and Political Rights (ICCPR), Article 1 of the European Convention on Human Rights (ECHR) and Human Rights Committee, 'General Comment 31 on Article 2 of the Covenant: The Nature of the General Legal Obligation Imposed on States Parties to the Covenant' (21 April 2004).

89 See the Guiding Principles on Internal Displacement (UN Doc.E/CN.4/1998/53/Add.2, Annex 1), Principles 1, para. 1 and 6, para. 1.

and access to justice in these States will often prove problematic. Even more problematic is the justifiability where it concerns economic, social and cultural (ESC) rights, a category of rights notably affected by climate induced displacement. In many jurisdictions, ESC rights cannot be invoked before a court of law.[90]

3.2.3 Protection for climate induced displacees across borders

If a person flees across a national border due to environmental factors, this person will be entitled to general human rights guarantees in a receiving State in line with the human rights obligations of that State.[91] A particularly relevant treaty in this respect is the International Convention on the Protection of the Rights of Migrant Workers and Members of their Families. However, as mentioned previously, climate displacement notably affects poorly developed regions of the world. In practice, therefore, the reliance on the receiving State to guarantee the human rights of CDPs often will prove illusory.

It is therefore essential also to examine whether States under international human rights law have extraterritorial obligations that might offer protection to people that have been displaced due to climate change. The human rights obligations of States under most human rights treaties are limited to those that are within their jurisdiction.[92] The International Covenant on Economic, Social and Cultural Rights (ICESCR), however, does not have such a restricting jurisdictional clause. Article 2 (1), provides that:

> Each State Party to the present Covenant undertakes to take steps, individually and through *international assistance and co-operation* [emphasis added, MG/NJ], especially economic and technical, to the maximum of its available resources, with a view to achieving progressively the full realization of the

---

[90] At the international level an optional protocol containing an individual right to complain about violations of the rights enshrined in the ICESCR, was adopted and opened for signature in 2009. The justifiability of ESC rights will greatly improve when this Optional Protocol enters into force, which will be the case after ten ratifications. At the time of writing, eight states have ratified the Optional Protocol.

[91] Certain human rights provisions may prohibit a State from returning the environmentally displaced. For more on this see *infra* the section on *non-refoulement*/complementary protection.

[92] See, for example, Article 2(1) ICCPR, which states that 'each party to the present covenant undertakes to respect and to ensure to all individuals within its territory and subject to its jurisdiction the rights recognized in the present covenant'.

> rights recognized in the present Covenant by all appropriate means, including particularly the adoption of legislative measures.

Consequently, ESC rights are generally considered to have an extra-territorial component. The obligation of States to respect and protect these rights is not restricted to a given's State jurisdiction.[93]

Moreover, Article 2(1) ICESCR explicitly refers to international cooperation in order to realize the rights in the Covenant.[94] According to Article 2(1) of the ICESCR, States are obliged to undertake joint and separate action to achieve the full realization of the rights in the Covenant.[95] Developed States have a particular responsibility and interest to assist the poorer developing States.[96] This is important as climate induced relocation will mostly affect poorer parts of the world. The Committee supervising the ICESCR has held that this obligation entails, inter alia, the obligation to 'take steps through international cooperation and assistance, depending on the availability of resources, to facilitate the fulfillment of human rights in other countries, *including disaster relief, emergency assistance, and assistance to refugees and displaced persons*'.[97] The ICESCR is particularly relevant for those that are displaced due to climate change as this will frequently imply a violation of ESC rights such as the right to housing, an adequate standard of living or the right to health. Furthermore, the ICESCR requires that States pay special attention to those that fall victim to natural disasters and people living in disaster-prone areas.[98] The duty of international cooperation can also be found in other international human rights instruments such as the Convention on the Rights of the Child,[99] the Convention on the Rights of

---

93 For more on the extraterritorial dimension of human rights see the 2011 Maastricht Principles on Extraterritorial obligations of States in the Area of Economic, Social and Cultural Rights; and Michal Gondek, *The Reach of Human Rights in a Globalizing World: Extraterritorial Application of Human Rights Treaties* (Intersentia 2009).

94 See also Articles 11(1), 15(4), 22 and 23 ICESCR.

95 See, inter alia, General Comment No. 19 (2008).

96 General Comment No. 3 (1990).

97 Emphasis added. See, e.g., General Comment No. 12 (1999) on the right to adequate food (Article 11); No. 13 (1999) on the right to education (Article 13); No. 14 (2000) on the right to the highest attainable standard of health (Article 12); and No. 15 (2002) on the right to water (Articles 11 and 12 of the Covenant).

98 General Comment No. 12, para. 13.

99 Articles 4 and 24(4).

People with Disabilities[100] and in the Declaration on Development.[101] As brought forward by the OHCHR, this complements the principle of 'common but differentiated responsibilities' laid down in the United Nations Framework Convention on Climate Change (UNFCCC).[102]

This obligation of international cooperation is crucial in the light of climate induced displacement. It is clear that climate change as a cause of displacement can only be addressed by means of international cooperation. Thus States are not only obliged to implement treaties within their jurisdictions but also to contribute, through international cooperation, to global implementation. Knox argues that the human rights obligation to cooperate 'requires the international community to establish a global polity for the purpose of addressing climate change'.[103] In light of the topic at hand it can be argued that the international human rights duty to cooperate obligates States to draw up a global instrument addressing climate induced displacement.

Notwithstanding the importance of the obligations of international cooperation, enforcing the international obligations that States have according to the international human rights treaties remains problematic.

3.2.4 Non-refoulement/complementary protection

As discussed previously, climate change does not provide people relocating to other countries with a right of entry based on the prime international document, the 1951 Refugee Convention. Nevertheless, those that are forcibly displaced from their homes to another State might find protection from other obligations under international human rights law that the receiving States might have. This is known as 'complementary protection', because it complements the Refugee Convention.[104]

Complementary protection follows from the principle of *non-refoulement*, which implies that asylum seekers already present within a jurisdiction cannot be removed by that State to a country where the person will face a real risk of persecution or exposure to torture, inhuman treatment or other serious human rights violations. The principle of *non-refoulement* has been enshrined in several human rights provisions such as the prohibition of torture and other inhuman or degrading

---

100 Article 32.

101 Articles 3, 4 and 6.

102 UN Doc.A/HRC/10/61, para. 87.

103 Knox, above note 82 at 495–496.

104 For a detailed study on complementary protection see, Jane McAdam, *Complementary Protection in International Refugee Law* (Oxford University Press 2007).

treatment (Article 3 EHCR, Article 7 ICCPR and Article 3(1) Convention Against Torture (CAT). Overall, it has been concluded that the level of complementary protection is lower than that provided for in the Refugee Convention.[105] Nevertheless, EDPs may find some protection from the obligations of receiving States that follow from the human rights provisions on *non-refoulement*. It has been argued that under certain circumstances returning a person who has been displaced due to climate change could amount to inhuman or degrading treatment in violation of *non-refoulement* provisions such as Article 3 ECHR.[106] 'Degrading treatment' may entail the denial of sufficient basic services necessary for a dignified existence. It is conceivable that an EDP will be confronted with such circumstances when returned. The receiving State returning such a person would then be in violation of its obligations under, for example, Article 3 ECHR and the displaced person could subsequently bring such a case to the attention of the supervisory body, the European Court of Human Rights.[107] However, as brought forward by McAdam, the chances of this complementary protection mechanism based on Article 3 ECHR becoming a meaningful protection mechanism is slim. She argues that it will only be under the most exceptional circumstances that a lack of resources will be found to be a reason prohibiting forced return.[108]

3.2.5 The responsibility to protect and climate induced displacement

From the above it becomes clear that the dependency on a single State to address climate induced migration is problematic in the face of this global phenomenon. Recently, a concept has been developed that rises above the single State approach and might prove beneficial to addressing the problems created by climate induced migration. There is widespread recognition that large-scale deprivation of basic human rights cannot be considered to belong to the domestic jurisdiction of a State. From this idea of 'responsible sovereignty'[109] the concept of the 'Responsibility to

---

[105] Jane McAdam, 'The European Qualification Directive, The Creation of a Subsidiary Protection Regime' (2005) 17 International Journal Of Refugee Law 461.

[106] Margit Ammer, *Climate Change and Human Rights. The Status of Climate Refugees in Europe* (Boltzmann Institute 2009) 58.

[107] Ibid. at 59.

[108] McAdam, above note 105.

[109] The then Secretary-General's Special Representative for Internally Displaced Persons, Francis Deng, is thought to be the first to coin the phrase

Protect (R2P)' has developed. It has been argued that it might offer an avenue for creating obligations for States *vis-à-vis* people that are displaced due to climate change.[110] R2P, first developed by the International Commission on Intervention and State Sovereignty (ICISS), focuses on the protection of every State towards its own population from certain threats.[111] If a State fails in living up to this responsibility to protect its population, the international community, especially through the UN Security Council, must bear the responsibility. It may be argued that climate change will lead to mass migration threatening regional and international stability. Based on this the UN Security Council would have the responsibility to use its Chapter VII powers to intervene. However, military intervention must be seen as a measure of last resort. R2P imposes a range of positive duties upon States to take steps to prevent victimization, statelessness and generally to address the grave circumstances of those living without any protection of their human rights.[112] Initially, the R2P concept explicitly mentioned disasters of a big magnitude as falling within the scope of R2P.[113] However, in 2005 the United Nations General Assembly[114] and, a year later, the Security Council[115] adopted a much narrower reading of R2P limiting the concept to cases in which a population is in need of protection from acts of genocide, ethnic cleansing, war crimes and crimes against humanity. Since then, it has been denied that R2P applies to environmental crises.[116] However, it has been argued that when a State deliberately causes serious injury to its

'responsible sovereignty'. Francis M Deng, 'Frontiers of Sovereignty: A Framework of Protection, Assistance, and Development for the Internally Displaced' (1995) 8(2) Leiden Journal of International Law 249–286.

[110] Ammer, above note 106 at 64.

[111] International Commission on Intervention and State Sovereignty, *The Responsibility to Protect* (ICISS 2001). R2P developed from the efforts to design an international system to protect IDPs. See Roberta Cohen, *Reconciling R2P with IDP Protection* (Global Responsibility to Protect 2010) 15–37.

[112] B Barbour and B Gorlick 'Embracing the Responsibility to Protect. A Repertoire of Measures Including Asylum for Potential Victims' (2008) 20(4) International Journal of Refugee Law 465.

[113] ICISS, above note 111 at 33.

[114] UN Summit Outcome Document 2005, UN Doc. A/Res/60/1.

[115] UNSC Resolution 1674 (2006) S/RES/1674 (2006) 28.04.2006, para. 4.

[116] See, for example, the speech by the UN Secretary General in 2008 warning that the concept of R2P should not be stretched beyond the 2005 consensus. Ban Ki-Moon, Address of the UN Secretary General at event on Responsible Sovereignty, International Cooperation for a Changed World, 15 July 2008.

civilian population by inaction this might invoke R2P.[117] It remains to be seen how useful the concept of R2P will be especially in the context of the slow onset of climate induced migration.

This section has shown that international human rights law does offer a limited degree of protection to people displaced as a result of climate change. Yet, the reliance on a single State to ensure that CDPs can enjoy their human rights is problematic. But the international human rights framework does contain some extraterritorial obligations that are relevant for the protection of CDPs. The duty of international cooperation, as has been laid down in several international human rights instruments, provides a potentially important foundation upon which to build an international framework addressing the effects of climate induced displacements. However, many legal gaps in protection remain, especially where it concerns cross-border displacement following slow-onset climate change.

3.3 Environmental Law Approach

Another way to approach the problem of climate induced displacement is through international environmental law, as it lays down the general responsibility of States for environmental damage and deals specifically with the issue of climate change. However, the question of responsibility under international environmental law, especially in the context of climate change-related effects, is a very controversial issue. There are ongoing debates whether responsibility towards the people induced to displacement by climate change can be based on the principles and rules of international environmental law. The relevance of these principles and rules for the problem of climate induced displacement will be discussed in the following section.

3.3.1 Principles of environmental law

One of the fundamental rules of international customary law is that every State shall be held responsible for transboundary environmental harm.[118] This principle is well established and was already expressed in 1941 in the conclusions of the *Trail Smelter* case:

> no State has the right to use or permit the use of its territory in such a manner as to cause injury by fumes in or to the territory of another or the properties

---

117 Anton and Shelton, above note 86 at 763 (quoting Roberta Cohen).

118 *Trail Smelter Arbitration, United States v Canada* [1938–1941] III RIAA [1905] 33.

> or persons therein, when the case is of serious consequence and the injury is established by clear and convincing evidence.[119]

In environmental law, this is reflected in the 'no harm principle', which is widely recognized and has been laid down in Principle 21 of the 1972 Stockholm Declaration and in Principle 2 of its successor, the 1992 Rio Declaration. Especially after the 1996 Advisory Opinion of the International Court of Justice (ICJ) on *The Legality of the Threat or Use of Nuclear Weapons* there can be no doubt that these provisions reflect a rule of international customary law.[120] Both Declarations establish the sovereign right of States to exploit natural resources, and their respective responsibility to ensure that the activities within their jurisdiction or control do not cause damage to the environment of other States or of areas beyond national jurisdiction.[121]

When a rule of customary international law, such as the 'no harm rule', is breached, States can be held responsible for the damages. To bring forth such a responsibility it has to be established that there is: (a) a wrongful act attributable to a State; (b) a causal link between the activity and damage; (c) a violation of either international law or a violation of a duty of care, which is (d) owed to the damaged State.[122] Yet, in case of responsibility for climate change, particularly in case of responsibility towards people displaced by climate change, proving these elements is highly challenging.

Even if the damaging activity causing climate change can be attributed to a particular State (as might be possible as mentioned earlier by identifying the top emitters of greenhouse gases (GHG)), the question of causation, as in the case with general causation between climate change and population movement, is likely to stay an insuperable obstacle. In this context, it is important to distinguish between general and specific causation. The first one concerns the general link between the amount of

---

119 Ibid.

120 Philippe Sands and Jacqueline Peel, *Principles of International Environmental Law* (3rd edn, Cambridge University Press 2012) 196.

121 Declaration of the United Nations Conference on the Human Environment (Stockholm Declaration) UN Doc. A/CONF/48/14/Rev.1 (16 June 1972), principle 21, Declaration of the United Nations Conference on the Human Environment (Rio Declaration) UN Doc. A/CONF/151/26/Rev.1 (1992), principle 2.

122 Richard S J Tol and Roda Verheyen, 'State Responsibility and Compensation for Climate Change Damages – a Legal and Economic Assessment' (2004) 32 Energy Policy 1111.

anthropogenic emissions of greenhouse gases and climate change damages. This type of causation has universal scientific support and is confirmed by a number of IPCC reports.[123] Specific causation however requires cogent evidence that a particular activity has caused a specific type of damage.[124] Legally speaking, it is problematic to establish that a certain State has caused damage to a particular individual or group of individuals by virtue of its greenhouse gas emissions.[125] All States to some extent have contributed to climate change. To determine that the emission of the particular State has caused the concrete climate related effect, which results in population movement, more likely will be impossible.

Among other reasons which hamper the establishment of causation between a damaging activity and harm, are the spatial and temporal factors. Firstly, States that have caused the harm are more likely to be geographically remote from those who will suffer the consequences. Namely, according to some scientific projections one of the effects of global warming may be the shift of the southern hemisphere cyclone belt to the south. If these prognoses come true the number of cyclones will increase enormously in Oceania, which will result in significant destruction and death. However, the causes of such grave effects will be geographically diffuse, as millions of people from different parts of the world have contributed to climate change. Moreover, people who are harmed will be remote in time from those who have harmed them. Certain Pacific islands, such as Kiribati and Tuvalu may be wiped off the face of the Earth in the twenty-first century, because of people's behavior during the nineteenth and twentieth centuries.[126]

Therefore, it seems that the problem of causation presents an insuperable obstacle in establishing the responsibility under the 'no harm rule' and does not allow for application of this legal principle in cases with EDPs.

Another principle of international customary law, which might be relevant for the protection of those displaced due to climate change, is the 'polluter pays principle'. This principle provides that:

---

[123] For example IPCC, above note 20.

[124] Tol and Verheyen, above note 122 at 1112.

[125] Jane McAdam, 'Environmental Migration Governance' (2009) 1 UNSW Law Research Paper 21.

[126] Stephen M Gardiner et al., *Climate Ethics: Essential Readings* (Oxford University Press 2010) 84.

> [N]ational authorities should endeavor to promote the internalization of environmental costs and the use of economic instruments, taking into account the approach that the polluter should, in principle, bear the costs of pollution, with due regard to the public interests and, without distorting international trade and investment.[127]

This principle has great relevance for CDPs. Developed countries have contributed the most to global warming. However, as discussed above, due to their geographical location and their poor adaptation capacity, developing countries are going to suffer the most of its consequences. The fact that developing countries are not blameworthy for climate change and usually have a limited capacity to deal with the problem makes the financial assistance from those who have caused the problem a logical and fare obligation. In fact in 2010 at the Conference on Climate Change in Cancun (COP16) it was decided to establish a Green Climate Fund, as an operating entity of the financial mechanism of the UNFCCC.[128] The main objective of the Fund is to promote the achievement of the UNFCCC's goals. This contribution is planned to be done through:

> providing support to developing countries to limit or reduce their greenhouse gas emissions and to adapt to the impacts of climate change, taking into account the needs of those developing countries particularly vulnerable to the adverse effects of climate change.[129]

Nevertheless, the 'polluter pays principle' has not received a lot of support and attention among States and in case law. Its legal content remains unclear and it is very doubtful that this principle has achieved the status of a generally applicable rule of customary international law, except perhaps in relation to States in the European Union, the United Nations Economic Commission for Europe (UNECE) and the Organisation for Economic Co-operation and Development (OECD).[130]

Overall, it can be concluded that principles of international environmental law do not provide a sufficient ground for dealing with people

---

127 Declaration of the United Nations Conference on the Human Environment (Rio Declaration) UN Doc. A/CONF/151/26/Rev.1 (1992) Principle 16.

128 Report of the Conference of the Parties on its sixteenth session, held in Cancun from 29 November to 10 December 2010, FCCC/CP/2010/7/Add.1, para. 102.

129 Green Climate Fund Report of the Transitional Committee, Draft decision -/CP.17, Annex, Objectives and guiding principles.

130 Sands and Peel, above note 120 at 229.

induced to displacement by climate change. Even though, the 'no harm rule' and 'polluter pays principle' prima facie seem to be relevant, in practice their application is hampered. With regard to the first rule, the issue of causation poses an obstacle to establishing responsibility, whereas the legal status of the 'polluter pays principle' raises a lot of doubts. There is no case law on this issue yet. Furthermore, as has been already emphasized, pre-emptive actions are crucial in case of climate induced displacement, as they help to avoid uncontrolled migration flights, conflicts and other instabilities. Yet, principles of international environmental law do not promote mitigation and adaptation to climate change, while these are crucial for people induced to relocate, as they can help to avoid displacement.

3.3.2 The UNFCCC and the Kyoto Protocol

Treaties in international environmental law are more concrete than the above-discussed principles and play a more significant role. The UNFCCC and the Kyoto Protocol[131] lay down rules which are better defined and provide for more differentiated obligations regarding implementation, control and enforcement of environmental rules and standards.[132]

The UNFCCC, which entered into force in 1994, is the prime international instrument that deals with the issue of climate change. The main goal of the Convention, as stated in Article 2 is:

> stabilization of greenhouse gas concentrations in the atmosphere at a level that would prevent dangerous anthropogenic interference with the climate system. Such a level should be achieved within a time-frame sufficient to allow ecosystems to adapt naturally to climate change, to ensure that food production is not threatened and to enable economic development to proceed in a sustainable manner.

Among the UNFCCC principles which can be relevant to climate induced displacement, is the requirement for the developed States Parties to take a lead in the protection of the climate system for the benefit of present and future generations, on the basis of equity and, in accordance with

---

[131] United Nations Framework Convention on Climate Change (New York, 9 May 1992), (1992) 31 ILM 849, entered into force 21 March 1994, and Kyoto Protocol to the United Nations Framework on Climate Change, UN Doc FCCC/CP/1997/7/Add. 1 (Kyoto, 10 December 1997), (1998) 37 ILM 22, entered into force 16 February 2005.

[132] Tol and Verheyen, above note 122 at 1114.

common but differentiated responsibilities and respective capabilities.[133] Another essential provision of the UNFCCC is that the specific needs of developing States Parties, especially those that are particularly vulnerable to the adverse effects of climate change and those that would have to bear a disproportionate burden under the UNFCCC, should be given full consideration.[134] Furthermore, the Parties are required to take precautionary measures to prevent or minimize the causes of climate change and mitigate its adverse effects. This principle also lays down that the 'lack of full scientific certainty should not be used as a reason for postponing precautionary measures' where there are threats of serious or irreversible damage.[135]

With regard to climate change mitigation and adaptation, the Kyoto Protocol and the UNFCCC contain a number of more specific provisions which potentially can be relevant for people induced to displacement by climate change. First of all, Article 4(1)(b) UNFCCC obliges all Parties to:

> formulate, implement, publish and regularly update national and, where appropriate, regional programmes containing measures to mitigate climate change (…) and measures to facilitate adequate adaptation to climate change[136]

This provision makes clear that adaptation is not a provisional obligation, but rather a binding rule for the Parties. Further, the UNFCCC requires all the developed countries (so-called Annex II countries) to assist the developing countries in meeting the costs of adaptation to the effects of climate change.[137] Thirdly, the Kyoto Protocol establishes the Adaptation Fund that aims to finance concrete adaptation projects and programs in developing countries which are parties to the Kyoto Protocol, and which are particularly vulnerable to the adverse effects of climate change. The Fund is financed from the revenues from operation of the Clean Development Mechanism and other sources of funding.[138]

However, if we analyze the wording of the UNFCCC, it becomes clear that it only foresees partial funding of adaptation measures by Annex II countries. The way, in which Article 4(4) is formulated, using the term 'assist', makes clear that there is no straight obligation for Annex II

---

133 Article 3(1).
134 Article 3(2).
135 Article 3(3).
136 Article 4(1)(b).
137 Articles 4(3), 4(4).
138 See extensively Chapter 2.

countries to bear the full costs of adaptation in all developing countries.[139] With regard to the responsibility for breaching the obligations of the UNFCCC and the Kyoto Protocol, there are also no clear grounds for bringing such a claim. Article 2 of the UNFCCC only establishes the objective of preventing any damage to the climate system, but does not prohibit the GHG emissions as such. As to the Kyoto Protocol, it also does not tackle the issue of damage, even though it sets legally binding targets. The Protocol presents in Article 2 the list of requirements which Parties are obliged to meet in order to reach reduction targets. However, the way to implement the policies stays largely at the States' discretion, and can be implemented in accordance with national circumstances. Therefore, while the European Union had chosen for more fixed and mandatory regulations, other States, such as Canada, Australia and some other Annex I Parties, have left themselves some space for deviation.[140] The policies and measures prescribed in Article 2 cannot be considered as being mandatory and, thus, establishing respective responsibility does not seem feasible.

Implicitly, the issue of climate induced migration was first mentioned in the UNFCCC assembly text in December 2008.[141] Since then, there have been heated discussions and a lot of research on the issue, which ultimately resulted in the Cancun Agreement – the outcome of the 2010 Conference on Climate Change in Cancun (COP16). The Agreement:

> invites all Parties to enhance action on adaptation ... by undertaking ... measures to enhance understanding, coordination and cooperation with regard to climate change induced displacement, migration and planned relocation, where appropriate, at national, regional and international levels.[142]

COP16 also created an Adaptation Committee and a Subsidiary Body for Implementation (SBI) work program on loss and damage where the issue

---

[139] Tol and Verheyen, above note 122 at 1115.

[140] Benjamin J Richardson, 'Kyoto Protocol to the United Nations Framework Convention on Climate Change' (1998) 2 New Zealand Journal of Environmental Law 253.

[141] Division of International Protection, *Climate Change Induced Displacement: Adaptation Policy in the Context of the UNFCCC Climate Negotiation* (UNHCR Legal and Protection Policy Research Series 2011) 4.

[142] Report of the Conference of the Parties on its sixteenth session, held in Cancun from 29 November to 10 December 2010, FCCC/CP/2010/7/Add.1.

of longer term foreseeable impacts such as sea-level rise and desertification could have implications for society, and established a new Green Climate Fund.[143]

Even though the issue of climate induced displacement was introduced to the UNFCCC agenda, the answer to the question whether the UNFCCC can be used for effective protection of people at risk of climate induced displacement, or as a basis for claiming certain States to be responsible for such protection, is still not positive. This is despite the shown fact that the UNFCCC recognizes the specific needs of developing countries, and that the principle of 'common but differentiated responsibilities and respective capabilities', which requires developed State Parties to 'take the lead in combating climate change and the adverse effects thereof' is a continuing theme throughout the UNFCCC.[144] When it comes to actual responsibility for damaging activities and especially responsibility towards CDPs, it appears too difficult to identify specific State obligations under the UNFCCC. The main reason for that lies in the fact that the UNFCCC sets up mainly common goals, shared principles and general interests of the international community and leaves States a lot of room for maneuver. It does not explicitly provide for specific State obligations and does not resolve the issue of State responsibility either for adaptation, or for damages. Another reason is that the Convention does not address the issue of climate induced displacement directly. Both, the UNFCCC and the Kyoto Protocol are silent about this problem and do not include any provisions concerning specific assistance or protection for those who will be directly affected by the effects of climate change.[145]

Ultimately, it can be concluded that at the current stage environmental law has too many limitations to be considered an effective legal framework for addressing the issue of climate induced displacement and does not offer sufficient protection to CDPs. Furthermore, neither international customary law, nor the provisions of the UNFCCC and the Kyoto Protocol contain any specific provisions on climate induced displacement or responsibilities towards its victims. Even though the 2010 COP16 made a positive contribution towards the recognition of the issue by the

---

143 Report of the Conference of the Parties on its sixteenth session, held in Cancun from 29 November to 10 December 2010, FCCC/CP/2010/7/Add.1. See extensively Chapter 2.

144 See Article 3(1).

145 Informal Group on Migration Displacement and Climate Change of the IASC, *Climate Change, Migration and Displacement: Who will be affected?* (IASC Working paper 2008).

international community, in order to make an actual improvement, further modifications of the environmental legal regime are required.

4 CONCLUSIONS AND OUTLOOK

Climate change and climate change related events have a horrific impact not only on the environment, but also on human inhabitants. The number of natural catastrophes increases as swiftly as the number of people who have to relocate because of them. Should the existing prognoses come true, we can expect between 50 and 350 million climate migrants by 2050. In the current chapter we have investigated the state of existing international law with regard to the issue of climate induced displacement and examined limitations and opportunities in providing legal protection to this group of people. The undertaken analysis allows us to draw the following conclusions.

First of all, as has been shown, the nature of the relationship between climate change and population movement is considered to be highly controversial. This results in numerous associated problems, such as the difficulty in conceptualizing people displaced by climate change and granting them a legal status. Nevertheless, the numerous prognoses made by the top scientific bodies and the UN agencies, and the analysis of the concrete impacts of climate change on certain regions show that there is enough evidence that some areas will become completely unsuitable for living as a result of changing climate. This will leave people no other choice than to relocate. Directly or indirectly, climate change will spur large-scale population displacement threatening regional and global stability. Therefore, it has been argued that instead of getting bogged down in theoretical disputes, policy-makers, politicians and scholars should set the priority on the search for ways to provide a timely response to the issue and protect CDPs.

In that light the current chapter has presented an overview of existing international legal frameworks which potentially can be applied to the issue of climate induced displacement.

The first regime that has been considered is international refugee law, as this is the main instrument at the international level that focuses on the protection of displaced populations. Yet, as has been shown, the attempt to apply refugee law to people induced to displacement by climate change is problematic. The central obstacle is that the definition of 'refugee' under the Refugee Convention does not extend to include people displaced by climate change. There are suggestions that this can be overcome by extending the conventional definition. Nevertheless, this solution seems to overlook some important points. One of them is that the amendment of the Refugee Convention will undermine the protection of political refugees. The UNHCR, which is the main institution with the mandate to assist and protect refugees, is already overloaded with work and lacks financial resources and capacity to assist those people for whom the Refugee Convention was initially created. Should people displaced by climate change be granted a refugee status, the entire refugee regime may collapse. Moreover, it should be kept in mind that some of the countries most severely affected by the issue of climate induced displacement, such as India and Bangladesh are not parties to the Refugee Convention. It is very doubtful that they will have any incentives to join the Convention in future. The underlying rule of the Refugee Convention is that the first host country has to accept refugees. For India the ratification of the Refugee Convention will mean the acceptance of unbearable obligations towards the growing flow of refugees from Bangladesh. Clearly, it is very unlikely that potential host States will be in favor of the extension of the Refugee Convention.

The second legal framework considered in this chapter is that of international human rights law. This is a potentially interesting framework in light of the problems caused by the issue of causality. Under international human rights law, States have the obligation to protect individuals against harm that affects the enjoyment of their human rights even if the State in question is not directly responsible for those threats. Human rights law thus provides a way to bypass the causality issue, which proves to be a significant stumbling-block to including CDPs in international legal frameworks offering protection. On paper, CDPs enjoy all human rights protection that the State they are in is obliged to ensure according to the international human rights treaties they have signed up to. In practice, however, this protection will often prove hard to invoke. The importance of the human rights framework for the protection of CDPs lies in the extraterritorial obligations that can be detected such as the international obligation to cooperate and provide assistance. This can be seen as a building block upon which a framework for the protection of CDPs can be built.

Another international legal framework which has been investigated as potentially relevant to the issue of climate induced displacement, is international environmental law, as it lays down the responsibility of States for environmental damage and deals specifically with the issue of mitigation and adaptation to climate change. Nevertheless, the current analysis has shown that neither customary international law nor the provisions of the UNFCCC and the Kyoto Protocol contain any concrete rules that can help protect people displaced by climate change. At the same time, the research has revealed certain positive features and initiatives towards the recognition of the issue of climate induced displacement within the international environmental legal framework. Therefore, some scholars argue that the problem can be tackled through the amendment of the UNFCCC with a separate protocol, which will focus specifically on the issue of climate induced displacement. The supporters of this solution state several reasons to justify its efficiency. First of all, the UNFCCC has a vast membership, and a lot of developing countries affected by climate change, such as Bangladesh, Tuvalu, Kiribati, and many of the affected African States, are currently among its Parties. The second important advantage arises from the fact that the principle of common but differentiated responsibilities is currently embedded into the UNFCCC, making it attractive for the States threatened by climate change to be supportive of such a protocol. Furthermore, the institutional regime under the UNFCCC is already established, functioning, and there is even an international arena, such as the Conference of the Parties, where this proposal can be introduced and discussed.

Lastly it should be mentioned that some scholars and policy-makers suggest creating a completely new convention, which will focus specifically on people displaced by climate change. An apparent advantage of this approach is its ability to reflect upon the major problems associated with climate induced displacement and to take into account all the previous relevant findings. However, the creation of the new legal instrument is, at the same time, a very time-consuming process. It requires bringing together large number of components, such as the political will, creation of respective institutions, and the sources of sufficient funding. Therefore, taking into account the urgency of the problem, negotiation of a new instrument might be not feasible.

Overall, it can be concluded that at the current stage, the issue of population displacement as a response to climate change is still associated with numerous uncertainties and remains debatable among scholars and politicians. The way to define and conceptualize this group of people remains unclear. This leads to the further confusions on whether the existing body of international law can regulate the issue of climate induced displacement and whether it can provide people with a sufficient level of protection. However, as the problem emerges at a dramatic speed, the international community cannot afford to further postpone addressing the problem of climate induced displacement and must decide on the framework suitable for the protection of people displaced by climate change without further delay.

5. Climate change adaptation and compensation

Michael Faure

1 INTRODUCTION

The various estimations concerning the consequences of climate change make clear that there is a serious probability that the number and intensity of disasters will increase.[1] Not only hurricanes and flooding, but also the mere risk of sea level rise may create substantial damage. This has obviously led to a debate on the necessity of adaptation, and the role that the law could play in that respect, the central focus of this book. The debate on adaptation often focuses on physical measures that public authorities could promote (e.g., building dikes and other infrastructural facilities) to adapt countries to the threat of climate change. Yet another aspect, which will be the central focus of this contribution, is how the legal system will have to deal with the damage that will undoubtedly result as a consequence of these disasters.[2] The advantage of approaching the topic in this way (by focusing on the compensation for victims of disasters) is that there is obviously already a wide debate on how victims of 'normal' disasters (hence, also irrespective of the climate change debate) should be compensated. This debate can also provide fruitful insights concerning the question how to compensate potential victims of climate change.

From a victim's perspective, it may be less interesting to know that a particular damage (e.g., resulting from flooding) was related to climate change or not. The only issue that is important in this respect is that the scientific evidence indicates that it is highly likely that as a result of

[1] See for further details J Spier and E De Jong, *Shaping the Law for Global Crisis: Thoughts about the Role the Law Could Play to Come to Grips with the Major Challenges of Our Time* (Eleven Intl Publ 2012) 13–14.

[2] There are other instruments insurers can use to promote adaptation, for example through measures to maintain the insurability of properties in risky areas: see Chr Courbage and W R Stahel, 'Insurance and Extreme Events' in Chr Courbage and W R Stahel (eds), *Extreme Events and Insurance: 2011 Annus Horribilis* (The Geneva Association 2012) 22.

climate change the number and intensity of natural disasters that would occur anyway (irrespective of climate change) will seriously increase.[3] That, hence, merits a careful analysis of victim compensation as one of the adaptation strategies.

There is some literature addressing explicitly how to compensate victims as a tool to adapt to climate change[4] but there is an overwhelming literature on compensation for victims of catastrophes. An early but very influential publication was Kunreuther (1968)[5] who already then made a case for comprehensive disaster insurance; he and his co-author Michel-Kerjan later published widely on the potential of mitigating disaster losses through insurance.[6] Literature also examined how victims of catastrophes are compensated in various legal systems[7] and which particular problems arise in case of compensating victims in developing countries.[8] This contribution will, to a large extent, build on this literature relating to the compensation for victims of disasters to ask the question to what extent the solutions advanced in that literature can also be used for the specific case of adaptation to climate change.

The remainder of this contribution is set up as follows: first, a few general principles concerning compensation will be sketched which can be useful in evaluating the various instruments that will subsequently be discussed (section 2); next, the possibilities and limits of liability rules

---

[3] In other words: global warming is likely to increase the catastrophe risk vulnerability, see Courbage and Stahel above note 2, 17.

[4] See, e.g., the interesting contribution by J McDonald, 'Paying the Price of Adaptation: Compensating for Climate Change Impacts' in T Bonyhady, A Macintosh, J McDonald (eds), *Adaptation to Climate change: Law and Policy* (The Federation Press 2010).

[5] H Kunreuther, 'The Case for Comprehensive Disaster Insurance' (1968) 11 Journal of Law and Economics 133–163.

[6] See, e.g., H Kunreuther, 'Mitigating Disaster Losses through Insurance' (1996) 12 Journal of Risk and Uncertainty 171 and E O Michel-Kerjan, 'Catastrophe Economics: The National Flood Insurance Programme' (2010) 24 Journal of Economic Perspectives 165, as well as H Kunreuther and E Michel-Kerjan, 'Insurability of (Mega-)Terrorism Risk: Challenges and Perspectives' in OECD, *Terrorism Risk Insurance in OECD Countries* (OECD 2005).

[7] M G Faure and T Hartlief (eds), *Financial Compensation for Victims of Catastrophes: A Comparative Legal Approach* (Springer 2006).

[8] J Linnerooth-Bayer and R Mechler, 'Disaster Safety Nets for Developing Countries: Extending Public-Private Partnerships' (2007) 7 Environmental Hazards 54; and for a focus on adaptation in developing countries see B J Richardson, Y Le Boutillier, H McLeod-Kilmurray and S Wood (eds), *Climate Change and Developing Countries: Legal and Policy Challenges for the World Economy* (Edward Elgar 2009).

will be evaluated (section 3) as well as the role of the potentially central instrument in compensation, being first party insurance (section 4). Then, to the extent that insurance may (for a variety of reasons) not be able to fully cover the damage, the potential role of compensation funds will be discussed (section 5) as well as possibilities to provide direct compensation by government (section 6). Section 7 will address a relatively new feature, being the intervention of government to support the functioning of insurance markets. Section 8 concludes.

The approach that will be taken in this contribution is to examine the various potential instruments that could be used and to analyze their potential strengths and weaknesses from a theoretical perspective. Moreover, to the extent that these instruments are implemented in some legal systems to compensate victims of disasters, examples will be provided and, of course, the potential of the instrument to finance compensation as an instrument of adaptation to climate change will be discussed as well. Hence, after discussing the general principles (section 2) I will address five possible compensation mechanisms (liability, insurance, compensation fund, direct compensation by government and reinsurance by government) whereby, to the extent possible, a similar structure will be followed: the potential (possibilities) of the instruments will be sketched, as well as the limits. Attention will be paid to the question to what extent the compensation mechanism is used in practice and whether it can be considered an appropriate instrument to compensate victims of climate change related damage.

2 GENERAL PRINCIPLES

2.1 Why Compensate?

A preliminary difficulty which could arise is why one should consider the compensation of damage caused by climate change at all. Using an economic perspective one could argue that compensation of victims is, as such, not a goal of the legal system, but rather welfare maximization through the minimization of accident costs. The question of how victims of catastrophes can be provided with appropriate compensation is, in that strict sense, not interesting from an economic perspective.[9] However, some have argued that there may be strong normative beliefs that

[9] R A Posner, *Economic Analysis of Law* (6th edn, Aspen 2003), 167–177.

providing disaster relief is one of the principal functions of government.[10] Moreover, leaving victims of catastrophes without any relief would probably be incompatible with the concept of the welfare state, at least as how this is conceived in most EU Member States.[11]

Indeed, a disaster or other damage resulting from climate change can lead to serious political and economic instability which hence justifies thinking about how to structure an efficient compensation for victims, including which role government should play in that respect. Economics can still be a useful instrument in indicating how a form of compensation can be provided at the lowest possible (administrative) costs and without unnecessary side effects (as far as affecting incentives for prevention is concerned) or negative distributional consequences.[12] Providing specific compensation for victims of catastrophes is indeed problematic from a distributional perspective for the simple reason that some victims (those that are hit by a catastrophe) may receive preferential treatment compared to other victims (hit by an individual accident). This different treatment is as such difficult to explain from the perspective of the equality principle. Some have indeed argued that it may be problematic if compensation would be guaranteed for victims of catastrophes only, whereas for victims of other accidents this generosity would not apply.[13]

From this it can be concluded that providing specific attention to victims of catastrophes (differently than to other victims) may indeed constitute a discrimination, but may on the other hand not only be unavoidable (given the pressure on politicians to act) but also desirable, given the potentially devastating effects of disasters and other damage resulting from climate change on society. From a policy perspective, there are hence good reasons to analyse how adequate compensation to victims could be provided.

---

[10] See G Priest,'The Government, the Market and the Problem of Catastrophic Loss' (1996) 12 Journal of Risk and Uncertainty 219, 235.

[11] See R Schwarze and G Wagner, 'In the Aftermath of Dresden. New Directions in German Flood Insurance' (2004) 29 Geneva Papers on Risk and Insurance 154.

[12] See R A Posner, *Catastrophe. Risk and Response* (OUP 2004).

[13] U Magnus, 'Germany' in M G Faure and T Hartlief, above note 7; and R Zeckhauser, 'The Economics of Catastrophes' (1996) 12 Journal of Risk and Uncertainty 113, 114.

2.2 Principles of Efficient and Fair Compensation

In subsequently analysing the various possible instruments of compensation, a few general rules of thumb could be used to analyse the strengths and weaknesses of the various instruments.

A first important principle is that no matter how compensation is organized, the incentives for preventing damage caused by climate change should always remain intact. This means that in principle the duty to compensate should rest upon the party that actually contributed to the risk, but equally that incentives are also provided to potential victims to mitigate the damage.

A second, related, principle is that a duty to contribute should also be related to the amount by which a specific activity or entrepreneur contributed to the risk. This remains important if a collectivisation of the compensation takes place. If such a risk differentiation is respected, the compensation will give incentives for prevention to contributors. This is not only important from a perspective of potential risk creators, but also for potential victims. This, hence, also means that in principle the individual seeking a particular protection will pay for this protection to the extent that he/she is exposed to the risk. Higher risks do, from this perspective, pay a higher contribution to a financing solution than lower risks. According to this principle, there should be no shifting of risks and costs to a collectivity.

The reason behind this principle is the idea that making individuals pay according to the risk they pose will make them aware of, for example, their exposure to a natural hazard and may hence have a positive impact on their behaviour, i.e., provide incentives for prevention. Hence the efficiency reason behind risk differentiation is to provide incentives for prevention, risk reduction and/or mitigation of damage.

Moreover, these principles are not only important from an efficiency point of view (providing optimal incentives for prevention to all stakeholders in the risk), but also include a fairness element. Indeed, if the principles were not followed, it would mean that good risks would have to pay for the bad risks as well and would therefore in fact subsidize bad risks. This negative redistribution should be avoided and therefore a compensation mechanism (no matter how it is organized) should in principle be financed by the parties actually contributing to the damage. But again, risk differentiation is also important from the victim's perspective. A solidarity on the basis of which all (tax payers) pay for those exposed to risk could imply a redistribution whereby those who accepted risk are rewarded by those who faced no risk. The distributional problem can be made clear on the basis of the following example:

suppose that a particular individual purchases a cheap villa in a flood prone area next to a river. An important reason why the nice villa is obviously cheap is its location in the flood prone area. If the individual could (when the flooding occurs) rely on compensation by government (thus from the general tax payers) a distributional problem could arise. The individual has (at least from an economic perspective) *ex ante* already been compensated for the loss (by purchasing the house at a lower purchase price), may not have taken out insurance and thus not have paid insurance premiums, not have taken the essential preventive measures and can subsequently shift the risk to the collectivity.

Such a solidarity may hence be at odds with efficiency and distributional principles. However, it is easier to apply this risk differentiation to potential injurers (to the extent that their contribution to the risk (i.e., climate change) can be established; it is much more difficult to apply this to potential victims). Indeed, in many developing countries individuals may simply not have another option than choosing to live in a hazard prone area. In such a case, a reverse argument could be made, that there would be a case in favour of providing compensation for that individual (who apparently had no alternative solution than to reside in the hazard prone area) from the collectivity. There is, in that particular case, nothing inherently unfair about redistribution.

A third general principle is that, to the extent possible, a solution should be introduced at the lowest administrative cost possible.

Fourth and finally, where possible, a competitive market solution may (if it is less costly) be preferred to a bureaucratic intervention by government. The market will usually be able to provide coverage at lower costs.[14] However, there may be some (perhaps exceptional) cases where a government monopoly in the provision of disaster insurance could provide better results than competitive markets.[15]

---

[14] See V Bruggeman, M G Faure and K Fiore, 'The Government as Reinsurer of Catastrophic Risks?' (2010) 35 Geneva Papers on Risk and Insurance Theory 381.

[15] This would more particularly be the case in Switzerland. For a discussion see W Emons, 'Imperfect Tests and Natural Insurance Monopolies' (2001) 49 Journal of Industrial Economics 247; Th Von Ungern-Sternberg, *Efficient Monopolies: The Limits of Competition in the European Property Insurance Market* (OUP 2004); Th Von Ungern-Sternberg, 'The Limits of Competition: Housing Insurance in Switzerland' (1996) 40 European Economic Review 1111.

3 ROLE OF LIABILITY RULES: POSSIBILITIES AND LIMITS

3.1 Possibilities

Whenever a liable tortfeasor can be identified, tort law has the main advantage that it can be used to provide a liable tortfeasor with adequate incentives for prevention. This was a point already made by Guido Calabresi in 1961.[16] The deterrent effect of liability rules will force a potential tortfeasor to follow efficient care.

The advantage of a liability rule is that a direct allocation of the social costs of accidents takes place to the source of the risk. The enterprise causing the particular risk will, via liability rules, also be confronted with the costs of its activity, which will precisely give incentives for prevention. Moreover, the tortfeasor, the one who creates the risk, pays the compensation as a result of which a perfect diversification of risk can take place. Liability rules can hence easily respect the principles of efficient and fair compensation since the costs are borne by the one who created the risk.

3.2 Limits

There are, however, various problems with applying liability rules to disasters, especially when the disaster would be related to climate change. A first and obvious problem is that a natural disaster will often be deemed an 'act of God'. Possibilities to apply tort law may be limited. One possibility to apply tort law in case of natural disasters is to argue that public authorities were at fault, for example, by failing to prevent the disaster or not taking adequate measures to mitigate the damage. It could, for example, be held that government failed to give adequate warnings, such as in case of a flooding. Liability questions that therefore arise after natural catastrophes are often linked to the liability of public authorities. This was, for example, also the case after hurricane Katrina.[17] However,

---

[16] See G Calabresi, 'Some Thoughts on Risk Distribution and the Law of Torts' (1961) Yale Law Journal 499; and G Calabresi, *The Costs of Accidents: A Legal and Economic Analysis* (Yale University Press 1970).

[17] See V Bier, 'Hurricane Katrina as bureaucratic Nightmare' in R J Daniels, D F Kettle and H Kunreuther (eds), *On Risk and Disaster: Lessons from Hurricane Katrina* (University of Pennsylvania Press 2006); and J Walters and D F Kettle, 'The Katrina Breakdown' in R J Daniels, D F Kettle and

many legislators have reacted with creating immunities for public authorities in order to limit the scope of governmental liability.

In the particular case of climate change one could of course question whether, for example, damage resulting from a sea level rise or a flooding is indeed the result of a natural disaster. The essence of climate change is that the resulting damage does not have a natural cause, but is precisely 'man-made'. In that case, it would qualify as a technological disaster in which case liability rules could in theory be applied.

3.3 Climate Change Liability

This brings us to the debate on climate change liability and its potential to compensate victims of climate change. For a long time, climate change was presented as the example where liability rules could never be applied. The reason was that it was held that the number of victims was potentially infinite, as well as the number of tortfeasors. Liability law would therefore be inherently inadequate to deal with a problem like climate change. However, inter alia as a result of changes both in scientific evidence as well as in liability law, climate change liability is now an issue which is seriously discussed.[18] Scientific evidence increasingly has possibilities to indicate a likely relationship between a particular natural phenomenon (such as a damaging heat wave) and anthropogenic emissions leading to climate change.[19] Moreover, a change in liability law is that many legal systems now accept a so-called proportional liability rule as a result of which a victim can claim compensation according to the likelihood that a particular tortfeasor contributed to his loss.[20] This may not result in compensation of the full damage, but of a proportion of it. Especially in cases where the amount of emitters is relatively large (and hence they contribute substantially to climate change), the feasibility of climate change liability should not, *ab*

H Kunreuther (eds), *On Risk and Disaster: Lessons from Hurricane Katrina* (University of Pennsylvania Press 2006).

[18] For an elaborate discussion see Spier and De Jong, above note 1 at 67–198.

[19] See inter alia M Allen, 'Attributing Extreme Weather Events: Implications for Liability' in Munich Re, *Liability for Climate Change? Experts' View on a Potential Emerging Risk* (Munich Re 2010); and Spier and De Jong, above note 1 at 17.

[20] See on the causation requirement in climate change liability also Spier and De Jong, above note 1 at 175–180.

initio, be totally ruled out.[21] Moreover, one could even argue that developed countries should be held liable to pay the adaptation costs of developing countries.[22]

However, even if the cynicism concerning the potential of climate change liability could be overcome, there are still substantial problems. For example, even if the causation hurdle could be met, another problem that would arise is that, at least in European legal systems, industry could claim that they already contributed to abatement via participation in the Emission Trading Scheme (ETS). Participation in the ETS would, from that perspective, have a pre-emptive effect in tort law. Even though this argument may not be watertight,[23] it may lead to a reluctance by judges, at least in the European context, to award liability suits against enterprises already participating in the ETS. That problem may not, however, arise in legal systems with a less developed legal framework addressing the mitigation of the emission of greenhouse gases, such as in the US at federal level. It is therefore not surprising that climate change liability cases have especially been brought in the US,[24] but its potential is examined for European legal systems as well.[25]

Another issue is that one could also question whether it is useful to use climate change litigation with the aim of compensating victims. Most climate change suits nowadays do not aim at compensation but rather at mitigation. Suing for compensation may, moreover, in times of financial crises, even lead to serious economic difficulties for enterprises and would from that perspective also not be very likely.[26] Another weakness in liability rules is, especially in developing countries, that victims may face large hurdles, for example, to go to court and effectively litigate. Legal aid, contingency fees or other instruments to lower the barriers to

---

[21] For the potential of a climate change liability suit, see the contributions in M G Faure and M Peeters (eds), *Climate Change Liability* (Edward Elgar 2011).

[22] See Spier and De Jong, above note 1 at 191–192.

[23] Spier and De Jong rightly argue that neither compliance with international agreements, nor compliance with national law (or with permits) exclude the possible application of liability law, above note 1 at 167–171.

[24] For a discussion of these cases, see M Haritz, *An Inconvenient Deliberation: The Precautionary Principle's Contribution to the Uncertainties Surrounding Climate Change Liability* (Maastricht University 2010).

[25] For the possibilities of climate change liability under English law, see, e.g., G Kaminskaité-Salters, *Climate Change Liability under English Law* (Maastricht University 2010).

[26] See Spier and De Jong, above note 1 at 182–187.

access to justice are often insufficiently available in developing countries.[27] Precisely given these hurdles for victims the use of liability rules may also have a rather ad hoc character as far as compensating victims of catastrophes is concerned. In other words: some may get (generous) compensation if they are successful in the 'tort law lottery', whereas others may receive no compensation whatsoever. Compensation via liability rules, therefore, has no structural nature and may also come at odds with the equality principle. A best liability law should be used, not to obtain compensation, but to require injunctive relief.[28]

This shows that, notwithstanding the increasing interest in the role of climate change liability, it is not very likely that this will provide a structural solution as far as the compensation of victims is concerned. Hence, other solutions will have to be examined.

4 FIRST PARTY INSURANCE

4.1 Possibilities

The type of insurance we refer to in this chapter is a so-called first party insurance coverage. This is a system whereby insurance coverage is provided and compensation awarded directly by the insurer to the victim. It is thus also the potential victim himself who buys this type of insurance coverage with the eye on possible future harm and corresponding damages. In a first party insurance the insurance coverage is triggered by damage, not by liability.[29] The main benefits of first party insurance are strongly advocated in law and economics literature. Priest has especially argued that first party insurance has the advantage that it allows an easier risk segregation than third party insurance (like liability insurance).[30] Particularly in the area of catastrophes, the ability of first party insurance to differentiate risks is quite important: first party insurance allows insurance companies to adapt the premium and policy

---

[27] For difficulties concerning access to justice see the contributions in M Tuil and L Visscher (eds), *New Trends in Financing Civil Litigation in Europe: A Legal, Empirical, and Economic Analysis* (Edward Elgar 2010).

[28] See Spier and De Jong, above note 1 at 194–199, who clearly argue 'compensation, all in all, is the wrong track in my view' (194).

[29] See generally on the use of first party insurance to cover catastrophes M Faure and V Bruggeman, 'Catastropic Risks and First Party Insurance' (2008) 15 Connecticut Insurance Law Journal 11–14.

[30] G Priest, 'The Current Insurance Crisis and Modern Tort Law' (1987) 968 Yale Law Journal 15–21.

conditions to the risk.[31] For the area of catastrophes this would mean that insurers could, for example, charge a higher premium for risky activities (such as building a house on a flood plain) or rewarding the good risks (e.g., those who construct their dwellings using materials that are less vulnerable to the impact of earthquakes). First party insurance therefore potentially has many advantages. One advantage is that victims in principle are able to purchase disaster coverage according to their own preferences; the other is that the insurer can (via the control of moral hazard) charge risk dependent premiums and hence positively affect the risk. Insurance has the main advantage that a correct pricing of the risk takes place.[32]

4.2 Limits

Notwithstanding these theoretical advantages of first party insurance as a mechanism to deal with catastrophes, the literature has indicated that for a variety of reasons in practice people do not use these theoretical advantages of first party insurance to the full extent.

A first problem is that empirical evidence has demonstrated that even in those countries where disaster insurance is widely available (such as in the US and in Europe), individuals tend not to make sufficient use of it, resulting in dramatic cases of underinsurance. This was shown, for example, after the 'flood of the century' of the river Elbe in Germany[33] as well as in the US, especially after Hurricane Katrina.[34] Several reasons are indicated for this low demand for disaster insurance. First, as a result of cognitive limitations, low probability events like natural disasters are systematically misjudged[35] resulting in a 'it will not happen to me' attitude.[36] Second, there is empirical evidence that people *ex ante* prefer

---

[31] See G Priest, above note 10.

[32] So strongly Kunreuther, above note 6.

[33] See for example A Endres, C Ohl and B Rundshagen, 'Land Unter! Ein Institutionenökonomischer Zwischenruf' (2003) 29 List Forum für Wirtschafts- und Finanzpolitik 294–294; U Magnus, 'Germany' in M Faure and T Hartlief (eds), *Compensation for Victims of Catastrophes: A Comparative Legal Approach* (Springer 2006) 119–144; and Schwarze and Wagner, above note 11.

[34] See R J Daniels, D F Kettle and H Kunreuther (eds), *On Risk and Disaster: Lessons from Hurricane Katrina* (University of Pennsylvania Press 2006).

[35] See P Slovic, H Kunreuther et al., 'Decision Processes, Rationality and Adjustment to Natural Hazards' in P Slovic (ed.), *The Perception of Risk* (Earthscan 2000) 1–31.

[36] Kunreuther, above note 6 at 175.

uncertain losses rather than the certain loss incurred by paying the premium. Insurance is considered an investment. The problem with disaster insurance is that a potential victim (like a house owner) is confronted with the certain loss of a premium, whereas there is a low expectation of a return on the 'investment' during a lifetime and hence a low demand.[37] Third, some literature indicates that *ex post* government relief (i.e., provided after a disaster) may also reduce incentives to purchase insurance coverage.[38]

A second problem is related to the supply side, more particularly to the 'difficult to predict' nature of catastrophes.[39] Many have indeed argued that the losses due to a catastrophe can be of such a nature that they can also endanger the financial viability of insurance companies.[40] Not only can the capacity of individual insurers be too limited for the large amount of losses caused by a catastrophe. The low probability, high loss character of a catastrophe makes it also difficult to insure.[41]

A third problem is related to the fact that insurance may provide a wonderful solution to catastrophes in the developed world, but will it really help for developing countries where many individuals lack the financial capacity to pay for the basic needs, let alone for an insurance premium? This, hence, comes down to the simple fact that even in situations where individuals are aware of the catastrophic risks they are exposed to and there would be sufficient supply by insurers, insurance coverage may simply not emerge for the reason that insurance coverage is not affordable because of a lack of funds.

---

37 P Slovic, B Fischhoff et al., 'Preference for Insuring against Probably Small Losses: Insurance Implications' (1977) 44 Journal of Risk and Insurance 237–258.

38 This is a point strongly made by R A Epstein, 'Catastrophic Responses to Catastrophic Risks' (1996) 12 Journal of Risk and Uncertainty 287–308; and by S E Harrington, 'Rethinking Disaster Policy' (2000) 23(1) Regulation 40–46.

39 For a more detailed analysis of problems on the supply side, see C Gollier, 'Some Aspects of the Economics of Catastrophe Risk Insurance' in OECD, *Catastrophic Risks and Insurance* (OECD Publishing 2005) 13–30; and the contributions in K A Froot, *The Financing of Catastrophic Risk* (University of Chicago Press 1999).

40 Attention is also increasingly paid to the consequences of climate change for insurance companies. See in that respect Spier and De Jong, above note 1 at 19–20 and see Courbage and Stahel, above note 2.

41 The basic problem is that the level of capital for insuring extreme events is often higher than can be collected via insurance premiums. See Courbage and Stahel, above note 2 at 24.

4.3 Remedies

Although the problems we have just indicated are absolutely serious, the literature has indicated that there are possible remedies to deal with these problems which may make insurance for catastrophes again possible. Looking at the first problem – the fact that there is a relatively low demand for insurance coverage – even if it would increase expected utility of individuals, an obvious solution would be to make disaster insurance mandatory. Many scholars and especially Kunreuther have argued in favour of compulsory (first party) insurance for property damage caused by all kinds of natural disasters.[42] An example of such a mandatory comprehensive disaster insurance constitutes the case of France. In that country, homeowners who purchase voluntary housing insurance are, as a result of a regulatory duty, automatically insured against damage caused by natural disasters.[43] This model, whereby the duty to insure against disasters is combined with an insurance against a high probability, low damage event, is also supported by behavioural literature. Individuals would prefer disaster insurance when it is combined with an insurance against high probability, low damage losses.[44] Thus it is argued that compulsory insurance plays an important role in improving hazard perception.[45] Even though such a compulsory insurance does not go totally undisputed (especially in those cases where it would also force individuals who run no risk at all[46] to take insurance

[42] It is an argument already made by Kunreuther in 1968, H Kunreuther, 'The Case for Comprehensive Dïsaster Insurance' (1968) 11(1) Journal of Law & Economics 133–163; and repeated regularly, inter alia, after Katrina, see H Kunreuther and M Pauly, 'Insurance Decision-making and Market Behaviour' (2005) 1(2) Foundations and Trends in Microeconomics 63–127.

[43] See O Moréteau, M Cannarsa et al., 'France' in M G Faure and T Hartlief (eds), *Financial Compensation for Victims of Catastrophes. A Comparative Legal Approach* (Springer 2006) 81–118; and O Moréteau, 'Policing the Compensation of Victims of Catastrophes: Combining Solidarity and Self-Responsibility' in W H van Boom and M Faure (eds), *Shifts in Compensation between Private and Public Systems* (Springer 2007) 199–218.

[44] See P Slovic (ed.), *The Perception of Risk* (Earthscan 2000) at 60–61 and 70–71.

[45] Slovic, Kunreuther et al., above note 35 at 25.

[46] Like the person living in an apartment on the tenth floor who barely has any risk of being flooded.

coverage), there seems to be support for such a regulatory policy in law and economics scholarship.[47]

As far as the failing of affordability of insurance premiums in developing countries is concerned, in this respect interesting proposals have also been formulated in the literature. Kunreuther held that it remains important that insurance premiums still reflect risk. If individuals in disaster prone areas lack capacity to pay insurance premiums, government could provide insurance vouchers or victims could be reimbursed by the government for a portion of the increased costs of insurance coverage. But the advantage would be that premiums would still reflect risk. Moreover, through the voucher system individuals could still be incentivized to adopt mitigation measures.[48] This relies on a study by Coate, who argues that if the government makes in-kind transfers of insurance to the poor they will not rely on disaster relief *ex post* in case of a loss.[49] This shows that through such a subsidization of insurance premiums the charity hazard could be avoided.[50] There is indeed a trade-off between on the one hand the efficiency requirement of charging risk dependent premiums and on the other hand the equity problem that risk dependent premiums may lead to problems for some individuals who are not in a position to reduce their risk exposure at a reasonable cost and who cannot afford the insurance premiums.[51] Nowadays, many interesting examples of insurance solutions are also provided in developing countries. Several countries have developed micro-insurance for natural

---

[47] Mandatory disaster insurance is, however, not the central focus of this chapter. For more details see Faure and Bruggeman, above note 29. See also S Levmore and K D Logue, 'Insuring Against Terrorism – and Crime' (2003) 102 Michigan Law Review 268, 304.

[48] H Kunreuther, 'Catastrophic Insurance: Challenges for the US and Asia' in Ch Scawthorn and K Kobayashi (eds), *Asian Catastrophe Insurance* (Risk Books 2008) 3, 13.

[49] S Coate, 'Altruism, the Samaritan's Dilemma and Government Transfer Policy' (1995) 85(1) American Economic Review 46–57.

[50] See in this respect also S E Harrington and G Niehaus, 'Government Insurance, Tax Policy and the Affordability and Availability of Catastrophe Insurance' (2001) 19(4) Journal of Insurance Regulation 591, and P Picard, 'Natural Disaster Insurance and the Equity-Efficiency Trade Off' (2008) 75(1) Journal of Risk and Insurance 17.

[51] If indeed risk reducing behaviour is not possible the insurance premium becomes analogous to a lump sum tax without any significant incentive effect, so Picard, ibid. at 19.

disasters.[52] For example, both in Indonesia and in Bangladesh private micro-insurance to deal with flood risks has been developed.[53]

4.4 Summary

Summarizing, insurance has a lot of potential in compensating victims for the damage resulting from climate change. The advantage is that insurance solutions can respect principles of efficient and fair compensation since insurers apply principles of risk differentiation and thereby also provide incentives for mitigation of damage (and hence for adaptation to climate change). However, given many behavioural biases, a voluntary scheme may not work since for a variety of reasons individuals may not *ex ante* purchase insurance coverage even if it would be beneficial (and even if, as is, e.g., the case in the US with the National Flood Insurance Plan, premiums are heavily subsidized). The type of mandatory coverage already proposed by Kunreuther since 1968 would then be the appropriate solution. However, as we also indicated, prescribing mandatory disaster coverage is one issue, making sure that individuals can really afford the coverage is a different one. That may especially be a problem in the case of developing countries where, hence, some form of subsidized insurance would have to be introduced. It is argued in the literature that if the government were to intervene to support insurance markets, providing a subsidy to low income individuals to purchase insurance (e.g., in the case of developing countries) would be more desirable than direct compensation by the government.[54] The reason is that insurance (even if subsidized) could still have an incentive effect of making individuals aware of differences in risk exposure (via differentiated premiums), whereas such an incentive effect (providing incentives for adaptation) would totally fail in case of direct compensation by the government.

However, there is one particular problem with insurance addressed above, being the fact that the scope of disasters can be such that insurance coverage may not be able to (totally) deal with the problem, that has not been addressed yet. That may, of course, especially in the

---

52 For an overview see Linnerooth-Bayer and Mechler, above note 8 at 10–11 and R Mechler, *Natural Disaster Risk Management and Financing Disaster Losses in Developing Countries* (Verlag für Versicherungswirtschaft 2004).

53 For details concerning the Bangladesh system, see S Akter et al., 'Exploring the Feasibility of Private Micro Flood Insurance Provision in Bangladesh' (2011) 35(2) Disasters 287.

54 To be discussed in section 6 below.

case of climate change, be a serious problem. Given some of the catastrophic scenarios sketched by the International Panel on Climate Change (IPCC), the damage resulting from climate change could potentially largely outweigh possibilities of the commercial insurance and reinsurance market. Hence, this still merits the question of whether (in addition to insurance) alternatives such as a compensation fund (section 5), direct compensation by government (section 6) or an intervention of government as reinsurer of last resort (section 7) would be indicated.

5 COMPENSATION FUNDS AND TAXATION

5.1 Possibilities

In the environmental field sometimes suggestions have been formulated that a compensation fund could be created to deal with environmental damage. The notion of a fund, however, can cover very different types of compensation mechanisms.[55]

Theoretically, a compensation fund could also be created *ex ante* to deal with damage resulting from climate change. Taking into account the principles of efficient and fair compensation mentioned in section 2, the question arises whether a compensation fund would be able to provide better protection than private insurance markets. In principle, it can be assumed that an insurer is better able to differentiate risks, being a specialist in risk differentiation and risk spreading. Insurers have the techniques to determine the share their insured contributed to the risk and if competitive insurance markets offer coverage, insurance may be the preferred option. A compensation fund could be used if the agency, acting as administrator of the fund, would have at least as good information on risks as an insurer.

5.2 Limits

If both insurance and a compensation fund are available, there seem to be no clear reasons why a fund would be the preferred option, and there may even be a preference for insurance. If the fund would be run by the government, the contributions paid to the fund could in fact amount to a tax on emissions. However, governments may lack information on the

---

[55] For an overview, see M G Faure and T Hartlief, 'Compensation Funds versus Liability and Insurance for Remedying Environmental Damage' (1996) 5(4) RECIEL 321.

particular situation of the various victims and thus opt for a lump sum payment of standardized amounts. In order to reduce moral hazard, compensation will not be full, but the incentivizing effects may be less than with differentiated insurance premiums under first party insurance.

A structural fund solution has, as far as disasters are concerned, also been criticized in economic literature. For example Gron and Sykes have argued that a structural fund may provide the wrong signal to the market. If market participants are aware that when a catastrophe occurs the financial consequences will be covered through government intervention. This will provide them with little incentive to develop financial solutions themselves.[56] A structural fund solution may also lead to the charity hazard[57] which occurs with any compensation by the government: it can dilute the incentives to self-insure by the victim or to take preventive measures in the framework of an effective risk management.[58]

5.3 A Climate Change Compensation Fund?

In the case of damage caused by climate change, there is not, *ab initio*, a reason to argue that insurance should do better than a compensation fund, on the condition that the compensation fund equally charges risk based premiums. A risk based contribution in case of a compensation fund for climate change damage would certainly be possible if one were to assume (which does not seem like such a strange idea) that the amount of greenhouse gas emissions of a particular industry would be a proxy for its contribution to climate change. Hence, the contribution to a climate change fund could be geared towards industries' greenhouse gas emissions, and hence made proportional and thus risk dependent. The advantage would then be (like in the case of insurance) that the contributions to the fund would provide incentives for mitigation. However, the administration of the fund should not only provide incentives to potential risk creators, but also to potential victims. Hence, the administrator of the fund would also have to take the individual situation of every victim into account and differentiate compensation payments on the basis of the victim's possibility to mitigate the damage. Insurers, specialized in risk differentiation, are probably better able to make such a differentiation than an administrator of a fund, especially in those countries where, for example, insurance for flooding risks is available.

---

56 A Gron and A O Sykes, 'A Role for Government' (2002) 25 Regulation 4.

57 To be discussed below in section 6.

58 Endres, Ohl and Rundshagen, above note 33 at 290.

For that reason, a compensation fund should, at best, only play a role to back up an insurance solution, for example, for amounts that are not coverable on insurance or reinsurance markets. However, in that case, one also has to realize that one would impose a tax (related to the amount of emissions) upon industry in addition to other regulatory solutions aiming at mitigation of emissions (like a command and control scheme or emissions trading). The political acceptability of such a climate change tax to finance a compensation mechanism that would be meant to back up insurance may be a difficult issue, taking into account the traditional resistance of industry against a carbon tax. Industry may feel that a contribution to such a compensation fund would hence reintroduce a carbon tax via the backdoor.

5.4 Practice

There are, in practice, some cases where compensation funds are used, for example, in the area of the oil pollution risk. The basic oil pollution risk is covered via the so-called protection and indemnity clubs which are based on a mutual risk sharing agreement between tanker owners.[59] A second layer of compensation is provided by an international oil pollution compensation fund, run by a specialized agency, where the maritime world has a large stake. In these cases, it could be argued that the operators themselves might be better suited than an insurance company to control moral hazard since they are better able to process information on particular risks.

There are, however, few general disaster funds in Europe. Belgium has a fund solution since 1976, but it applies a deductible and has standardized limited amounts. It is completely financed through the public purse and the amounts of compensation are hence relatively modest. In 2003, the Belgian system was dramatically changed and Belgium turned to the French solution of additional mandatory coverage on voluntary insurance policies. The Belgium fund solution obviously had the disadvantage that since it was funded by the public purse, there was no incentive whatsoever on risk creators (which was in that case pretty obvious since the goal was mainly to compensate victims from natural disasters). Another European country that has a structural compensation fund is

---

[59] See T G Coghlin, 'Protection and Indemnity Clubs' (1994) Lloyds Maritime and Commercial Law Quarterly 403.

Austria.[60] However, the Austrian fund only covers a relatively limited part of the damage. It seems, thus, that only few countries in Western Europe still have structural disaster funds and that insurance seems to be the preferred solution today.[61]

5.5 Summary

There may be some potential for a climate change compensation fund. However, the financing of the fund should always be structured in such a way that risk dependent contributions can be charged in order to provide appropriate incentives for emission reductions. Insurance companies may be better able to differentiate risks than the manager of a compensation fund. Moreover, contributions to a climate fund would in practice often amount to a tax on emissions of which the political feasibility is not very high.

Moreover, a compensation fund might not only lead to perverse incentives on the side of emitters (if no risk dependent premiums are charged), but could equally lead to a charity hazard on the side of victims. In that respect, it suffers from the same flaws as direct compensation by government.

6 DIRECT COMPENSATION BY GOVERNMENT

6.1 Possibilities

One possibility to organize the compensation for victims of climate change is obviously to simply assume that governments will pay compensation as a form of charity. This looks like the compensation fund discussed above, but the major difference is that there would be no structural solution *ex ante* (defining precisely the conditions under which compensation would be paid), but compensation would be *ex post* and ad hoc. In that case, the government (assuming the national government) would *ex post* decide, on the basis of the scope and nature of the damage, how much of what particular type of relief would be provided. Usually, it

---

[60] D Hinghofer-Szalkay and B A Koch, 'Austria' in M G Faure and T Hartlief (eds), *Financial Compensation for Victims of Catastrophes. A Comparative Legal Approach* (Springer 2006) 7–36.

[61] For an overview see M Faure, 'Financial Compensation for Victims of Catastrophes: A Law and Economics Perspective' (2007) 29 (3) Law & Policy 339.

involves *ex post* relief through lump sum payments to victims that are financed by the general tax payers.

Although there are obvious (economic) criticisms that can be formulated (and have been in literature) on this type of government charity there are some positive aspects as well. A positive aspect of government intervention is that the prospects of large-scale payments in the aftermath of a disaster might encourage the government to take cost–benefit justified precautions long before disasters strike.[62] This argument may be strong in the case of damage caused by climate change. Climate change may cause substantial damage, especially when the damage does not occur gradually but is related to a disaster. Often, this can lead to a total disruption of society. Providing *ex post* relief may then help to restore public trust. Moreover, government may be better able to determine adaptation strategies necessary to react against climate change. Think of the example of large infrastructural works which would be necessary to prevent catastrophes (e.g., building dikes against tsunamis). These are typically the public goods that would not be provided through private action and may hence require government intervention.

A related argument in favour of government intervention would be that the government has the capacity to diversify the risks over the entire population and to spread past losses to future generations, thus creating a form of cross-time diversification, which the market could not achieve. The argument can be made that the government can be in a better position to adapt to climate change than individuals.[63] Hence, in some cases the government may be in the best position to prevent disasters. Government intervention would from this perspective provide incentives to politicians to invest in preventive measures. One can think about building dikes against the risk of flooding or sea level rise resulting from climate change or effective zoning and planning decisions to avoid locating residences in flood prone areas.

Moreover, one could also make the (too easy) argument that the government can (in principle) compensate without limits. If the damage would exceed the current budgetary possibilities of the government, the already mentioned cross diversification over time and future generations could in principle take place.

---

[62] See Levmore and Logue, above note 47 at 310.

[63] This argument has been strongly made by Kunreuther and Michel-Kerjan, above note 6.

6.2 Limits

Notwithstanding the theoretical advantages of government compensation, the literature also points at high costs and disadvantages of this type of government intervention.

A first major disadvantage of the lump sum payment under government relief is that no incentives are provided to potential victims to take effective preventive measures. Since the payments under government relief usually do not relate to risk, they offer no incentives for taking preventive measures. Of course it will very much depend upon the nature of the disaster whether it is realistic that victims do take preventive measures. Preventive measures that need great infrastructural *ex ante* disaster management will, for the public good reason discussed above, be primarily undertaken by the government. Taking structural measures for example, to protect a country against the consequences of climate change, such as sea level rise, hurricanes or flooding, will primarily be a government task. However, there are certainly measures individual potential victims can take, not so much to prevent a (natural) disaster, but rather to limit the impact and damage. For example, as far as flood protection is concerned, an important factor relates to the location decision. Damage could obviously be prevented by not locating in flood prone areas. But again the government may be the best situated to know where these flood prone areas are and thus, for example via zoning and refusal of building permits, prevent location in these types of areas. However, even when exposed to particular risks victims can to some extent take preventive measures (e.g., in case of flooding not installing the most valuable objects in the cellar or on the ground floor). These measures may be relatively limited compared to the impact of infrastructural works that governments could undertake, but they remain important. The problem is that *ex post* relief by government may not provide incentives to take those appropriate preventive measures.[64] In that respect, the literature argues that competitive insurance markets are better able to deal with moral hazard and adverse selection via risk differentiation.[65]

Another problem is that victims may be counting on government compensation, which may exactly create an incentive not to purchase

---

[64] For further examples see V Bruggeman, M G Faure and M M Haritz, 'Remodelling Reparation: Changes in the Compensation of Victims of Natural Catastrophes in Belgium and the Netherlands' (2011) 35(4) Disasters 766.

[65] Priest, above note 10.

insurance.[66] The problem of government provided compensation is indeed that it may dilute incentives to purchase insurance since victims could simply free ride on the State.[67] In the words of Gollier: 'Solidarity kills market insurance'.[68] Coate has identified the lack of insurance resulting from the generosity of the government.[69] The problem has been referred to as the 'charity hazard'.[70] Also an experimental study on crop insurance in the Netherlands showed that the willingness of producers to purchase private insurance (supported by the government) was significantly negatively influenced by the producer's belief about the availability of disaster relief in the future.[71] The study concluded that if governments continue to provide free ad hoc disaster relief, an important incentive to participate in a crop insurance programme would be severely undermined.[72] A similar conclusion was also recently reached based on an empirical research comparing compensation mechanisms available in different countries after the August floods of 2005 in Austria, Switzerland and the German state of Bavaria.[73] The study argued that there was a substantial charity hazard leading to a lower purchase of insurance in Austria, where there was a disaster fund which pays (low) amounts of compensation. Participation in flood insurance was higher in Bavaria, although it was the highest in the Swiss canton of Grison, which had a public monopoly on insurance and mandatory participation.[74]

---

[66] Levmore and Logue, above note 47 at 281 and L Kaplow, 'Incentives and Government Relief for Risk' (1991) 4(2) Journal of Risk and Uncertainty 167.

[67] Gron and Sykes, above note 56; A Gron and A O Sykes, 'Terrorism and Insurance Markets: A Role for the Government as Insurer?' (2003) 36 Indiana Law Review 447; Endres, Ohl and Rundshagen, above note 33 at 290.

[68] Gollier, above note 39 at 25.

[69] Coate, above note 49.

[70] More particularly by P Raschky and H Weck-Hannemann, 'Charity Hazard – Real Hazard to Natural Disaster Insurance' (2007) 7 Environmental Hazard 321.

[71] See with respect to a crop insurance programme in the Netherlands M van Asseldonk, M Meuwissen and B Huirne, 'Belief in Disaster Relief and the Demand for a Public-Private Insurance Programme' (2002) 24(1) Review of Agricultural Economics 196.

[72] Ibid.

[73] See P Raschky, R Schwarze, M Schwindt and H Weck-Hannemann, *Alternative Financing and Insurance Solutions for Natural Hazards: A Comparison of Different Risk Transfer Systems in Three Countries – Germany, Austria and Switzerland – affected by the August 2005 Floods* (AlpS GmbH Centre for Natural Hazard and Risk Management 2009).

[74] Ibid.

A related problem is that there may be negative distributional effects since some victims (who probably purchased houses at low prices in flood prone areas) may free ride on other individuals (the general tax payers) who finance the *ex post* relief.

Ex post government compensation is, therefore, generally seen as problematic by law and economics scholars, which is nicely expressed by the title of a contribution by Epstein. He qualifies *ex post* relief as 'catastrophic responses to catastrophic risks'.[75] If one would weigh these disadvantages against the potential benefits (in providing incentives to the government to take measures aiming at the prevention of disasters), the disadvantages seem to be overwhelming. It is, indeed, not clear to what extent the payment of *ex post* relief effectively provides incentives to a government for taking preventive measures. Moreover, as Depoorter recently showed, the foresight of having to pay *ex post* relief in the future apparently does not at all provide incentives to politicians to invest in disaster prevention mechanisms for the simple reason that the political benefits may be different. The result is, as Depoorter showed, precisely an oversupply of *ex post* relief and undersupply of disaster preparation measures.[76]

6.3 Practice

This is related to the fact that the positive incentive effects on politicians from having to pay compensation may largely be absent. Indeed, Depoorter has argued that politicians may receive too little reward from *ex ante* disaster management policies as a result of which they may be undersupplied. Also, political benefits of planning activities may be shared with other levels of government. The political reward for *ex post* compensation may, to the contrary, be very strong as a result of which *ex post* relief is likely to be oversupplied.[77] This has also been confirmed in other literature. Public choice theory shows that governments will always tend to intervene when the number of victims is large.[78] Potential victims are, after all, voters and the government will try to seduce the largest number of electors in order to gain power or maintain its existing power.

---

[75] Epstein, above note 38.

[76] B Depoorter, 'Horizontal Political Externalities: The Supply and Demand of Disaster Management' (2006) 56 Duke Law Journal 101.

[77] Ibid. Depoorter also provides empirical evidence that this may have been the case after Katrina.

[78] See inter alia Kaplow, above note 66; Epstein, above note 38; Gron and Sykes, above notes 56 and 67.

A political (and strategic) interest might thus be hidden behind the argument of victims' protection. Public choice theory also demonstrates that public intervention is costly. The organization of the financing of compensation by the State (especially through a public insurance pool) can be analysed analogically as a bureaucracy. In that respect, Tullock shows that this type of scheme might suffer from financial waste and cost-inefficiencies.[79] This waste appears to be due mainly to internal rent-seeking activities and in the case of a public insurance pool, to the capture of a monopolistic profit.[80] Indeed, Gerrett and Sobel showed that disaster expenditures by the Federal Emergency Management Agency (FEMA) are mostly politically motivated. States that are politically important to the President have a higher rate of disaster declaration by the President and disaster expenditures are higher in states having congressional representation on FEMA oversight committees.[81] Hence, they argue that nearly half of all disaster relief is motivated politically rather than by need. Finally, we can also point to the fact that from the victims' perspective ad hoc compensation by government still remains ad hoc, in other words uncertain. Some victims may, after a (large and politically important) disaster, receive generous compensation, whereas if the disaster would receive less attention in the media and thus be politically less sensitive, government compensation could be absent. The ad hoc nature of charity by the government hence creates inequality.

Moreover, the assumption that governments can in principle always compensate without limit is certainly not correct either. Some small island states may be subject to large catastrophes of which the costs can easily outweigh their budgetary possibilities. This is a problem which is recognized by the World Bank, as a result of which they have also examined the possibility to develop sovereign natural disaster insurance for developing countries in order to solve the short-term liquidity need.[82]

---

79 G Tullock, 'The Welfare Costs of Tariffs, Monopolies, and Theft' (1967) 5(3) Western Economic Journal 224.

80 Obviously, the criticisms partially depend upon the nature of the arrangement. To the extent that government relief can be structured in such a way that some incentives can still be provided to victims (e.g., by including deductibles or by differentiating the amount of compensation taking into account preventive measures taken) and that the financing can to some extent still be risk-based (although that is seldom the case) the criticisms may be less serious.

81 Th A Gerrett and R S Sobel, 'The Political Economy of FEMA Disaster Payments' (2003) 41(3) Economic Inquiry 496.

82 See F Ghesquiere and O Mahul, 'Sovereign Natural Disaster Insurance for Developing Countries: A Paradigm Shift in Catastrophe Risk Financing', (2007) The World Bank Policy Research Working Paper WPS 4345.

For example, for a country like Indonesia the compensation resulting from disasters amounted in 2006 to USD 1,278 million. This shows that more particularly for developing countries the assumption that governments will solve all compensation problems is simply not realistic.

6.4 Summary

It is therefore not difficult to argue that charity by the government may not be the most appropriate way to provide compensation to victims. In developed countries, this could create a charity hazard, reducing incentives to develop market solutions (such as insurance) and providing little incentive to potential victims to adapt to climate change. In developing countries, the charity hazard may not be as serious as in the developed world, but there, to the contrary, the problem may arise that governments would be lacking resources to provide the compensation needed. Hence, the question still arises whether it is not possible to come up with a smart solution for government intervention in case commercial insurance markets cannot provide full compensation for the damage caused by climate change.

7 REINSURANCE BY GOVERNMENT

7.1 Possibilities

In various legal systems an interesting model has been developed whereby the government acts as reinsurer of last resort. Under this approach of government involvement, the State assumes at least part of the risk for losses from catastrophes. Even though government intervention is required (since the private insurance market cannot provide adequate catastrophe insurance coverage), the underlying philosophy of this approach is that private insurance should keep on playing a significant role in allocating compensation for victims of catastrophes. This option then usually takes the form of a multi-layered insurance programme. Such a programme is normally administered by private insurance companies, meaning that they sell insurance, collect premiums and pay claims.

There can indeed be arguments to favour such a reinsurance by the government; assuming that capacity on the private insurance market is indeed severely falling behind, it can be held that without State intervention, insurance coverage for disasters would simply not have developed.[83] Reinsurance by the State can then be considered as an adequate method to resolve the uninsurability problem.[84] A condition is, of course, that the government charges an actuarially fair premium for its intervention.[85] This type of government intervention has, moreover, the advantage that *ex post* relief sponsored through the public purse can be avoided. Where the government acts as reinsurer, this at least has the advantage that a premium can be paid by those who actually cause or run the risk. It can thus facilitate market solutions, still provide incentives for prevention to potential victims and avoid the negative redistribution discussed above.[86] Thus a State intervention as reinsurer may avoid the 'catastrophic responses to catastrophic risks'.[87]

Kunreuther and Michel-Kerjan also argue in favour of this type of government provided reinsurance. They argue that one advantage is that the government has the capacity to diversify the risks over the entire population and to spread past losses to future generations, thus creating a form of cross-time diversification which the private market could not achieve.[88] On the other hand, they argue that, especially as far as terrorism is concerned, government participation in insurance programmes is crucial since the risk of terrorist attacks is partly in the government's control and the government can have more information on on-going terrorist groups' activities through intelligence services.[89]

7.2 Limits

Both in law and economics, as well as in legal scholarship, criticisms have been formulated on the facilitative role of a government stimulating insurance markets. For example, Gron and Sykes argue in several papers

---

[83] This is a point strongly made by Kunreuther, above note 6 at 180–183; Harrington, above note 38; as well as by Schwarze and Wagner, above note 11.

[84] These public-private initiatives to cover extreme risks are also supported by OECD recommendations. See A Monti, 'Public-Private Initiatives to Cover Extreme Events' in Chr Corbage and W R Stahel (eds), *Extreme Events and Insurance: 2011 Annus Horibilis* (The Geneva Association 2012) 27.

[85] Faure, above note 61 at 358.

[86] See above section 2.

[87] Epstein, above note 38. See in this respect also Kunreuther and Pauly, above note 42 at 113 (arguing that this government's role in assisting the supply side allows avoiding the inefficiencies and inequities associated with disaster assistance).

[88] See Kunreuther and Michel-Kerjan, above note 6 at 210.

[89] Ibid.

that it would be unjust for the government to provide (re)insurance at a lower price than the market price.[90] This would give a wrong signal to the market as far as stimulating insurability is concerned. The authors would prefer ad hoc solutions whereby compensation is provided to accident victims on an *ex post* basis. This would avoid the situation that market participants are aware that the government will anyway guarantee compensation.

It is striking that most of this criticism is not addressed against the intervention of the government as such, but is based on the assumption that the government will not ask premiums that reflect market prices. It is a criticism shared by Levmore and Logue, who argue that such a regime (of acting as reinsurer of last resort) only has its desired effect of encouraging the purchase of commercially provided terrorism coverage when it involves a substantial subsidy.[91] They are sceptical of these types of interventions in the market (for terrorism insurance) arguing that, without government intervention 'the market would likely have been able to provide the necessary coverage'.[92]

Apparently, the arguments against government intervention are based on the assumption that the government will not ask competitive reinsurance premiums, thereby subsidizing catastrophe insurance. Moreover, without this government support, insurance coverage could have probably developed anyway (at least for terrorism events). Those points can of course only be validated on the basis of empirical research.

7.3 Practice

In the developed world there are many examples of governments acting as a reinsurer of last resort with the goal of promoting capacity for disaster risks.[93] This is, for example, the case as far as terrorism coverage is concerned. The current schemes for terrorism insurance in France (GAREAT), Germany (Extremus), the UK (Pool Re), the Netherlands (the Dutch Terrorism Risk Reinsurance Company, NHT) and the US (the Terrorism Risk Insurance Act, TRIA) all have features of a government

---

[90] See Gron and Sykes, above notes 56 and 67.

[91] Levmore and Logue, above note 47 at 304 (arguing that otherwise disaster insurance would still not be 'available').

[92] Ibid. at 311.

[93] For an overview of the different forms of public intervention in disaster insurance schemes see Monti, above note 84.

acting as a reinsurer of last resort, at least in general terms.[94] For natural hazards a well-known example is the Caisse Centrale de Réassurance or CCR as it functions in France. Insurers can reinsure the risk of natural hazards with this CCR, which benefits from a State guarantee in the event that the CCR exhausts its resources.

Also in many developing countries insurance solutions against natural disasters have been developed.[95] A variety of insurance schemes exist in developing countries which often rely on (more or less large) government support to make the risk insurable.

7.4 Application to Climate Change?

The arguments presented above in favour of an intervention of the government as reinsurer of last resort can certainly also be advanced with respect to the compensation for damage caused by climate change. Insurers rightly argue that damage caused by climate change may have such an impact, not only as far as predictability is concerned but also as far as the magnitude of the damage – they also hold that this may lead to serious problems of insurability on commercial insurance and reinsurance markets. These problems are, however, not a reason to leave the insurance model with all of its advantages,[96] or to move to either a compensation fund or government charity (with its negative effects on incentives), but rather to look for a smart solution whereby governments would rather intervene to stimulate the functioning of insurance markets and thus increase the insurance capacity for damage caused by climate change. The examples, such as the CCR (in France) but also with the California Earthquake Authority (CEA) and the National Flood Insurance Plan (NFIP) in the US, are interesting in that respect. However, it is important to structure this government intervention in such a way that it carefully mimics the market solution and has the least distortive effects as possible. Three important conditions would be that there currently is an absence of a market solution that would cover the risk; that insurers are also provided with the option (but not the obligation) to reinsure with the State; and that the State intervention would have a temporary

---

94 See D M Jaffee, *Report on the Role of Government in the Coverage of Terrorism Risks* (OECD 2004) 25–31.

95 For an overview see Linnerooth-Bayer and Mechler, above note 8.

96 Extensively discussed above in section 4.

character, thus stimulating the development of market solutions. Moreover, the State intervention should not be provided for free and, to the extent possible, risk based premiums should be charged.[97]

7.5 Summary

Providing reinsurance for climate change related damage certainly looks like a promising solution, backing up first party insurance. There is some experience with this model in providing compensation for victims of disasters.

It can certainly not be argued that the current cases in which the government acts as reinsurer of catastrophic risks are all complying with these economic principles. In some cases it is not clear that a market solution failed; in other cases the government does not charge any premium and the reinsurance by the government hence, in fact, amounts to State aid to the insurance industry. However, these institutional design failures can be remedied. Indeed it should not be forgotten that the model of government acting as reinsurer of last resort has many advantages, not only compared to the alternatives (more particularly a compensation fund or government charity) but also as a stimulant to the adequate compensation of victims of climate change related damage, the central focus of this contribution.

8 CONCLUDING REMARKS

This chapter discussed adaptation from one particular angle. The starting point was that scientific evidence indicates a high likelihood that serious damage may result from climate change. Some of this damage may have a gradual nature (like sea level rise) but some may have an accidental nature, like an increasing frequency and magnitude of disasters such as hurricanes and flooding. Of course not all continents or even countries will be hit in the same way by climate change, but many undoubtedly will, which hence justifies the crucial question of this book, being how an institutional framework can be designed to adapt to the consequences of climate change. One of the aspects of this institutional framework relates to the question how victims could be compensated for the damage resulting from climate change.

---

[97] For further details on these conditions see Bruggeman, Faure and Fiore, above note 14 at 376–379.

The chapter started by arguing that this is a policy issue that cannot simply be ignored given the potentially catastrophic nature of climate change related damage. Given the potentially disruptive effects of this damage, society will have to work out some institutional framework to guarantee compensation to victims in order also to provide a stabilization of society after disasters hit.

Notwithstanding the increasing literature on climate change liability and the fact that a few real climate change cases are now being litigated, it was argued that climate change liability itself is not a good candidate for providing adequate compensation to victims. Not only are there legal uncertainties surrounding climate change liability claims (causation and justificative effect of following prior regulatory solutions) that may constitute the problem, but also most of the debate concerning climate change liability sees litigation as a means to force emitters towards mitigation measures rather than as an adequate tool of compensation. Liability rules may, moreover, only help a few victims and not provide compensation in a structural manner.

The most promising compensation tool seems to be first party insurance. The advantage of this tool is that it does not explicitly focus on climate change related damage, but, for example, on damage resulting from heavy rain, flooding or hurricanes. Of course, precisely the fact that these events may increase in frequency and magnitude may cause serious problems for the insurance industry. Moreover, various biases on the side of the victims may inhibit them from purchasing adequate insurance coverage even where this is available (and sometimes even subsidized). Given these informational and behavioural limits there may be a strong case in favour of comprehensive mandatory coverage against the consequences of natural disasters. The main advantage of the use of the insurance instrument is, moreover, that insurers can use their experience with risk differentiation to force victims to implement adaptive strategies via premium differentiation. It is precisely at that point that insurance has a major advantage compared with direct compensation by the government. Government intervention basically uses tax payers' money to bail out victims, does not have a positive incentive effect and has potentially negative distributional effects. If insurance markets need some support from the government, a more effective solution may be a reinsurance by government as last resort, whereby the government should, of course, follow insurance market principles and hence also charge (differentiated) contributions for its intervention instead of merely providing State aid.

This shows that even though first party insurance may be the most desirable option, the real focus will be on the question how an optimal mix of various instruments can be employed to compensate victims of climate change related damage. One possibility may still be, in a first layer, to use liability rules were possible. It may have the advantage of still providing additional incentives for prevention to those who could influence the risk (of course, primarily emitters of greenhouse gases but also public authorities who would have failed to take appropriate adaptation strategies). First party insurance (probably with appropriate deductibles to cure moral hazard) would provide victim compensation, provided by commercial insurance and reinsurers, supplemented (where necessary given market failure) by government acting as reinsurer of last resort.

Another important question is, of course, the political feasibility of any solution one proposes. Insurance solutions that largely rely on potential victims financing the intervention may be politically feasible as there will not be a large resistance from industry. One can also expect cooperation from the insurance world especially in cases where for large risks governments would intervene as reinsurer of last resort. Such a regime seems to have a greater likelihood of getting accepted than, for example, a compensation fund to be paid by taxes on greenhouse gas emissions. Industry may feel that this amounts to a carbon tax as a result of which industry would pay twice (after, e.g., also having purchased emissions via an auction in an emission trading scheme).

Of course, the (optimal combination of) solutions to be designed may well strongly depend upon particular (political and economic) features of the region where a compensation system will be implemented. The solutions for developing countries like Bangladesh or Indonesia may, for obvious reasons, be different than for Europe. It is, however, clear that it is time that, also within the EU, increasing attention is paid to the tasks governments, including the EU itself, could fulfil in promoting an adequate compensation for victims of climate change related damage. In this respect it should not be forgotten that without adequate stimuli there is a large danger that nothing happens and hence *ex post* (inefficient) solutions will have to be designed once the damage has occurred. In this respect it should not be forgotten that, for example, even in a seemingly civilized country like the Netherlands since 1955 until today (2012) flooding insurance for ordinary households is not yet available. A binding decision of the insurance industry prohibiting its members from providing flooding insurance has, under pressure from the EU cartel authorities, been abrogated, but the market still has not developed an adequate policy offering flooding insurance. This shows that it is indeed high noon[98] to examine to what extent today victims of climate change related damage can call on insurance or other financial instruments to obtain compensation and to what extent government policy should aim at stimulating the development of financial mechanisms to provide adequate compensation to victims of climate change related damage. That issue should now urgently be addressed.

[98] Paraphrasing the words of Jaap Spier, 'High Noon: Prevention of Climate Change as the Primary Goal of Liability?' in M Faure and M Peeters (eds), *Climate Change Liability* (Edward Elgar 2011) 47.

6. Climate change adaptation and disaster law

Dewald van Niekerk[1]

1 INTRODUCTION

Climate change has been used as the bogeyman, the scapegoat, the *el Hombre de la Bolsa* if you may, to deflect attention from underlying issues which contribute to vulnerability in society such as poverty, social deprivation, lack of resources and poor education. Political advocates of climate change use the 'unknown' to drive agendas and, sometimes, misguide the layperson to achieve their own objectives. One such message is that we urgently need new action to adapt to and mitigate climate change. However, there is more than enough evidence to suggest that communities have been adapting to a changing climate for centuries. An undeniable fact, however, is that we are experiencing quicker and more unforeseen dramatic changes in climate. This is no truer in any other focus than in the realm of disaster risk reduction, where vulnerability reduction to hazard exposure is the main focus.

Climate change adaptation, as with disaster risk reduction, is a transdisciplinary issue. Moreover, scholars and practitioners in the domain of climate and disaster risk science see these two concepts as largely complementary and inseparable. It is acknowledged that solutions to reduce the risk of disasters occurring also greatly solve climate change related issues. One undeniable fact of the impact of climate change is that the frequency and magnitude of hydrometeorological hazards is on the increase. This has a profound effect on the way in which communities and governments need to conduct their business. Governments will do well to acknowledge the interplay between actions for climate change adaptation and disaster risk reduction. In the past decade, a number of countries have revisited, changed and/or promulgated disaster risk reduction law. Similarly, new legislation governing climate change and adaptation issues has seen the light. However, governments have generally

[1] The author would like to acknowledge the research undertaken by Leandri Hildebrandt on the various case studies in this chapter.

been weak in explicitly stating and acknowledging the relationship between these two focuses and their interplay. In most instances, the failure to make this necessary connection has been more political than practical in nature.

This chapter will look into disaster risk reduction as a form of climate change law. Firstly a theoretical grounding for disaster risk reduction will be established. Secondly, the linkages between climate change adaptation and disaster risk reduction will be alluded to. The chapter will, thirdly, aim to show the numerous similarities in actions towards reducing disasters and adapting to climate change. Fourthly, a description will be given of the application of disaster risk reduction as a transdisciplinary field within other disciplines and government planning. Throughout the chapter a number of case studies will be used to illustrate examples in practice.

2 INTERACTION OF TERMINOLOGY

Before one can talk about disaster risk reduction it is important to understand what is meant by a *disaster*. This ambiguous term has been used to describe various phenomena, as shown by the work of Quarantelli and others.[2] However, over the past two decades (1990 to 2010) scholars and countries alike have gravitated towards a common understanding of the term. This is not only aimed to facilitate a common focus but to provide a basis for action, which in many countries is legislated through disaster risk management or disaster management law. In this vane, the United Nations International Strategy for Disaster Reduction (UNISDR) now defines a disaster as: 'A serious disruption of the functioning of a community or a society involving widespread human, material, or environmental losses and impacts which exceeds the ability of the affected community to cope using only its own resources.'[3] Three elements from this definition must be highlighted. Firstly, a disaster is linked to an event or events which cause a serious disruption in a given society. These events are referred to as natural or anthropogenic (human made) hazards (such as floods, earthquakes, fires, cyclones and the like). Secondly, the impact of a hazard must be on humans and/or the various

---

[2] E L Quarantelli, *What is a Disaster?: A Dozen Perspectives on the Question* (Routledge 1998); Ronald W Perry and E L Quarantelli, *What Is A Disaster?: New Answers to Old Questions* (Xlibris 2005).

[3] United Nations International Strategy for Disaster Reduction, *Terminology for Disaster Risk Reduction* (United Nations 2009).

systems on which humans depend (such as the environment or infrastructure). Thirdly, the impact of the event must exceed the ability of the affected community to cope when only relying on their own resources. Linking the above to a climate change argument, evidently more hydrological and meteorological hazards of varied intensities increasingly cause disasters. It would, however, be wrong to argue that these hazards are disasters. Hazards are natural phenomena with an inherent ability to cause harm. Without a direct impact on humans and/or the systems on which they rely, one can hardly talk about a disaster. For example, an earthquake in a desolate desert area void of any humans or natural resources is not considered a disaster. Only when vulnerable human conditions meet hazardous phenomena, to the extent that these phenomena exploit humans beyond their own capacities, can we define an event as a disaster. However, not all hazards affect different communities in the same way. Communities have varied capacities, skills, knowledge and capabilities which either increase or decrease their *disaster risk* profile.

The UNISDR defined disaster risk as: 'The potential disaster losses, in lives, health status, livelihoods, assets and services, which could occur to a particular community or a society over some specified future time period.'[4] This definition thus has a future perspective. If a community is continually exposed to hazardous conditions and their ability to manage such conditions is not addressed (through building their own capacities), then the risk of a disaster occurring increases. However, the varied types of potential losses must be acknowledged, although it may not always be possible to qualify these. Thus, in order to determine disaster risk three variables should be considered. The following notation tries to clarify this relationship:

$$\text{Disaster risk} = \frac{\text{Hazard x Vulnerability}}{\text{Capacity}}$$ [5]

Disaster risk can thus be determined by considering the various types, frequency, intensities and impacts of hazards on a given system taking into account the vulnerabilities to these hazards and the existing coping capacities to offset the types of vulnerability (e.g., social, economic, political, physical, legal).

---

4 Ibid.

5 United Nations, *Living with Risk: A Global Review of Disaster Reduction Initiatives 2004* (United Nations 2004).

How would one then go about addressing such seemingly perilous conditions? The answer lies within a joint focus on disaster risk reduction, climate change and adaptation. Disaster risk reduction refers to the 'practice of reducing disaster risks through systematic efforts to analyse and manage the causal factors of disasters, including through reduced exposure to hazards, lessened vulnerability of people and property, wise management of land and the environment, and improved preparedness for adverse events'.[6]

The key to the above definition is the focus on the causal factors of disasters (such as the lack of capacities and vulnerability). The Intergovernmental Panel on Climate Change (IPCC) defines climate change as:

> a change in the state of the climate that can be identified (e.g., by using statistical tests) by changes in the mean and/or the variability of its properties, and that persists for an extended period, typically decades or longer. Climate change may be due to natural internal processes or external forcings, or to persistent anthropogenic changes in the composition of the atmosphere or in land use.[7]

The United Nations Framework Convention on Climate Change (UNFCCC) defines climate change as: 'a change of climate which is attributed directly or indirectly to human activity that alters the composition of the global atmosphere and which is in addition to natural climate variability observed over comparable time periods'.[8] The above definitions link with disaster risk reduction through the increase of the causal factors which leads to disasters or heightens the risk of disasters occurring. If we agree with the definition of the IPCC (which seems less restricted than that of the UNFCCC) then the change in climate due to human and natural activities over a given period of time is key.

The argument can thus be made that there is a direct correlation between the risk of a disaster occurring and the increase in climate change conditions. However, as has been argued above, disaster risk is not the final product of *only* hazardous conditions, but one needs to consider how communities adapt, and build their own resilience. To this

---

6 United Nations International Strategy for Disaster Reduction, above note 3.

7 M L Parry et al. (eds), *Climate Change 2007: Impacts, Adaptation and Vulnerability: Contribution of Working Group II to the Fourth Assessment Report of the Intergovernmental Panel on Climate Change* (Cambridge University Press 2007).

8 United Nations Framework Convention on Climate Change (New York, 9 May 1992), (1992) 31 ILM 849, entered into force 21 March 1994, Art. 1(2).

end, adaptation can be defined as 'the adjustment in natural or human systems in response to actual or expected climatic stimuli or their effects, which moderates harm or exploits beneficial opportunities'.[9] Adaptation in the broader sense also relates to hazards caused by non-climatic factors (such as soil erosion, earthquakes and tsunamis). Adaptation can also occur in an autonomous fashion which could include better government policies or plans, global market change or societal conscience. Disaster risk reduction therefore can directly contribute to addressing increases in hydrometeorological hazards (more natural science approach) and the aspects of adaptation and vulnerability (more social science approach).

Considering all of the above, clearly there is an undeniable direct relationship between climate change adaptation and disaster risk reduction.[10] Many disaster risk reduction measures can directly contribute to better adaptation. These can either be direct (such as policies, law and plans) or indirect (such as a better understanding by communities of their own disaster risk profile and possible impact of climate change on their livelihoods and security). However, an alarming notion within national government structures is to view and manage disaster risks and climate change issues as two totally separate issues. Evidence shows the development of climate change, climate change adaptation and disaster risk reduction policies and legislation in fair isolation of each other (see some of the case studies in this chapter). Climate change is mostly seen as the responsibility of the ministry of the environment, whereas many governments are still stuck in disaster response mode when it comes to managing their disasters (and risks), and this is addressed by a totally different ministry. Little to no interaction between these various ministries occurs. Mercer believes that this is largely due to the unit of analysis associated with climate change and disaster risks.[11] Climate change is mostly addressed in policies and law through a national and sometimes sub-national level (for example National Adaptation Programmes for

---

9 Parry et al., above note 7 at 6.

10 K O'Brien et al., 'Disaster Risk Reduction, Climate Change Adaptation and Human Security', Report prepared for the Royal Norwegian Ministry of Foreign Affairs by the Global Environmental Change and Human Security Project, (2008) 3 GECHS Report.

11 Jessica Mercer, 'Disaster Risk Reduction or Climate Change Adaptation: Are We Reinventing the Wheel?' (2010) 22 Journal of International Development 247.

Action).[12] Disaster risks, however, have enjoyed decades of attention, which has become much more community based in its focus and application.[13] There is thus a disconnect between the national focus of climate change (by governments) and the local implementation of disaster risk reduction (mostly by civil society).[14] The challenge is to reconcile these two, and a challenge it is. The Philippines provides us with an insightful case study as to the integration of these two agendas,[15] although this country has also enjoyed its fair amount of criticism.[16] This integration challenge is highlighted by the development of separate disaster risk reduction and climate change and adaptation policies in various countries such as Colombia, South Africa, India and Mozambique.

BOX 6.1 CLIMATE CHANGE AND DISASTER LAW IN INDIA

India is a democratic republic that functions under a parliamentary system, with the President as the constitutional head of the government.[1] The handling of different disasters in the past years in India is a testimony to an inadequate and ill-equipped disaster management system. Legislation allocates responsibilities in legal form, to ensure uniformity of approach and provide for the protection of people's rights.[2] India addresses disaster risk reduction and climate change issues mainly through two

[12] Walter Leal Filho (ed.), *Experiences of Climate Change Adaptation in Africa* (Springer 2011).

[13] Global Network of Civil Society Organisations for Disaster Reduction, *Clouds but little rain…* (Global Network of Civil Society Organisations for Disaster Reduction 2009).

[14] Willemien Faling et al., 'Rhetoric or Action: Are South African Municipalities Planning for Climate Change?' (2012) 29 Development Southern Africa 241.

[15] C Benson, *Mainstreaming Disaster Risk Reduction into Development: Challenges and Experience in the Philippines* (International Federation of Red Cross and Red Crescent Societies/the ProVention Consortium 2009); Republic of the Philippines, Climate Change Act No. 9729 of 2009 (Government Printer 2009).

[16] Arnold Padilla et al., *On the Road to Disaster. Philippine Climate Change Policies* (Center for Environmental Concerns Philippines 2011). See also Box 6.3.

documents. The first is The Disaster Management Act of 2005,[3] which is the legislation put in place to address disaster risk reduction and management in India, and the second is the National Action Plan on Climate Change (NAPCC).[4] It is the goal of the NAPCC to identify the measures that promote India's development objectives, while also yielding co-benefits for addressing climate change effectively. This plan outlines steps to advance India's development and climate change related objectives of adaptation and mitigation. The core of the NAPCC consists of eight national missions, which represent strategies for achieving key goals in the context of climate change. It is also the purpose of this document to create awareness in the public, different departments of the government and the community about the threat of climate change and the proposed steps to counter it.[5] In this plan the terms 'disaster management' or 'disaster risk reduction' are mentioned to a great extent. Section 4.3 especially focuses on the disaster management response to different climate change events. The Disaster Management law is concerned with the enforceable right of a disaster victim, but the existing law, government schemes and policies in India are not known to anyone, until a disaster strikes.[6]

Notes:
1 Dewald van Niekerk et al., *An Alternative Financing Model for Disaster Risk Reduction in South Africa* (Financial and Fiscal Commission 2011).
2 Jacob P Alex, *Disaster Management: Towards A Legal Framework* (The Indian Institute of Public Administration 2006).
3 Government of India, The Disaster Management Act No. 53 of 2005 (Government Printer 2005).
4 Government of India, National Action Plan on Climate Change (Government Printer 2008).
5 Government of India, National Action Plan on Climate Change (Government Printer 2008).
6 Alex, above note 2.

3 INTERNATIONAL PROGRESS IN DISASTER REDUCTION AND DISASTER RISK LAW

Civil protection (also called civil defence) finds its basis in the norms which regulate armed conflict – *ius in bello*. Civil protection was initially linked to all non-military means to protect civil population and mitigate harm. At first, these mostly referred to the means, methods and conduct of war, or the Law of the Hague. Following the atrocities committed during World War II (WWII), a new branch emerged which mainly concerned itself with what actions were permissible and what obligations the belligerents had in relation to persons not taking part in hostilities.[17] This later became known as the Law of Geneva, which consists of the Geneva Conventions and Protocols.

It was, however, prior to and following WWII, that a number of countries came to promulgate formally civil protection legislation. The need for structured protection of domestic civilians was heightened during the Cold War era especially within Western states. After the collapse of the USSR the focus on military threat shifted towards so-called natural and human made (anthropogenic) disasters. Although the United Nations' various agencies have to a greater and lesser degree been involved in responses to disasters since its inception after WWII, no one agency was dedicated to focusing purely on issues relating to disaster reduction until the late 1980s. The inadequacies of the international relief and humanitarian system to address continuous losses due to natural disasters brought with it a realization that alternative interventions were needed.[18] In 1987 an international decade on natural disasters was first proposed at the Eighth International Congress of Earthquake Engineering.[19] This initiative was met favourably by the international scientific community. In December 1987, the UN General Assembly adopted Resolution 42/169,[20] which proclaimed the years 1990 to 1999 as the International Decade for Natural Disaster Reduction, or IDNDR.[21] During this decade a concerted international effort was attempted to reduce the loss of life, property, livelihoods, and social and economic disruption caused by the violent impact of nature on vulnerable conditions. The aim

---

17 University of Macedonia, *NATO and Civil Protection: Future Perspectives* (University of Macedonia 2012).

18 D Van Niekerk, 'From Disaster Relief to Disaster Risk Reduction: A Consideration of the Evolving International Relief Mechanism' (2008) 4 The Journal for Transdisciplinary Research in Southern Africa 355.

19 M F Lechat, 'The International Decade for Natural Disaster Reduction: Background and Objectives' (1990) 14 Disasters 1.

20 United Nations, General Assembly Resolution 42/169, 11 December 1987.

21 C S Melching and P J Pilon, *Comprehensive Risk Assessment for Natural Hazards* (World Meteorological Organization 2006); Keith Smith, *Environmental Hazards: Assessing Risk and Reducing Disaster* (5th edn, Routledge 2009); United Nations International Strategy for Disaster Reduction, *Natural Disasters and Sustainable Development: Understanding the Links between Development, Environment and Natural Disasters* (United Nations 2002); Lechat, above note 19.

of this decade was to ensure a shift in the reactive approach towards natural disasters to that of proactive planning and prevention.[22]

The emphasis that the decade placed on scientific solutions, as well as the transfer of hazard-mitigation technologies to developing countries, was capital-intensive most of the time and did not take the capacities of these countries into consideration.[23] Although the concept of development/disaster integration was not new at the times of planning the IDNDR,[24] the decade failed to recognize the importance of the inclusion of risk reduction initiatives within programmes of sustainable development, thus creating certain development dilemmas.[25] Even so, during this era discussion on climate change issues reached prominent international importance, yet did not adequately find its way into the natural disaster management agendas until much later.

Another significant event which shaped the disaster risk reduction focus in the 1990s was the World Conference on Natural Disaster Reduction held in Yokohama, Japan from 23 to 27 May 1994, and the subsequent adoption of the Yokohama Strategy and Plan of Action for a Safer World. The Yokohama Strategy and Plan (1994) stressed that each country has the sovereign responsibility to protect its citizens from natural disasters and that priority must be given to developing countries, in particular the least developed, land-locked countries and the small island developing states. It further emphasized the importance of 'developing and strengthening national capacities and capabilities and, where appropriate, national legislation for natural and other disaster prevention, mitigation and preparedness, including the mobilisation of non-governmental organisations and participation of local communities'. Lastly, the Strategy pointed to the importance of promoting and strengthening sub-regional, regional and international cooperation in prevention, reduction and mitigation of natural and other disasters. The absence of any mention of the climate change agenda was evident; however, one can

---

[22] George W Housner, 'An International Decade of Natural Disaster Reduction: 1990–2000' (1989) 2 Natural Hazards 45; Lechat, above note 19; K Smith, *Environmental Hazards: Assessing Risk and Reducing Disaster* (5th edn, Routledge 2009); Van Niekerk, above note 18.

[23] P Blaikie et al., *At Risk: Natural Hazards, People's Vulnerability, and Disasters* (Routledge 1994).

[24] Phil O'Keefe et al., 'Taking the Naturalness out of Natural Disasters' (1976) 260 Nature 566.

[25] D A McEntire, 'Reflecting on the Weaknesses of the International Community during the IDNDR: Some Implications for Research and its Application' (1997) 6 Disaster Prevention and Management 221; U Rosenthal et al., *Managing Crisis: A Global Perspective* (Thomas Publishers 2001).

argue this is mostly due to the fact that an official international stance on curbing climate change was only formulated in 1997 with the Kyoto Protocol.[26]

As the successor to the IDNDR, the International Strategy for Disaster Reduction (ISDR) proceeded with the emphasis on the protection against hazards, reducing vulnerability and building resilient communities.[27] However, the connection was made between the need for natural hazard mitigation and the impact of climate change. The most significant progress that the ISDR aimed to make, vis-à-vis the ideals of the IDNDR and the Yokohama Strategy and Plan of Action for a Safer World,[28] is the cultivation of a multi-disciplinary approach to disaster reduction within the broader context of sustainable development.[29] The ISDR took a global approach to disaster reduction inculcating a culture of risk avoidance behaviour through the fostering of partnerships at community level, and solutions grounded in development interventions.

Subsequently, a number of international policies and frameworks have been developed since the 2000s, these include the Hyogo Framework for Action: Building the Resilience of Nations and Communities (2005–2015),[30] the African Regional Disaster Risk Reduction Framework and its Plan of Action,[31] the ASEAN Agreement on Disaster Management and Emergency Response,[32] the Saint Marc Plan of Action for Disaster Reduction of the Association of Caribbean States[33] as well as the draft Southern African Development Community Disaster Risk Reduction

---

[26] Kyoto Protocol to the United Nations Framework on Climate Change, UN Doc FCCC/CP/1997/7/Add. 1 (Kyoto, 10 December 1997), (1998) 37 ILM 22, entered into force 16 February 2005.

[27] United Nations International Strategy for Disaster Reduction, above note 21.

[28] United Nations, *Yokohama Strategy and Plan of Action for a Safer World* (United Nations 1994).

[29] United Nations, *Living with Risk: A Global Review of Disaster Reduction Initiatives 2004* (United Nations 2004).

[30] United Nations, *Hyogo Framework for Action 2005–2015: Building the Resilience of Nations and Communities to Disasters* (United Nations 2005).

[31] African Union Commission and United Nations International Strategy for Disaster Reduction, *Extended Programme of Action for the Implementation of the Africa Regional Strategy for Disaster Risk Reduction (2006–2015) and the Declaration of the 2nd African Ministerial Conference on Disaster Risk Reduction 2010* (International Strategy for Disaster Reduction 2010).

[32] Association of Southeast Asian Nations, *ASEAN Agreement on Disaster Management and Emergency Response* (ASEAN 2005).

[33] Association of Caribbean States, *Saint-Marc Plan of Action for Disaster Reduction* (Association of Caribbean States 2008).

Framework. The above are examples of how disaster risk reduction has become a policy priority for governments and regions world-wide. Such policies can well be seen as the first stepping stones towards sound disaster risk governance.

A core function of disaster risk reduction governance is ensuring that the necessary support exists within government to drive the disaster risk reduction agenda. There is wide international consensus that governments as the administrative entity must ensure that disaster risk reduction becomes a priority.

BOX 6.2 DECENTRALIZING ACTIONS FOR DISASTER RISK REDUCTION AND CLIMATE CHANGE ADAPTATION IN COLOMBIA

Colombia is the only South American country with coastlines on both the Pacific Ocean and the Caribbean Sea, which for obvious reasons makes it particularly vulnerable to climatic fluctuations. Colombia is also the place with the most complex tectonic systems. The severe disaster impacts that affected Colombia in the past motivated the authorities to promote a more effective coordination among different sectors involved in disaster risk reduction.

The National Platform for Disaster Risk Reduction in Colombia is called the National System for Disaster Prevention and Response (SNPAD).[1] Law 46, whose main goal was to define responsibilities and functions of all stakeholders involved in disaster risk reduction, disaster risk management and recovery issues, created the mentioned national platform in 1988. It was only in 2004 that the SNPAD was officially designated as a National Platform of Colombia, anticipating the world conference of Disaster Reduction in Kobe 2005. The National Plan for Prevention and Attention of Disasters in Colombia (PNPAD) sets decentralization as a fundamental principle. This meant that territorial entities became responsible for disaster risk reduction and related policies, leaving the general definition and co-ordination actions for disaster risk reduction to the national government.[2] The PNPAD calls for harmony, consistency and coherence among all the territorial instances involved.

Colombia has submitted only one National Communication to the UNFCCC, in December 2001.[3] This laid out the actions already taken by the government and the analytical basis for its policy response to climate change and its commitment to take future actions within an official international framework. This communication thus established the first National Greenhouse Gas Inventories for 1990 and 1994, and also vulnerability and adaptation studies for the coastal area, water resources, and agricultural sector and land management. Colombia has developed an Integrated National Adaptation Project (INAP) which was designed with the purpose of supporting Colombia's efforts to define and implement prevention and adaptation measures to climate change. In all of these project/plans/platforms they aim to address both climate change adaptation and disaster risk reduction. For example, in the INAP mention is not specifically made to the term disaster risk reduction, but reference is made to general hazards and risks caused by climate change. Colombia is widely considered as a leader in instituting a policy and legal framework that enables a comprehensive, multi-sectoral approach to local climate change adaptation and disaster risk reduction.

Notes:

1 Luz Amanda Pulido, *Colombia: Integrating Disaster Risk Reduction at the Local Level* (Dirección de Prevención y Atención de Desastres 2006).

2 Ibid.

3 Svetlana Edmeades, *Country Note on Climate Change Aspects in Agriculture* (World Bank 2009).

4 TRANSDISCIPLINARY NATURE OF DISASTER RISK REDUCTION AND CLIMATE CHANGE ADAPTATION

Disaster risk reduction and climate change adaptation are both transdisciplinary issues. Transdisciplinarity is a term being used, moreover, within a number of disciplines to indicate the interaction with other disciplines in identifying and solving real world problems through the joint adoption, incorporation and application of various perspectives, theories and methodologies (merging 'soft' and 'hard' sciences).[34]

[34] Johann Tempelhoff, 'Editorial comment' (2005) 1 The Journal for Transdisciplinary Research in Southern Africa 1.

In contrast to the one-dimensional reality of classical thought, transdisciplinarity acknowledges multidimensionality.[35] Transdisciplinarity also takes into consideration non-traditional creators and custodians of knowledge (such as communities and practitioners). A transdisciplinary orientation to studying or acting to resolve a defined real life problem (such as climate change and increased societal vulnerability) necessarily also seeks to engage a range of contributing interests, skills and abilities, but the various relationships involved are intended to become synthesized and more holistic in their merged, rather than shared approach.[36] Transdisciplinary research blurs disciplinary boundaries. The combined knowledge or experience brought to bear on a problem is thus seen to be greater than the sum of its parts.[37]

Transdisciplinarity provides an ideal vehicle for jointly solving disaster risk reduction and climate change adaptation real world problems. It aims to cross the traditional boundaries of thought which has come to dominate our view of disaster risk reduction and climate change adaptation as two separate entities. The complexities associated with climate change and disaster risks require a new way of understanding their driving factors. Therefore, through defining a common real world problem (for example: how should a community adapt their livelihood model considering a changing climate/or increases in disasters) the focus is removed from traditional isolated disciplinary research to an open and combined endeavour, which would need the knowledge of a number of traditional disciplines.

When considering the transdisciplinary merit of disaster risk reduction and climate change adaptation their interaction becomes clear. Such a transdisciplinary focus removes the emphasis on either of these inherent problems (disaster risk or climate change), but places the critical issue on a real life problem which is shared by both focuses (such as societal resilience). In doing so the interaction with development driven problems/solutions comes into play. Thus the real world problem derives from human/systems lack and inabilities which can be bridged through development oriented interventions, and so one will find that the actions needed to address climate change and disaster risk issues become significantly intertwined and similar.

---

35 Julie Thompson Klein, 'Prospects for Transdisciplinarity' (2004) 36 Futures 515.

36 Ibid.

37 B Blattel-Mink and H Kaslenholz, 'Transdisciplinarity in Sustainability Research: Diffusion Conditions of an Institutional Innovation' (2005) 12 International Journal of Sustainable Development and World Ecology 1.

5 DISASTER RISK REDUCTION AND CLIMATE CHANGE ADAPTATION: TWO SIDES OF THE SAME COIN?

Climate change has become a politically loaded term and governments are trying to optimize their benefits of climate change and adaptation. Similarly, disaster risk reduction is enjoying much more international attention (specifically following a number of major international disasters since 2000 such as: the Gujarat earthquake in India in 2001; the Dresden floods in 2002; the 2003 heat waves in Europe; the Ban earthquake in December 2003; the 2004 Indian Ocean tsunami; unprecedented and record rainfalls in India in July 2005; Hurricanes Katrina, Rita and Wilma in 2005; the 2005 earthquake in Kashmir; the Java earthquake in October 2005; Cyclone Nargis in 2008; the devastating Sichuan earthquake in China in 2008; followed by the Haiti earthquake in January 2010;[38] and then the Japan earthquake and ensuing tsunami of 2011).

Managing disasters and, more recently, engaging in actions aimed towards disaster risk reduction is not new. Various disciplines have studied disasters and their consequences for decades.[39] It is now widely acknowledged that hydrometeorological hazards are increasing in frequency and intensity. There also exists adequate consensus among climate scientist that the increase in these hazards can be attributed to a changing and variable climate. Linked to aspects such as societal vulnerability, food insecurity and diminishing local capacities, it is only logical to assume more future losses and heightened disaster risk profiles of these communities. Although significant research has been undertaken in understanding disaster risks as well as the impact of climate change on vulnerability communities, very little thought has gone into the linkages between climate change and disaster risk reduction. Birkmann's research into the use of common terminology is a clear indication of the misconceptions which exist and can occur between the disaster risk reduction and climate change scientific community.[40] It is thus prudent to first establish a theoretical grounding for the terms which constitute disaster risk reduction and climate change and adaptation, in order to put the two concepts within context.

---

38 Sean Deely et al., *World Disasters Report 2010: Focus on Urban Risk* (Renouf 2010).

39 Faling et al., above note 14.

40 J Birkmann, *Measuring Vulnerability to Natural Hazards: Towards Disaster Resilient Societies* (United Nations University Press 2006) 9–54.

6 CONVERGENCE OF AGENDAS

Kelman and Giallard are of the opinion that the sterling work undertaken by many disaster risk scientists over the past decades has not been acknowledged by the climate change sciences. This in essence means that climate change policy and actions are missing out on the long experience from dealing with climate related disasters. There seems to be a general consensus within the disaster risk reduction arena that there needs to be greater convergence between disaster risk reduction, climate change and adaptation.[41] Furthermore, disaster risk reduction has an existing set of community based tools which, in essence, relate to similar strategies for achieving climate change adaptation objectives.[42] King believes that much of disaster risk reduction programming comes under the rubric of natural resource management.[43] Some organizations do much of their disaster risk reduction work on soil and water conservation, conservation agriculture, healing environmental 'hot-spots', drought mitigation, livestock asset protection, irrigation, drought resistant seeds and agroforestry. Given that these types of interventions are specifically focused on increasingly scarce natural resources, funding can be easily linked with climate change initiatives.[44]

On a policy level, it is valuable to understand the interaction between disaster risk reduction and climate change adaptation. Figure 6.1[45] aims

---

[41] M Aalst and T Mitchell, *Convergence of Disaster Risk Reduction and Climate Change Adaptation: A Review for DFID* (IDS Report 2008). O'Brien et al., above note 10; M B King and K Bouchard, *Examining Linkages between Disaster Risk Reduction and Livelihoods* (Feinstein International Center 2011).

[42] Gideon van Riet and Dewald van Niekerk, 'Capacity development for participatory disaster risk assessment' (2012) Environmental Hazards 1; Rajib Shaw, *Community-Based Disaster Risk Reduction (Community, Environment and Disaster Risk Management)* (Emerald Group Publishing 2012); Robert Chambers, *Participatory Workshops: A Sourcebook of 21 Sets of Ideas and Activities* (Routledge 2002).

[43] King and Bouchard, above note 41 at 8.

[44] J Birkmann, 'Integrating Disaster Risk Reduction and Climate Change Adaptation: Key Challenges – Scales, Knowledge, and Norms' (2010) 5(2) Sustainability Science 171.

[45] A Gero, K Méheux and D Dominey-Howes, *Disaster Risk Reduction and Climate Change Adaptation in the Pacific: The Challenge of Integration* (University of New South Wales, Australian Tsunami Research Centre-Natural Hazards Research Laboratory 2010) 16. See also A Gero, K Méheux and D Dominey-Howes, 'Integrating Disaster Risk Reduction and Climate Change Adaptation in the Pacific' (2011) 3 Climate and Development 310; A Gero, K Méheux and D

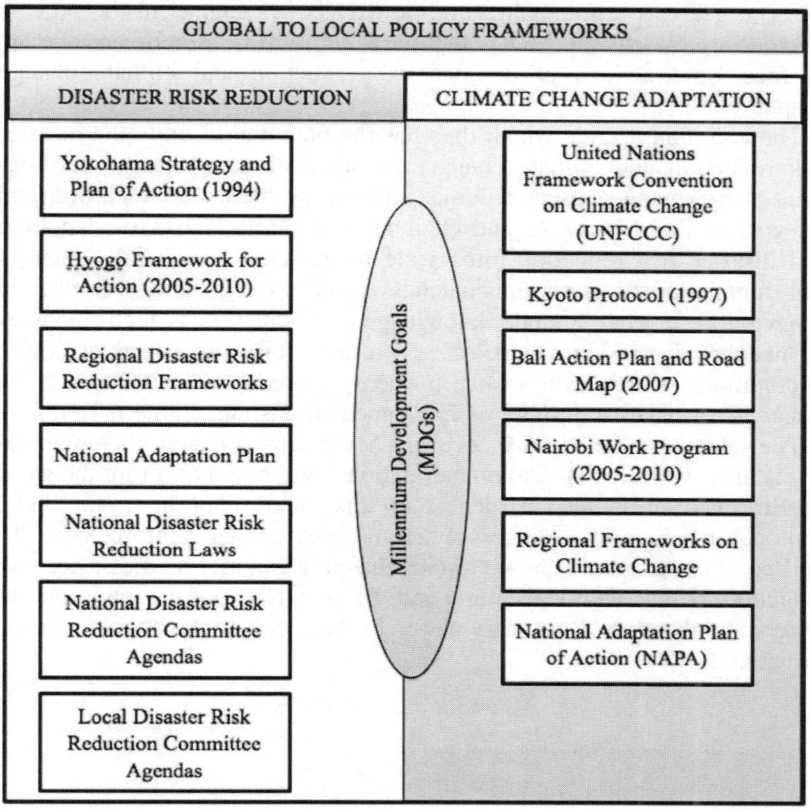

Figure 6.1 Comparative global to local policy framework for disaster risk reduction and climate change adaptation

to show the interaction of the various levels and the proposed policies and strategies which are, or should be, in place. In both cases sustainable development, and in this case the Millennium Development Goals (MDGs), are used as the benchmark for measuring success. The argument is thus that disaster risk reduction and climate change adaptation

Dominey-Howes, 'Integrating Community Based Disaster Risk Reduction and Climate Change Adaptation: Examples from the Pacific' (2011) 11 Natural Hazards and Earth System Sciences 101.

must be geared towards achieving the MDGs. Of course, in the process of working towards the MDGs there are, and will be, significant overlap of focus and actions of disaster risk reduction and climate change adaptation.

The common agenda which thus has the potential to unite the disaster risk reduction and climate change and adaptation, is that of development.[46] Birkmann's research looks into the problem of integration and categorizes the differences and challenges of climate change adaptation and disaster risk reduction into 'scale mismatches'[47] (spatial, temporal and functional),[48] 'norm mismatches' (such as legislation, culture or behavioural norms),[49] and 'knowledge mismatches' (which includes mismatches due to knowledge of certain areas, climatic effects or failure to communicate scientific results to the correct audience).[50] Through his research he makes a number of recommendations (see Table 6.1[51]).

The integration of the two focuses should thus occur at all functional levels, that is local, state/provincial, national and regional. From the work of Birkmann it is also evident that this should not be a top-down approach but a parallel process which ensures mutual reinforcement. To find common ground in how climate change adaptation and disaster risk reduction policy and legislation can be integrated, it is important to understand the transdisciplinary nature of these two development oriented focuses.

---

46 Mercer, above note 11. J Mercer et al., 'Reflections on Use of Participatory Research for Disaster Risk Reduction' (2008) 40(2) Area 172; W Solecki et al., 'Climate Change Adaptation Strategies and Disaster Risk Reduction in Cities: Connections, Contentions, and Synergies' (2011) 3(3) Current Opinion in Environmental Sustainability 135; Birkmann, above note 40.

47 Ibid.

48 F Thomalla et al., 'Reducing Hazard Vulnerability: Towards a Common Approach between Disaster Risk Reduction and Climate Adaptation' (2006) 30(1) Disasters 39.

49 Ibid.

50 I Kelman, 'Climate Change and Small Island Developing States: A Critical Review' (2009) 5(1) Ecological and Environmental Anthropology 1; E Bryan et al., 'Adaptation to Climate Change in Ethiopia and South Africa: Options and Constraints' (2009) 12 Environmental Science & Policy 413.

51 Birkmann, above note 40 at 178.

Table 6.1 *Birkmann's recommendations of integrating disaster risk reduction and climate change adaptation*

Recommendation	Target communities
Adoption of a cross-sectoral, multi-scale and integrative approach to link disaster risk reduction and climate change adaptation and to mainstream both into other activities on sustainable development	UNFCCC, ISDR, national governments, humanitarian and development communities
Development of standardized methods and quality criteria for impact and needs assessment as well as vulnerability, capacity and adaptation assessment	IPCC, ISDR, humanitarian and development communities
Translation of guiding principles, such as resilience and adaptive societies into more precise goals in order to be useful for practical actions and strategies on the ground	UNFCCC, ISDR, humanitarian and development communities
Coordination of actors, institutions and organizations to build on existing capacities and explore synergies	UNFCCC, ISDR, humanitarian and development communities, national governments
Creation of flexible funding schemes that shift from short-term and project-oriented financing to the support of forward-oriented strategies that ultimately lead to long-term sustainability	UNFCCC, national governments, other donors
Consideration of disasters as windows of opportunities that allow for change and progress if appropriate measures are taken, and a long-term perspective is adopted	National governments
Creation of structures and instruments that improve social learning and memory	National governments, humanitarian and development communities
Provision of necessary information, such as scientific data, local knowledge or experimental knowledge accumulated in institutional and personal memory	IPCC, UN University, scientific community, national governments, humanitarian and development communities
Development of a comprehensive and internationally accepted framework that could serve as a conceptual and practical tool when integrating disaster risk reduction and climate change adaptation in practice	UNFCCC, ISDR, IPCC

BOX 6.3 POLICY AND LEGISLATION INTEGRATION IN THE PHILIPPINES

The Republic of the Philippines is one of the most disaster-prone countries in the world.[1] The Philippines is vulnerable to almost all types of natural hazard, due to its geographical location. According to Duque, the Philippines is located within the Circum-Pacific belt of fires and along cyclone paths.[2] Its geographic location exposes the Philippines to a number of natural hazards. The UNISDR ranked the Philippines twelfth on their Mortality Risk Index (MRI), among the 200 countries that were analysed with respect to climatic events.[3] As in the case of South Africa and Indonesia, change in the Philippines came about as a result of rising concern about increasing disasters in the country.[4] Between 1999 and 2009, dozens of bills were submitted to the Philippines' Congress with the aim of changing the primarily reactive 1978 disaster management legislation to a more proactive, preventive law. However, as Benson indicates, progress in passing a disaster risk reduction bill in the Philippines was thwarted in part by lack of coordination among the many, often conflicting bills, as well as additional submissions for piecemeal change around specific issues.[5] The Philippine Disaster Risk Reduction and Management Act of 2010, addresses both climate change and disaster risk reduction. According to the Philippine Disaster Risk Reduction and Management Act of 2010[6] it is the policy of the State to: adopt a disaster risk reduction and management approach that is holistic, comprehensive, integrated and proactive in lessening the socio-economic and environmental impacts of disasters including climate change, and promotes the involvement and participation of all sectors and all stakeholders concerned at all levels, especially the local community. The Act further addresses disaster risk reduction and climate change in section 2(e), (g) and (j) and also in section 3(d), (n) and (o). The Philippines Disaster Management System (PDMS) is carried out at various administrative levels through the National Disaster Coordinating Council (NDCC), 80 Provincial disaster coordinating councils, 113 City disaster coordinating councils, 1,496 Municipal disaster coordinating councils, 41,956 Barangay disaster coordinating councils, and 17 Regional disaster coordinating councils.

The Philippines' Climate Change Act of 2009 is intended to integrate disaster risk reduction measures into climate change adaptation plans, development and poverty reduction programmes.[7] This Act also addresses disaster risk reduction. According to the Climate Change Act of 2009 it is the policy of the State to recognize the vulnerability of the Philippines and its local communities and to cooperate with the global community in the resolution of climate change issues, including disaster risk reduction. The Climate Change Act of 2009 addresses the challenge of inter-sectoral government collaboration by creating a commission to be chaired by the President and attached to the President's office, thus ensuring highest-level political support for collaborative implementation of the law. The commission is composed of the secretaries of all relevant departments as well as the 'Secretary of the Department of National Defense, in his capacity as Chair of the National Disaster Coordinating Council',[8] and representatives from the disaster risk reduction community. The main functions of the Commission are to '[e]nsure the mainstreaming of climate change, in synergy with disaster risk reduction, into the national, sectoral and local development plans and programs and to create a panel of technical experts, consisting of practitioners in disciplines that are related to climate change, including disaster risk reduction'.[9]

Some critics of the climate change and disaster risk reduction legislation believe that 'at best, the law reiterates what has been recognized by the international community more than a decade ago: that climate change is a pressing global concern that needs to be addressed collectively at local and national levels. However, it also reflects how the Philippine government adheres to specific development frameworks which have historically been unable to resolve the ecological and socio-economic crisis the country has faced.'[10] Although well intended, it seems as if the policies and laws for climate change and disaster risk reduction assume a functioning sustainable development environment which is not present.[11] Padilla et al. are further critical of the bias towards climate and disaster risk mitigation strategies which are driven by international funding institutions, trans-national corporations and developed countries.[12] That being said, on a pure policy and legislation level, the Philippines serves as a good example for disaster risk reduction and climate change adaptation integration.

Notes:
1 United Nations International Strategy for Disaster Reduction, *Global Assessment Report on Disaster Risk Reduction* (United Nations 2009).
2 P P Duque, *Disaster Management and Critical Issues on Disaster Risk Reduction in the Philippines* (National Disaster Coordinating Council 2005).
3 United Nations International Strategy for Disaster Reduction, above note 1.
4 C Benson, *Mainstreaming Disaster Risk Reduction into Development: Challenges and Experience in the Philippines* (International Federation of Red Cross and Red Crescent Societies/the ProVention Consortium 2009).
5 Ibid.
6 Republic of the Philippines, Philippine Disaster Risk Reduction and Management Act of 2010.
7 Philippines, Climate Change Act No. 9729 of 2009.
8 Ibid.
9 Ibid.
10 Arnold Padilla et al., *On the Road to Disaster: Philippine Climate Change Policies* (Center for Environmental Concerns Philippines 2011).
11 Benson, above note 4.
12 Padilla, above note 10.

7 LINKING THE DISASTER RISK REDUCTION AND CLIMATE CHANGE ADAPTATION AGENDAS: LESS WORK WITH FEWER RESOURCES

One aspect which still has to be addressed is that of integration. Perusal of literature shows a three-way argument: integration of disaster risk reduction into climate change adaptation[52] and vice versa,[53] and the linkage with development.[54] Most of the scholars writing on the topic

[52] SVRK Prabhakar et al., 'Climate Change and Local Level Disaster Risk Reduction Planning: Need, Opportunities and Challenges' (2009) 14 Mitigation and Adaptation Strategies for Global Change 7; I Kelman and J C Gaillard, 'Placing Climate Change within Disaster Risk Reduction' (2008) 1 Disaster Advances 3.

[53] Lisa Schipper and Mark Pelling, 'Disaster Risk, Climate Change and International Development: Scope for, and Challenges to, Integration' (2006) 30 Disasters 19; Mercer, above note 11.

[54] Geoff O'Brien et al., 'Climate Change and Disaster Management' (2006) 30 Disasters 64; E Schipper, 'Climate change adaptation and development: Exploring the linkages' (2007) 107 Tyndall Centre Working Paper; Solecki et al., above note 46; M Munasinghe, 'Exploring the Linkages between Climate Change and Sustainable Development: A Challenge for Transdisciplinary Research' (2001) 5(1) Conservation Ecology 14.

agree that disaster risk reduction and climate change adaptation have been working in isolation.[55]

Mercer says that given the history of disaster risk reduction and its link to development it would make logical sense to view climate change as one among many drivers of vulnerability and disaster risk.[56] Kelman and Giallard, however, take a bit of a more strategic stance.[57] They argue that the 'scapegoat' status of climate change provides an ideal opportunity for the disaster risk reduction community to access different funding streams and gain more political prominence. Venton and La Trobe also believe that the difference between disaster risk reduction and climate change lies within the political importance and widespread recognition which climate change enjoys, as opposed to disaster risk reduction.[58] O'Brien et al. are of the opinion that the sometimes short-term focus on rapid onset hazards and the more longer-term focus on the impacts of climate change, need to be merged with development planning, with capacity development and resilience being key (as has been seen in the example of the Philippines).[59]

Disaster risk reduction and climate change adaptation share many common goals with the development (or sustainable development) agenda. The main driver of development in the future will be the ability of humans to adapt, change and remain/become resilient in the face of climate uncertainties and heightened disaster risk. Integral to sustainable development is: 'how future development patterns will affect climate change; how climate change impacts, adaptation, and mitigation will affect future sustainable development prospects; and, how climate change responses might be better integrated into emerging sustainable development strategies'.[60]

---

55 Paul Venton and Sarah La Trobe, *Linking Climate Change Adaptation and Disaster Risk Reduction* (Tearfund 2008); Thomalla et al., above note 48; Philip Ireland, 'Climate Change Adaptation and Disaster Risk Reduction: Contested Spaces and Emerging Opportunities in Development Theory and Practice' (2010) 2 Climate and Development 332; Birkmann, above note 40; Kelman and Gaillard, above note 52; Solecki at al., above note 46.

56 Mercer, above note 11.

57 Kelman and Gaillard, above note 52 at 3.

58 Venton and La Trobe, above note 55.

59 O'Brien et al., above note 54.

60 Munasinghe, above note 54.

BOX 6.4 SOUTH AFRICAN POLICY RESPONSE TO CLIMATE CHANGE

The main planning document at local municipal level is the 'Integrated Development Plan' (IDP). The aim of the IDP is to provide a five-year planning guide for municipalities to implement development oriented projects.[1] The IDP is directly linked to the budget of the municipality as well as the annual allocation of the equitable share by government. The logic is that no project should be funded if it is not within the IDP of a municipality. The process of compiling an IDP is very participatory in nature.[2] A required part of any IDP must be an integrated disaster risk reduction plan.[3] The disaster risk assessment of the municipality must be contained in the disaster risk reduction plan and must inform all development planning. However, as the research of Hoogstad and Kruger shows, in reality the above is not always achieved.[4] In contrast to the 2002 Disaster Management Act, the Government of South Africa only published an official response to climate change in 2010, through a Green Paper called the National Climate Change Response Green Paper.[5] As with so many other climate change policies at national government level, this Green Paper is very silent on local actions for adaptation and the link to development planning. Although the Green Paper contains a section focusing on disaster risk management, the focus remains reactive to climate stimuli without creating a link to the reduction of risks associated with climate related hazards. Development and varied development focuses are mentioned in the Green Paper; however, it is difficult to judge exactly how the South African Government will address climate change related issues. Several authors suggest that the most effective way of disaster risk reduction and climate change adaptation integration must be through the development mechanism at local government level.[6] However, Falling et al. show that very little is currently being done at municipal level in South Africa to address issues of climate change adaptation in a coherent and integrated manner.[7]

Notes:

1 D Van Niekerk, *A Comprehensive Framework for Multi-Sphere Disaster Risk Reduction in South Africa* (Unpublished PhD thesis North-West University 2005).

2 G Van Riet, 'Qualitative and Quantitative Community Level Approaches: Disaster Risk

Assessment in the Fezile Dabi District Municipality' (2010) 39(2) Africanus: Journal of Development Studies 31.
3 Republic of South Africa, Disaster Management Act (Government Gazette 29 January 2003).
4 Willem Hoogstad and Christi Kruger, *The Link between Development Planning and Disaster Risk Management in Selected Municipalities* (South African National Disaster Management Centre 2008).
5 Republic of South Africa, *National Climate Change Response Green Paper* (Government Printer 2010).
6 Gideon van Riet, 'Disaster Risk Assessment in South Africa: Some Current Challenges' (2009) 40 South African Review of Sociology 194; Gideon van Riet and Dewald van Niekerk, 'Capacity Development for Participatory Disaster Risk Assessment' (2012) Environmental Hazards 1; Global Network of Civil Society Organisations for Disaster Reduction, *Clouds but little rain* …' (Global Network of Civil Society Organisations for Disaster Reduction 2009).
7 Willemien Faling et al., 'Rhetoric or Action: Are South African Municipalities Planning for Climate Change?' (2012) 29 Development Southern Africa 241.

The issues of scale, norms and knowledge across these three domains must be addressed. As has been shown, this chapter is more concerned with the norms mismatches and in particular the legislative and policy perspectives on climate change adaptation and disaster risk reduction. The sections to follow aim to argue the need and use of disaster (risk reduction) law as a method for climate adaptation and addressing climate change drivers.

8 MITIGATING CLIMATE CHANGE THROUGH DISASTER RISK LAW

The Global Assessment Report of the ISDR indicates that since the signing of the Hyogo Framework for Action (HFA) in 2005, many countries have made significant strides in developing and implementing disaster risk reduction law.[61] Most of the post-2005 laws are aligned with the HFA and its priority areas. One such priority area is ensuring that disaster risk reduction gains heightened political interest. To this end a number of national multi-sectoral disaster risk reduction platforms have been established with the aim of ensuring a coordinated approach to addressing disaster risk factors. One therefore finds that an institutional mechanism is firmly in place where multiple role-players address a common problem, which is mostly legislated. However, one still finds a duplication of structures in the name of climate change adaptation and

---

[61] United Nations International Strategy for Disaster Reduction, *Global Assessment Report on Disaster Risk Reduction* (United Nations 2009).

climate mitigation, which brings about significant wastage of time and resources within public sectors.[62] Disaster risk law in the majority of the cases already provide for institutional structures. These structures are enforceable and span the various tiers of governments. It would therefore only make logical and practical sense to utilize such structures for climate change adaptation issues as well.

The prior discussion in this chapter makes a clear argument for the connection between disaster risk reduction and climate change and climate change drivers. More unknowns linked with climatic hazards heightens a country's disaster risk profile. It is widely accepted that a global disaster risk driver is climate change. Therefore, from disaster risk reduction a connection must be made to climate change as a causal factor of vulnerability which in turn drives poverty, under-development, fragile rural livelihoods and declining ecosystem health. Climate change magnifies the interactions between disaster risk and poverty. The development of a disaster risk profile is a common requirement in all disaster risk laws. These disaster risk assessments must therefore consider the above global factors as well, and linkages to wider interdependent systems must be made.[63] Gero et al. argue that disaster risk reduction policies and laws can address both 'hard' and 'soft' solutions to shared climate change and disaster risk problems.[64] For example, hard solutions might be related to the construction and maintenance of infrastructure such as sea walls, whereas a soft solution might be changing behaviour or attitudes through education, awareness and training. In a sense, Artur and Hilhorst believe that the successes and failures of adaptation and policy imperatives depend highly on the cultural and political realms of societal perceptions and the sensitivity of institutions.[65]

---

[62] Mercer, above note 11; Kelman and Gaillard, above note 52; United Nations International Strategy for Disaster Reduction, *At the Crossroads: Climate Change Adaptation and Disaster Risk Reduction in Asia and the Pacific* (Asia Pacific Secretariat, International Strategy for Disaster Reduction 2011).

[63] United Nations, above note 61.

[64] Gero, Méheux and Dominey-Howes, above note 45.

[65] Luis Artur and Dorothea Hilhorst, 'Everyday Realities of Climate Change Adaptation in Mozambique' (2012) 22 Global Environmental Change 529.

9 BARRIERS TO DISASTER RISK LAW IN THE CONTEXT OF CLIMATE CHANGE ADAPTATION

In practical terms, the adoption of an overarching policy framework for disaster risk reduction should provide a vehicle for a closer integration of existing climate change and development policy and strategy instruments such as Poverty Reduction Strategy Papers, National Adaptation Programmes/Plans of Action and other action plans to implement the HFA, thus improving coherence and gaining synergy. In turn this will be facilitated if the plethora of planning, reporting and funding mechanisms at the international level is streamlined.[66]

However, disaster risk reduction alone is not the silver bullet to our climate change woes. Recognition is needed of the fact that disaster risk reduction also focuses on other possible disaster risks unrelated to climate change. Similarly, there are certain very specific climate change adaptation actions needed which do not in themselves directly reduce the risk of hazards turning into disasters.

The research of Artur and Hilhorst on Mozambique as a case illustrates the above. Mozambique is highly vulnerable to natural hazards.[67] A detailed analysis carried out by the National Government of Mozambique shows that the frequency and magnitude of natural hazards have steadily increased over the last 50 years and in effect they will increase further because of climate change.[68] Following the 2000 floods in Mozambique, the Government embarked on a process to strengthen its disaster risk management institutional arrangements. This process provided an ideal opportunity to also consider issues of climate change and adaptation in parallel. The National Disaster Management Institute (INGC) is currently developing a detailed catalogue of concrete climate change adaptation measures. This will outline new areas of activity, such as setting up flood early warning systems. Such systems are particularly targeted at urban coastal areas, as the rising sea level is likely to increase damages and losses. Besides this, the introduction of adapted agricultural practices in rural areas is considered an important step in reducing climate related

---

66 Elizabeth Burleson, 'Climate Change Consensus: Emerging International Law' (2010) 34 William and Mary Environmental Law and Policy Review 543. See also Figure 6.1.

67 Artur and Hilhorst, above note 65.

68 Deutsche Gesellschaft für Internationale Zusammenarbeit (GIZ), *Mozambique: Disaster Risk Reduction as the Basis for Climate Change Adaptation* (GIZ 2011).

disaster risks.[69] According to Göhl, the emphasis is placed on the vital responsibility of governments to develop strategies and policies through the fostering of local capacities and also to reduce vulnerabilities to hazards to promote economic and social development.[70] The government of Mozambique has also acknowledged a commitment to follow this path.

There is, in fact, no legislation for Disaster Risk Management or Climate Change in Mozambique, but they do have many established projects or programmes that address both of these agendas.[71] Some of the projects or programmes that are already established in the vulnerable or affected areas of Mozambique are:[72]

- The Mozambique Flood Emergency Recovery Project;
- Mozambique IDA/GEF Zambezi Valley market-led smallholder development;
- Malawi and Mozambique mainstreaming disaster risk reduction;
- flood early warning systems in Mozambique;
- drought mitigation strategies in Mozambique;
- effective flood management in Mozambique; and
- integrating disaster risk management into rural development in Central Mozambique.

However, Artur and Hilhorst found that the emergency discourse prevailing internationally around climate change does not resonate with any of the actors in Mozambique.[73] They also observed that the stakes of climate change are high in two ways. Climate change is said to result in increased flooding which affects Mozambique. Secondly, the international attention to climate change has opened up an arena of contestation over ideas and resources. The authors found a remarkable cross-cutting tendency that institutions and people alike tend to seek continuity by incorporating the climate change agenda into their lifestyles and institutional interests. In Mozambique there is also evidence of clashes or contradictions within and between the domains of governance, international interventions and people (i.e., government policy is torn between

---

69 Ibid.; Artur and Hilhorst, above note 65.

70 Sandra E Göhl, Local Governance and Disaster Risk Management in Mozambique (Thesis, University of the Western Cape 2008).

71 Disaster Preparedness Programme DIPECHO, *Community-Based Best Practices for Disaster Risk Reduction* (UNDP 2010).

72 GIZ, above note 65.

73 Artur and Hilhorst, above note 65.

addressing disaster risk reduction, climate change adaptation and development). In many instances there is a disconnect between the scientific understanding of climate change and how it is used to drive political policy agendas and reinforce social order. Lastly, Artur and Hilhorst believe that technocratic responses to the challenge of climate change adaptation take place in a social context of significant inequality and differential power.[74]

Thus, various authors emphasize that certain barriers exist in only relying on disaster risk reduction laws to address climate change related issues. The response to climate related hazards are complex. It needs to be understood within certain historical, cultural as well as political processes.

10 CONCLUSION

This chapter aimed to argue the convergence of the disaster risk reduction and climate change adaptation agendas. It showed that the legal path followed to institutionalize disaster risk reduction is well established compared to climate change. The uncertainties of climate change, as well as the inability of governments to act, bring disaster risk reduction more and more into play. Disaster risk reduction serves as a type of insurance scheme for development, where climate change now threatens our hard earned development. This is even truer in the developing world. Although the hype surrounding climate change has somewhat voided its meaning to the layperson, we still find many national policies being developed. The challenge remains to enhance rather than duplicate the disaster risk reduction and climate change actions.

Climate change, climate change adaptation and disaster risk reduction remain complex issues. So complex, that humankind almost finds itself paralysed towards actions. One universal truth when one considers climate change is that in the years to come it will only contribute, more and more, to devastating disasters, increase disaster risk, exploit vulnerabilities and impact on urban as well as rural livelihoods. However, the actions needed to mitigate the effects of climate change and those needed to address disaster risks are very similar. By joining forces, the climate change adaptation and disaster risk reduction agendas can be merged with obvious benefits. The implementation of disaster risk reduction laws in so many countries already provide a foundation from which climate

---

[74] Ibid.

change can be mitigated and adaptation supported. A remaining major obstacle is political conscience, understanding and action. Addressing climate change and enhancing adaptation methods is a moral just as much as it is a legal obligation of governments.

7. Climate change adaptation and public health law

*Lindsay F Wiley**

1 INTRODUCTION: A SHIFT IN PERSPECTIVE

In 2002, Oxford philosopher Benito Müller diagnosed 'a clear North-South divide' in climate change policy. In the 'ecological view' of the industrialized North, he wrote, 'Climate change is perceived as a problem of polluting the environment, of degrading the eco-system[;] … its essence is seen to be that of a wrongful act against "Nature". … The reality in the South is quite different: climate change has primarily come to be seen as a human welfare problem … The harm is against humans, it is largely other-inflicted, and it is not life-*style*-, but *life*-threatening'.[1] In the intervening decade, the southern view has gained global traction. The global health community has made great strides in its efforts to 'put a human face on climate change'. Epidemiological studies have shed light on the current and anticipated impacts of climate change on infectious diseases, cardiovascular and pulmonary diseases, natural disasters and mental health. The documentation of these impacts has lent weight to arguments for climate change mitigation – policies aimed at reducing atmospheric greenhouse gas concentrations as a means to prevent climate change from happening. At the same time, growing awareness of the enormous human health burden associated with climate change raises important questions about adaptation – policies aimed at addressing the impacts of climate change and reducing their severity. Given the significance of the health impacts associated with climate change, global public health infrastructure should be a crucial component of any adaptation strategy.

This chapter addresses the important linkages between climate change adaptation policy and the rapidly growing fields of public health law and global health law. As defined by Lawrence Gostin in an influential treatise, public health law is concerned with 'the legal powers and duties

[1] Benito Müller, *Equity in Climate Change: The Great Divide* (Oxford Institute for Energy Studies 2002) 2.

of the state, in collaboration with its partners … to ensure the conditions for people to be healthy'.[2] At the international level, public health law becomes global health law – a field defined by Gostin as encompassing 'the legal norms, processes, and institutions needed to create the conditions for people throughout the world to attain the highest possible level of physical and mental health'.[3] Both fields are distinct from the law of health *care*. Rather than being concerned with the medical treatment of individuals, public health law and global health law focus on protecting and promoting health at the population level by ensuring access to healthy living conditions. This chapter will address the two fields in combination by using the term 'global public health law'.

Climate change will make the work of ensuring access to healthy living conditions considerably more challenging for global public health advocates. For the most part, climate change will act as an intensifier of existing threats to health. Thus, an important starting point for health adaptation is examination of existing law and policy tools for addressing these threats. Unfortunately, climate change is expected to intensify many health threats that are currently neglected by global public health law frameworks. For a variety of reasons, problems such as diarrheal illness (94% of which is attributable to modifiable environmental factors) and vector-borne disease (nearly half of which is attributable to modifiable environmental factors)[4] have not historically received adequate attention from policymakers even though they are among the greatest sources of global disease burden. We understand the mechanisms by which environmental factors contribute to the global disease burden and we have readily available interventions for combating these threats to health. The problem of health adaptation to climate change is not so much a need for innovative technologies (though significant resources are being deployed to develop new tools, such as a vaccine for malaria[5] or cheaper, lower energy water filtration[6]). The problem is that we lack the political will

---

[2] Lawrence O Gostin, *Public Health Law: Power, Duty, Restraint* (2nd edn, University of California Press 2008) 4.

[3] Lawrence O Gostin and Allyn L Taylor, 'Global Health Law: A Definition and Grand Challenges' (2002) 1 Public Health Ethics 53, 55.

[4] A Prüss-Üstün and C Corvalán, *Preventing Disease through Healthy Environments* (World Health Organization 2006) 9.

[5] PATH Malaria Vaccine Initiative, 'About Us', www.malariavaccine.org, accessed 11 October 2012.

[6] Oxfam America, 'The New Adaptation Marketplace: Climate Change and Opportunities for Green Economic Growth' (2009) 2, http://www.oxfamamerica.org/files/the-new-adaptation-marketplace.pdf-1, accessed 11 October 2012.

and effective legal mechanisms to translate existing public health approaches into widespread practice.

In the next section, this chapter describes the current and anticipated impacts of climate change on human health and provides examples of the kinds of interventions that have been developed to address them. Section 3 assesses the usefulness of current global public health law frameworks as potential tools for health adaptation to climate change. Section 4 describes the current level of focus on the climate-health nexus within the United Nations Framework Climate Change Convention (UNFCCC) and the World Health Organization (WHO). Section 5 highlights three crucial challenges that must be addressed by health adaptation planners at the international, regional, domestic, and local levels: investment in basic public health and sanitation infrastructure; integration across sectors; and a shift from security to equity as the justification for prioritizing public health.

2 THE HEALTH IMPACTS OF CLIMATE CHANGE AND INTERVENTIONS FOR ADDRESSING THEM

Environmental factors are already responsible for a considerable portion of global disease burden – a statistical measure that aggregates loss of healthy years of life due to disease or disability, measured in disability-adjusted life years, or DALYs. Nearly one-quarter of the total global disease burden and as much as one-third of the disease burden among children under age 14 is believed to be attributable to modifiable environmental conditions such as lack of adequate water for drinking and sanitation, exposure to disease-carrying insects and rodents, and air pollution.[7] The impact of unhealthy environments is felt disproportionately by children (who are especially vulnerable to complications and death from these illnesses), particularly in the developing world. The WHO estimates that more than four million child deaths each year are attributable to environmental causes, and that the infant death rate from environmental causes is 12 times higher in developing countries than in developed countries.[8] To address environmental threats to health, policymakers in many countries have developed plans and legal frameworks. However, these plans are largely threat-specific and fail to address the

---

[7] Prüss-Üstün and Corvalán, above note 4 at 9.
[8] Ibid.

upstream determinants of health. There are several well-developed examples of country-level approaches to mosquito control, heat wave planning, and natural disaster response. But some of the most concerning health impacts continue to be neglected, especially in the area of water-related diarrheal illness.

The climate change impacts discussed in previous chapters (including water scarcity, diminished agricultural capacity, ecosystem degradation, and severe weather events) are potentially devastating for human health. Climate change is expected to intensify existing environmental threats to health, including diarrheal illness, vector-borne diseases, and cardio-respiratory diseases. It is also likely to redistribute some of those threats to new areas by bringing malaria or dengue fever to new areas, for example. Finally, climate change may create fertile conditions for the potential emergence of new infectious pathogens, by altering the ways in which animal species interact with each other and with humans, for example. In recent decades, health scientists have begun to document the current effects of anthropogenic climate change on human health and to predict effects that are likely to be seen in the future.[9] The health sector has also begun to assess the resources that will be needed to respond to these threats and public health policymakers have intensified their planning and response efforts, at least in some areas.[10]

2.1 Heat Waves and Hazardous Air Quality

The 2003 European heat wave, which caused thousands of deaths,[11] played an important role in drawing public attention to the health dimension of climate change. Extreme heat can complicate cardiovascular, respiratory, and renal disease, as well as central nervous system disorders and diabetes.

---

[9] A McMichael and A Githeko, 'Human Health' in James L McCarthy et al. (eds), *Climate Change 2001: Impacts, Adaptation, and Vulnerability: Contribution of Working Group II to the Third Assessment Report of the Intergovernmental Panel on Climate Change* (Cambridge University Press 2001).

[10] Anthony Costello et al., 'Managing the Health Effects of Climate Change' (2009) 373 The Lancet 1693.

[11] S Vandentorren et al., 'August 2003 Heat Wave in France: Risk Factors for Death of Elderly People Living at Home' (2006) 16 European Journal of Public Health 583.

Thus, heat waves are particularly dangerous for the elderly.[12] Air pollution – especially ground-level ozone[13] and particulate matter[14] – is exacerbated by higher temperatures, with resulting effects on cardiovascular and respiratory illness.[15] In response to the 2003 heat wave, several European countries developed heat wave plans. Many, such as those for France,[16] Belgium,[17]

---

12 Eric Klinenberg, *Heat Wave: A Social Autopsy of Disaster in Chicago* (1st edn, University of Chicago Press 2002); J Diaz et al., 'Heat Waves in Madrid 1987–1997: Effects on the Health of the Elderly' (2002) 75 International Archives of Occupational & Environmental Health 163; N Gouveia et al., 'Socio-economic Differences in the Temperature-Mortality Relationship in Sao Paulo, Brazil' (2003) 32 International Journal of Epidemiology 390; R S Kovats and others, 'Contrasting Patterns of Mortality and Hospital Admissions During Hot Weather and Heat Waves in Greater London, UK' (2004) 61 Occupational & Environmental Medicine 893; J Schwartz, 'Who is Sensitive to Extremes of Temperature? A Case-Only Analysis' (2005) 16 Epidemiology 67.

13 Ground-level ozone (the primary constituent of smog) is formed by chemical reactions between certain air pollutants and sunlight. It is distinct from the upper-level atmospheric ozone that protects us from harmful UV rays. Susan M Bernard, 'The Potential Impacts of Climate Variability and Change on Air Pollution-Related Health Effects in the United States' (2001) 109 Environmental Health Perspectives 199.

14 Particulate matter ($PM_{2.5}$), which includes all airborne particles that are less than 2.5 micrometers in diameter, is also formed through atmospheric reactions (which are influenced by rising temperatures) among various pollutant gasses. United States Environmental Protection Agency, 'Fine Particle ($PM_{2.5}$) Designations', http://www.epa.gov/pmdesignations/basicinfo.htm, accessed 11 October 2012.

15 Ulisses Confalonieri et al., 'Human Health' in M L Parry et al. (eds), *Climate Change 2007: Impacts, Adaptation, and Vulnerability: Contribution of Working Group II to the Fourth Assessment Report of the Intergovernmental Panel on Climate Change* (Cambridge University Press 2007).

16 R Sari Kovats and Gerd Jendritzky, 'Heat-waves and Human Health' in Bettina Menne and Kristie L Ebi (eds), *Climate Change and Adaptation Strategies for Human Health* (Steinkopff 2006) 86.

17 Belgian National Climate Commission, 'Climate Change: Belgium's Fifth National Communication under the United Nations Framework Convention on Climate Change [UNFCCC]' (Belgian Federal Public Service: Health, Food Chain Safety and Environment) 119, http://unfccc.int/resource/docs/natc/bel_nc5_en_final.pdf, accessed 11 October 2012; Belgian Federal Public Service: Health, Food Chain Safety and Environment, 'What does Belgium do?', http://www.health.belgium.be/eportal/Environment/Airetchangementclimatique/Ozone/WhatdoesBelgiumdo, accessed 11 October 2012; Belgian National Climate Commission, 'Belgian National Climate Change Adaptation Strategy' (2010), http://www.lne.be/themas/klimaatverandering/adaptatie/nationale-adaptatie-strategie, accessed 11 October 2012.

Italy,[18] and the United Kingdom,[19] include tiered levels of warnings and responses, depending on the forecast temperatures. Response plans typically involve a combination of media warnings and direct contact with high risk individuals and facilities. This awareness-raising approach has effectively reduced mortality associated with subsequent heat waves in Europe.[20] Similar plans have been adopted in the US.[21] While the plans are largely similar, some have unique characteristics. The heat wave plan for England, for instance, requires local governments to encourage high risk individuals and their families to develop personal plans. It also requires that every nursing home and residential facility provide a 'cool room' where the temperature does not exceed 26ºC.[22] France's plan includes provision of air conditioning units for the elderly.[23] Some countries also include response plans for hazardous air quality events in their heat wave plans.[24] Belgium's

---

[18] Italian Ministry for the Environment, Land and Sea, 'Fifth National Communication under the UNFCCC: Italy' (2009) 6–21, http://unfccc.int/resource/docs/natc/ita_nc5.pdf, accessed 11 October 2012; Italian Ministry of Health, 'Heat Waves', http://www.salute.gov.it/emergenzaCaldo/emergenzaCaldo.jsp, accessed 11 October 2012.

[19] National Health Service, England, *Heatwave Plan for England: Protecting Health and Reducing Harm from Extreme Heat and Heatwaves* (26 May 2011), United Kingdom's National Health Service, 'Heatwave Plan for England: Protecting Health and Reducing Harm from Extreme Heat and Heatwaves' (United Kingdom's Department of Health 2011), http://www.dh.gov.uk/prod_consum_dh/groups/dh_digitalassets/documents/digitalasset/ dh_127235.pdf, accessed 11 October 2012.

[20] Anne Fouillet et al., 'Has the Impact of Heat Waves on Mortality Changed in France Since the European Heat Wave of Summer 2003? A Study of the 2006 Heat Wave' (2008) 37 International Journal Epidemiology 309.

[21] Maryland Commission on Climate Change, 'Health Impact Assessment of the Climate Change Action Plan' (2008) (Maryland's CCAP requires coordination across health and safety agencies, including the Department of Mental Health, local health departments, the MD Emergency Medical Association, and the MD Departments of Agriculture and Environment); Laurence S Kalkstein et al., 'The Philadelphia Hot Weather-Health Watch/Warning System: Development and Application, Summer 1995' (1996) 77 Bulletin of the American Meteorological Society 1519 (the city of Philadelphia's Hot Weather-Health Watch/Warning System (PWWS) coordinates weather forecasting systems with decisions by the Philadelphia Department of Health about dissemination of media messages advising the public regarding how to minimize heat exposure).

[22] 'Heatwave Plan for England', above note 19 at 27.

[23] Kovats and Jendritzky, above note 16 at 85.

[24] Dianne Lowe et al., 'Heatwave Early Warning Systems and Adaptation Advice to Reduce Human Health Consequences of Heatwaves' (2011) 8 International J of Environmental Research and Public Health 4631–4633.

plan, for example, establishes alerts for high concentrations of ground-level ozone, triggering media alerts when a defined threshold is reached.[25]

2.2 Natural Disasters as Public Health Emergencies

Weather-related natural disasters such as tropical storms, tornadoes, floods, and wildfires cause physical injuries and deaths. Many people are exposed to infectious diseases and hazardous chemicals through contaminated floodwaters or unsanitary living conditions following an event.[26] The displacement caused by natural disasters can lead to a dangerous disruption in health care for those suffering from chronic diseases, which require regular access to medications and treatments such as HIV/AIDS, diabetes, and cardiovascular disease. The effects of natural disasters on mental health have gained attention in recent years, with some suggesting that they make up the largest portion of disease burden attributable to such events.[27]

Although disaster law (addressed in a previous chapter) encompasses many provisions aimed primarily at addressing property loss and long-term rebuilding, recent reform efforts have emphasized the importance of the public health emergency. In particular, the International Federation of Red Cross and Red Crescent Societies (IFRC) has promoted a comprehensive vision of disaster response law that addresses everything from protecting volunteer health workers from legal liability to facilitating the transportation of needed health care supplies.[28] The IFRC has also

---

[25] 'What Does Belgium Do?', above note 17.

[26] United States Centers for Disease Control and Prevention, 'Morbidity Surveillance after Hurricane Katrina – Arkansas, Louisiana, Mississippi, and Texas' (2005) 55 Morbidity & Mortality Weekly Report 727.

[27] K L Middleton et al., 'Natural Disasters and Posttraumatic Stress Disorder Symptom Complex: Evidence from the Oklahoma Tornado Outbreak' (2002) 9 International J of Stress Management 229; C V Russoniello and others, 'Childhood Posttraumatic Stress Disorder and Efforts to Cope After Hurricane Floyd' (2002) 28 Behavioral Medicine 61; P Verger et al., 'Assessment of Exposure to a Flood Disaster in a Mental-health Study' (2003) 13 Journal of Exposure Analysis and Environmental Epidemiology 436; C S North, et al., 'The Course of PTSD, Major Depression, Substance Abuse, and Somatization After a Natural Disaster' (2005) 192 The Journal of Nervous and Mental Disease 823; J G I Barbee and M H Townsend, 'Mental Health and Recovery in the Gulf Coast after Hurricanes Katrina and Rita' (2006) 296 Journal of the American Medical Association 585.

[28] International Federation of Red Cross and Red Crescent Societies, 'Introduction to the Guidelines for the Domestic Facilitation and Regulation of International Disaster Relief and Initial Recovery Assistance' (2011), http://www.

integrated psychosocial support, to address the enormous mental health burden associated with disasters, into its basic emergency response operations.[29]

2.3 The Devastating Impact of Water Scarcity on Public Health

Increasing aridity in some regions and periodic drought and flood in others will impede regular access to safe, adequate water supply for drinking and sanitation, which can be devastating for health.[30] Water-washed diseases, such as rotavirus, are affected by hygiene practices and are therefore highly sensitive to water scarcity.[31] Water-borne diseases, such as cholera, cryptosporidiosis, and campylobacteriosis, are affected both by droughts (which concentrate pathogens in pools) and floods (which increase runoff, resulting in microbial contamination of water supplies).[32]

Sanitation systems that provide for the safe disposal of human and animal waste are crucial to public health. Research suggests that investment in sanitation infrastructure leads to significant financial returns associated with improved health. Pit latrines in rural areas are associated with a one to five return on investment. Off-site treatment plans in urban areas are associated with a one to four return on investment.[33] Globally, the return on investment in sanitation has been estimated to be as high as

---

ifrc.org/PageFiles/41203/1205600-IDRL%20Guidelines-EN-LR%20(2).pdf, accessed 11 October 2012.

[29] International Federation Reference Centre for Psychosocial Support, 'Psychosocial Interventions: A Handbook' (2009), http://psp.drk.dk/graphics/2003referencecenter/Doc-man/Documents/docs/Psychosocial%20interventions%20A%20handbook%20LowRes.pdf, accessed 11 October 2012.

[30] World Health Organization 'Safer Water, Better Health' (2008), http://whqlibdoc.who.int/publications/2008/9789241596435_eng.pdf, accessed 11 October 2012; Corrine J Schuster-Wallace et al., 'Safe Water as the Key to Global Health' (United Nations University International Network on Water, Environment, and Health 2008); Bettina Menne and Roberto Bertollini, 'The Health Impacts of Desertification and Drought' (2000) 14 Down to Earth: United Nations Convention to Combat Desertification Newsletter 4.

[31] Ibid.

[32] McMichael and Githeko, above note 9 at 408.

[33] Water and Sanitation Program, 'The Economic Returns of Sanitation Interventions in the Philippines' (2011), http://www.wsp.org/wsp/sites/wsp.org/files/publications/wsp-philippines-economic-returns.pdf, accessed 11 October 2012.

one to nine.[34] Despite the obvious benefits of prioritizing sanitation infrastructure, one in three people worldwide still lacks access to basic sanitation. In sub-Saharan Africa, that number may be as high as two in three.[35] The Millennium Development Goal (MDG) – aimed at halving by 2015 the number of people who lack sustainable access to safe drinking water and basic sanitation – has been described as the 'MDG orphan'.[36] The goal is not simply in danger of going unmet; it may actually be that previous gains are *lost* during the MDG timeframe:[37]

> [I]n many countries, policies and programmes have far too little emphasis on ensuring adequate financial and human resources to both sustain the existing infrastructure and expand access to sanitation, drinking-water and hygiene services. The danger of slippage against the MDG target is a real one.[38]

Nearly 80 percent of countries recognize a human right to water and nearly 50 percent recognize a human right to sanitation; most countries have made political and/or financial commitments to ensuring access to sanitation and drinking water, typically at the ministerial level.[39] The problem is one of translating those commitments into sustainable infrastructure.

2.4 Infectious Diseases

Vector-borne infectious diseases (such as malaria, dengue fever, West Nile Virus, and Lyme Disease), zoonotic diseases (such as Hantavirus carried by rodents or H5N1 influenza carried by birds), and food-borne illnesses (such as Salmonellosis and ciguatera poisoning) are all highly sensitive to changing environmental conditions. Climate change threatens

---

[34] Tom Kemeny, 'Sanitation and Economic Development: Making an Economic Case for the MDG Orphan' (WaterAid 2007), http://www.wateraid.org/documents/sanitation_and_economic_development.pdf, accessed 11 October 2012.

[35] World Health Organization and UNICEF, 'Meeting the MDG Drinking Water and Sanitation Target: The Urban and Rural Challenge of the Decade' (2006), http://www.who.int/water_sanitation_health/monitoring/jmpfinal.pdf, accessed 11 October 2012.

[36] Kemeny, above note 34.

[37] World Health Organization, 'GLAAS 2012 Report: UN-Water Global Analysis and Assessment of Sanitation and Drinking-Water: The Challenge of Extending and Sustaining Services', http://whqlibdoc.who.int/publications/2012/9789241503365_eng.pdf, accessed 11 October 2012.

[38] Ibid.

[39] Ibid. at 10.

to alter the survival, persistence, habitat range, and transmission of a variety of species and pathogens in ways that may bring them into greater contact with humans.[40] Abrupt ecosystem changes can lead to the introduction of illnesses into areas that do not have sufficient surveillance and response infrastructure in place. Health care workers, for example, may not be prepared to evaluate patients for illnesses that have not previously been prevalent in their communities. It is also possible that abrupt, climate related ecosystem changes will make the emergence of new human diseases more likely. Novel infectious diseases are particularly concerning for their epidemic potential, because the human population will not have developed any natural immunity.

The impact of climate change on malaria and dengue fever, the vector-borne illnesses with the greatest disease burden, is particularly concerning. Changing rainfall and temperature patterns can alter the geographic distribution of vector mosquitoes. As mosquitoes carrying malaria and dengue move to higher latitudes and altitudes, they will threaten human populations that have not previously been exposed to these diseases and thus are particularly vulnerable to complications.[41] The public health sector has a long (and occasionally controversial) history of intervention to control and prevent vector-borne disease. Many countries combine their malaria control programs with dengue control into integrated mosquito control programs. Plans typically involve a combination of vector control (killing the insects, limiting breeding sources), transmission prevention (distribution of insecticide-treated nets, mosquito-proofing dwellings), and drug treatments for infected individuals. In India, for example, where the government is implementing its

---

[40] Confalonieri, above note 15 at 408; Ying Zhang et al., 'Climate Change and the Transmission of Vector-Borne Diseases: A Review' (2008) 10 Asia-Pacific Journal of Public Health 64; D J Rogers and S E Randolph, 'Climate Change and Vector-Borne Diseases' (2006) 62 Advances Parasitology 345; P Gale et al., 'Predicting the Impact of Climate Change on Livestock Diseases in Great Britain' (2008) 162 The Veterinary Record 214; John S Browstein et al., 'A Climate-Based Model Predicts the Spatial Distribution of the Lyme Disease Vector Ixodes Scapularis in the United States' (2003) 11 Environmental Health Perspectives 1152; R S Kovats et al., 'Early Effects of Climate Change: Do They Include Changes in Vector-Borne Disease?' (2001) 356 Philosophical Transactions of the Royal Society 1057; Simon Hales et al., 'El Niño and the Dynamics of Vectorborne Disease Transmission' (1999) 107 Environmental Health Perspectives 99.

[41] Ibid.

eleventh five-year malaria plan,[42] the government's approach combines integrated vector control with case surveillance.

For adult mosquitoes, the Indian health service employs indoor residual spraying using DDT and alternative pesticides in high risk areas (those where more than 10% of the population tests positive for malaria) and priority spraying in other risky areas (those where more than 5% of the population tests positive for malaria). In coordination with local government agencies and non-governmental organizations (NGOs), the Indian government also seeks to control mosquito larvae through introduction of larvae-eating fish into high risk areas.[43] In addition to direct government efforts to control mosquitoes, many countries enlist community support in reducing the prevalence of potential breeding sites. In Cambodia, for example, the government sprays for mosquito larvae during an active outbreak of dengue, while also reaching out to communities to eliminate breeding sources, particularly containers where water might accumulate.[44] In Vanuatu, the government encourages vector control by educating the population on the dangers of standing water and by distributing large bags for residents to dispose of tin drums and discarded tires in which water may accumulate and larvae develop.[45] These and other vector control efforts require significant coordination between federal and local governments and among government agencies, funders, NGOs, and community leaders.[46]

Many countries also focus on the use of long lasting, insecticide-treated bed nets to protect high risk populations from exposure, mostly though mass distributions funded partially by international partners.[47] Tanzania, assisted with funds from the Global Fund to Fight Aids, Tuberculosis, and Malaria, created the Tanzania National Voucher Scheme (TNVS), which uses a two-pronged approach to increase access

---

42 Directorate of National Vector Borne Disease Control Program, 'Strategic Action Plan for Malaria Control in India 2007–2012' (date of publication unknown) 24–40, http://nvbdcp.gov.in/Round-9/Annexure-2%20%20Strategic%20action%20plan.pdf, accessed 11 October 2012.

43 Ibid. at 40.

44 Sokrin Khun and Lenore Manderson, 'Community Participation and Social Engagement in the Prevention and Control of Dengue Fever in Rural Cambodia' (WHO 2008) 32 Dengue Bulletin 145.

45 George Taleo et al., 'Dengue Control in Vanuatu: Towards an Integrated Vertical and Horizontal Control Programme' (WHO 2000) 24 Dengue Bulletin.

46 Indian Ministry of Health and Family Welfare, 'National Vector Borne Disease Control Programme', http://nvbdcp.gov.in, accessed 11 October 2012.

47 'Strategic Action Plan for Malaria Control in India', above note 42 at 36–37.

to insecticide-treated nets.[48] The TNVS provides a voucher for nets to pregnant women and children (who are considered most vulnerable to complications arising from malaria infection) and also provides support for the distribution and manufacture of insecticide-treated nets. Similar bed net programs were found to be effective in Malawi, where the government promoted nets through subsidization and distribution via public health facilities.[49]

Improving diagnosis and treatment of existing malaria infections is also an important strategy, particularly given the growing risk of drug resistance. Treatment-oriented programs generally focus on improving the accuracy of diagnosis (over-diagnosis can contribute to drug resistance); distributing more effective treatments through simplified blister packs; and surveillance to track the prevalence of specific strains of the disease. In Cambodia, village malaria workers – laypeople trained to administer rapid diagnostic tests, provide treatment with combined drug therapy blister packs, and refer patients to a physician where appropriate – have effectively improved diagnosis and treatment in rural areas.[50] India also has a program to increase blood tests for malaria and use drug combination therapy in areas where drug resistance is a problem.[51]

Many countries have developed public education programs to raise awareness of the risks posed by ticks and mice and to identify the signs of infection. In response to outbreaks of tick-borne encephalitis, the Czech Republic has instituted a tick warning system.[52] Warnings about tick activity are distributed via a website maintained by the Czech Hydrometeorological Institute, the State Health Institute, and the Ministry of Health. Similarly, the European Centre for Disease Prevention and Control publishes guidelines to prevent the transmission of Hantavirus,

---

48 Jo-Ann Mulligan et al., 'Costs and Effects of the Tanzanian National Voucher Scheme for Insecticide-treated Nets' (2008) 7 Malaria Journal 32.

49 Warren Stevens et al., 'The Costs and Effects of a Nationwide Insecticide-treated Net Programme: the Case of Malawi' (2005) 4 Malaria Journal 22.

50 Junko Yasuoka et al., 'Assessing the Quality of Service of Village Malaria Workers to Strengthen Community-based Malaria Control in Cambodia' (2010) 9 Malaria Journal 109.

51 'Strategic Action Plan for Malaria Control in India', above note 42 at 29–31.

52 Ministry of the Environment of the Czech Republic, 'Fifth National Communication of the Czech Republic on the UNFCCC Including Supplementary Information Pursuant to Article 7.2 of the Kyoto Protocol' (2009) 135, http://unfccc.int/resource/docs/natc/czenc5.pdf, accessed 11 October 2012; Bill Lehane, 'Hikers Warned of High Tick Risk' (*Prague Post*, 15 June 2011).

which is carried by rodents.[53] These programs primarily counsel avoidance of contact (by wearing protective clothing) and self-monitoring for exposure and symptoms of illness.

Higher temperatures may also lead to an increase in food-borne illness.[54] Salmonellosis has been shown to be particularly sensitive to higher ambient temperatures.[55] Higher ocean surface water temperatures may lead to an increased incidence of food poisoning through contamination of shellfish and reef-fish. Harmful algae blooms are promoted by higher temperatures and can lead to ciguatera poisoning when fish contaminated with toxins produced by marine microalgae are consumed.[56] Climate change also increases cyanobacteria (blue green algae) blooms, which are potentially toxic.[57] In Scotland, the government publishes local warnings and awareness to prevent exposure to the toxins released by the cyanobacteria.[58] Regional governments in the United States and Australia also publish warnings about blue green algae.[59] These warnings and information are designed to encourage people to avoid exposure.

---

53 European Centre for Disease Prevention and Control, 'Factsheet for Health Professionals [Hantavirus]' (2012), http://ecdc.europa.eu/en/healthtopics/hantavirus/basic_facts/Pages/factsheet_health_professionals.aspx, accessed 11 October 2012.

54 Ibid.

55 Ibid.

56 James L McCarthy et al. (eds), *Climate Change 2001: Impacts, Adaptation, and Vulnerability: Contribution of Working Group II to the Third Assessment Report of the Intergovernmental Panel on Climate Change* (Cambridge University Press 2001) 400.

57 Hans W Paerl and Jef Huisman, 'Climate Change: A Catalyst for Global Expansion of Harmful Cyanobacterial Blooms' (2009) 1 Environmental Microbiology Reports 27; John Whittington, 'Blue Green Algal Blooms: A Preventable Emergency?' (1999) 14 Australian Journal of Emergency Management 21.

58 Health Protection Scotland, 'Cyanobacteria – Blue Green Algae', http://www.hps.scot.nhs.uk/enviro/cyanobacteria-bluegreenalgae.aspx, accessed 11 October 2012; The Scottish Government, 'Actions for Sector', http://www.scotland.gov.uk/Topics/Environment/climatechange/scotlands-action/adaptation/AdaptationFramework/SAP/Health/Actions, accessed 11 October 2012.

59 Wisconsin Department of Natural Resources, 'Blue Green Algae', http://dnr.wi.gov/lakes/bluegreenalgae, accessed 11 October 2012; Queensland Government, Natural Resources and Water, 'Blue Green Algae' (Department now called the Department of Environment and Resource Management, Queensland 2006), http://www.derm.qld.gov.au/factsheets/pdf/water/w3.pdf, accessed 11 October 2012.

3 EXISTING GLOBAL PUBLIC HEALTH LAW FRAMEWORKS ARE NOT ADEQUATE FOR CLIMATE CHANGE ADAPTATION

The threat-specific programs described above – in the areas of heat wave planning, vector control, and disaster preparedness – provide useful models for climate change adaptation, but they are hanging on at the margins of global public health law. They have mostly developed in a piecemeal, reactionary fashion and do little to meet the need for comprehensive and deliberate health adaptation planning. Interventions in many of these areas are sorely underfunded and not prepared to withstand climate related intensification. Even more concerning, the most dire environmental threats to health – those associated with water scarcity and lack of access to improved sanitation – are among the most neglected by current frameworks.

The dominant focus of global public health law is on emergency preparedness and response. The fields of international health law and domestic public health law have experienced rapid growth since the early 2000s, primarily in response to threats of terrorism and novel infectious disease pandemics (such as SARS and avian influenza). At first blush, it may seem that the substantial legal infrastructure being developed for public health emergency preparedness would be well suited to the goals of climate adaptation. But because the climate related threats with the highest disease burden are associated with gradual processes rather than rapidly emerging events, the emergency preparedness focus of global public health law may create a false sense of security. Assessment of existing frameworks reveals missed opportunities for integration of climate change considerations within the broader work of the WHO, even as the WHO is working to integrate health considerations into the work of the UNFCCC.

3.1 International Health Law

In 2005, the state parties to the World Health Assembly adopted a major revision of the International Health Regulations (IHRs), the primary instrument of international health law. In their earlier form, the IHRs (previously known as the International Sanitary Regulations) were limited to controlling the cross-border spread of a short list of highly communicable diseases: cholera, plague, yellow fever, and smallpox. Minor amendments were adopted from time to time – removing smallpox from the list of covered diseases upon its global eradication and altering provisions regarding cholera – but the basic approach remained the same. The Regulations' narrow applicability, the fact that they were largely ignored by state parties even when they did apply, and the failure of amendments to keep pace with scientific advances,[60] caused them to lapse into a long period of irrelevance.[61] Beginning in 1995, driven by concerns about terrorism and newly emerging infectious diseases, policymakers turned their attention to modernizing the Regulations in ways that dramatically expanded the WHO's role in coordinating a response to public health emergencies, broadly defined.

The revised IHRs' purpose is 'to prevent, protect against, control and provide a public health response to the international spread of disease in ways that are commensurate with and restricted to public health risks and which avoid unnecessary interference with international traffic and trade'.[62] Most radically, the revised IHRs apply to all 'public health emergencies of international concern' (PHEICs), including not only infectious disease outbreaks, but also chemical and radio-nuclear events, and perhaps other threats to health as well.[63] A PHEIC is defined as 'an extraordinary event which is determined, as provided in these Regulations: (i) to constitute a public health risk to other States through the international spread of disease; and (ii) to potentially require a coordinated international response'.[64] 'Disease' is defined quite broadly as 'an illness or medical condition, irrespective of origin or source, that presents or could present significant harm to humans'.[65]

In addition to imposing obligations on member states to notify the WHO of a potential PHEIC, the IHRs also provide for a process whereby the WHO Director-General has authority to officially declare a PHEIC, in consultation with an Emergency Committee.[66] The IHRs further provide for WHO and other international assistance throughout the evaluation process as well as during a declared PHEIC.[67] Upon declaration of a

---

60 Boris Velimirovic, 'Do We Still Need International Health Regulations?' (1976) 133 Journal Infectious Diseases 478 ('Each revision was obsolete on the very day when it was published').

61 David P Fidler, 'From International Sanitary Conventions to Global Health Security: The New International Health Regulations' (2005) 4 Chinese Journal of International Law 325, 333–338.

62 WHA58.3, Article 2 (World Health Assembly 58 2005).

63 Ibid. Article 13 para. 1.

64 Ibid. Article 1.

65 Ibid.

66 Ibid. Article 12 para. 1.

67 Ibid. Article 13 para. 4.

PHEIC, the Director-General is authorized to issue temporary recommendations regarding appropriate health measures to be taken to prevent the international spread of disease, while avoiding unnecessary interference with traffic and trade.[68]

Although notification and reporting obligations are in some ways the centerpiece of the IHRs, the revised Regulations are perhaps even more important for the domestic capacity-building obligations they impose. Member states are required to establish a National IHR Focal Point and develop surveillance and response capacities for health emergencies. The minimum capacities required under the IHR are set forth in considerable detail in Annex 1. State parties are obligated 'to establish, operate and maintain a national public health emergency response plan, including the creation of multidisciplinary/multisectoral teams to respond to events that may constitute a public health emergency of international concern'.[69] They must ensure their surveillance capacity to 'detect events involving disease or death above expected levels for the particular time and place in all areas within the territory of the State Party'.[70] States must also ensure their response capacity to institute control measures, including 'specialized staff, laboratory analysis of samples (domestically or through collaborating centres) and logistical assistance (e.g. equipment, supplies and transport)'.[71]

The expanded application of the IHRs to any 'public health event of international concern' and the capacity-building obligations the Regulations impose would seem to make the revised IHR quite useful in the face of climate related public health threats. But in practice, the application of the Regulations is more limited than the broad definition of PHEIC might suggest. According to the Decision Instrument for Assessment and Notification of Events Which May Constitute a Public Health Event of International Concern (Annex 2), an event detected by a national surveillance system triggers an obligation to notify the WHO under the IHRs primarily based on four criteria: (1) the seriousness of the public health impact; (2) whether the event is unusual or unexpected; (3) whether there is a significant risk of international spread; and (4) whether there is a risk of international restrictions on trade or travel.[72] These factors are not weighed in a balancing test. Rather, they are

---

[68] Ibid. Article 17.
[69] Ibid. Annex 1 para. 6(g).
[70] Ibid. Annex 1 para. 4(a).
[71] Ibid. Annex 1 para. 6(b).
[72] Ibid. Annex 2.

applied through a very specific decision tree.[73] The seriousness of the public health impact alone is not sufficient to trigger the IHRs. Even if the public health impact of an event is serious, if it is not also unusual or unexpected, it triggers obligations under the IHRs only where there is a threat of international spread or a risk of international restrictions (presumably based on the misperception of a threat of international spread). Health threats associated with environmental degradation, especially those that are gradually emerging, are generally not likely to be deemed 'unusual or unexpected'. Many of them – such as increased incidence of malaria or exacerbation of respiratory diseases – also will not pose a significant risk of cross-border spread. Even cholera, a disease that prompted some of the earliest efforts at international public health cooperation, is no longer likely to trigger the IHRs.[74]

Presently, the public health infrastructure enjoyed in wealthy countries is almost entirely sufficient to protect them from cross-border spread of cholera and other diseases of poverty that are likely to be exacerbated by climate change. Like dengue fever,[75] cholera is no longer a disease that 'knows no borders'. During an outbreak, a few cases of infection are typically identified in the border regions of neighboring countries, but significant spread is typically halted at the borders between high-resource and low-resource populations. The WHO Director-General did not declare a PHEIC in response to the cholera outbreak in Zimbabwe in 2007–2008, in spite of several cases identified across Zimbabwe's borders with Mozambique, South Africa, Botswana, and Zambia.[76] These limited cases (which were primarily among Zimbabwean migrants) did

---

73 Ibid.

74 World Health Organization Global Task Force on Cholera Control, 'Cholera: Mechanism for Control and Prevention', http://www.who.int/cholera/technical/resolution/en/index.html, accessed 11 October 2012.

75 Joan Marie Brunkard et al., 'Dengue Fever Seroprevalence and Risk Factors, Texas-Mexico Border, 2004' (2008) 13 Emerging Infectious Diseases (Assessing dengue prevalence in two cities directly across the US–Mexico border from each other and finding that Mexican residents were more than three times as likely as US residents to test positive for recent dengue infection and nearly twice as likely to test positive for any past infection; also noting that although dengue is present in the US, it has been limited to five south Texas border counties and has not encroached further into the US).

76 Center for Public Health and Human Rights, 'Zimbabwe Health System Crisis' (Johns Hopkins Bloomberg School of Public Health 2012), http://www.jhsph.edu/research-and-centers/center-for-public-health-and-human-rights/locations/africa/Zimbabwe_Health_System_Collapse.html, accessed 10 June 2012.

not pose a threat of significant encroachment beyond two border provinces in South Africa, because public health infrastructure in South Africa was adequate to support an effective response and halt further spread of the disease.[77] Other neighboring countries did not fare nearly as well. In Mozambique, for example, cholera spread to ten of eleven provinces.[78]

The WHO also refrained from applying the revised IHRs to the 2010–2011 cholera outbreak in Haiti. In spite of cases identified across the border in the Dominican Republic and the fact that the UN has acknowledged its own role in causing the outbreak,[79] the Haitian cholera outbreak was treated as a matter for ad hoc, charity-based assistance, rather than triggering binding international obligations. Of course, failure to invoke the IHR does not amount to total neglect of a public health crisis. The WHO, the Pan-American Health Organization (PAHO), and several additional international agencies partnered to develop a strategy to combat the spread of cholera.[80] Notably, PAHO's primary recommendation was to urge governments to invest in water and sanitation services.[81] This urgent recommendation was made even as the WHO and its regional bodies were devoting considerably more attention to urging member states (including developing countries) to commit resources to

---

[77] Barbara Hogan, 'Statement on the Outbreak of Cholera in Zimbabwe and South Africa' (Department of Health, South Africa, 26 November 2008), http://www.info.gov.za/speeches/2008/08112711451003.htm, accessed 11 October 2012.

[78] World Health Organization (2009) 2(8) Africa Weekly Emergency Situation Update (23 February 2009).

[79] A 2011 UN report found that the precipitating cause of the outbreak in Haiti was improper disposal of human fecal matter at a camp for UN peacemakers who came to Haiti to aid earthquake recovery efforts. Alejandro Cravioto et al., 'Final Report of the Independent Panel of Experts on the Cholera Outbreak in Haiti' (WHO 2011) 4, http://www.un.org/News/dh/infocus/haiti/UN-cholera-report-final.pdf, accessed 11 October 2012.

[80] UN Office for the Coordination of Humanitarian Affairs, 'Cholera Inter-Sector Response Strategy for Haiti Nov. 2010 – Dec. 2011' (2010), http://reliefweb.int/sites/reliefweb.int/files/resources/8B2E07AB1A31BABEC12577D8004FE057-Full_Report.pdf, accessed 11 October 2012.

[81] UN News Centre, 'UN Urges Investing in Water and Sanitation Services to Combat Cholera' (Haiti, 11 January 2012).

meeting the rapid detection and response capacities required under the IHRs.[82]

3.2 Domestic Public Health Law

The same concerns about terrorism and novel infectious disease pandemics that have motivated legal reform at the international level have also led to an explosion in domestic public health law reform. Domestic public health law concerns the public health authorities and activities of national and sub-national levels of government, including health inspections and standards, vector control, quarantine, isolation, and mandatory vaccination and treatment. It also regulates the collection, use, and disclosure of information relevant to public health surveillance and monitoring. A central issue in public health law is finding a balance between the needs of a society to conduct disease surveillance and control against the rights of individuals to privacy and liberty. Most states have some form of public health or sanitation code for addressing routine public health needs, and many states have undertaken reform efforts in the last decade or so. However, in other states, these laws are still woefully out-of-date. Like the pre-revision IHRs, public health laws from the colonial era tend to be organized around particular, identified disease threats and nuisances. They focus predominantly on outmoded legal authorities to impose 'command and control' quarantine and isolation measures.

Public health law reform efforts have focused on striking a better balance between civil liberties and public health needs while also incorporating public health innovations. One such innovation, social distancing, suggests temporarily closing schools and other gathering places to reduce opportunities for disease transmission. Such compromises are considered less coercive alternatives to quarantine and isolation. Reform efforts have also adopted a more flexible 'all hazards' approach that provides flexibility to address threats that are newly emerging or evolving. The work of ensuring healthy living conditions and adequate preparedness for public health emergencies depends on legal frameworks in sectors outside those that have traditionally been seen as the province of health agencies. As noted in the WHO guidance materials regarding legislative implementation of IHR obligations, other areas affected by IHR implementation (and by public health more generally)

---

[82] World Health Organization, *Protocol for Assessing National Surveillance and Response Capacities for the International Health Regulations (2005)* (WHO 2010).

include: regulation of international ports, airports, and ground crossings; environmental regulation; customs law; food safety; agriculture; radiation safety; chemical safety; and transportation.[83]

As part of its IHR implementation efforts, the WHO is actively engaged in promoting domestic public health law reform aimed to ensure public health emergency surveillance and response capacity.[84] In addition to efforts aimed directly at IHR implementation, the WHO is aiming to provide guidance on modernization of public health law more generally via a Public Health Law Manual currently in development.[85] The WHO Regional Offices and individual donor states are also playing an important role in supporting and shaping national public health law reform efforts.[86] The WHO maintains a searchable database of national and sub-national health laws of member states, the WHO International Digest of Health Legislation (IDHL), available online at www.who.int/legislation.

Model public health laws may prove useful, particularly when tailored to specific regions with shared concerns. For example, in 2008 the Australian Agency for International Development funded a project aimed at developing a model public health law for the Asian Pacific Region.[87] In the United States, where public health law is primarily a matter of

---

[83] World Health Organization, *International Health Regulations (2005): A Brief Introduction to Implementation in National Legislation* (WHO 2009).

[84] Ibid. at 10.

[85] World Health Organization, 'Health and Human Rights: Projects', http://www.who.int/hhr/health_law/projects/en/index.html, accessed 11 October 2012.

[86] World Health Organization Regional Office for the Western Pacific, 'Workshop on Public Health Law for Pacific Island Countries, Auckland, New Zealand, 12–15 February 2007: Report' (2007); World Health Organization Western Pacific Region, 'Expert Consultation on Public Health Law: 5–6 May 2011, Manila, Philippines' (2011), http://www.wpro.who.int/health_services/documents/expert_consultation_public_health_law.pdf, accessed 11 October 2012 (WPRO Secretariat is developing a tool that Member States could use to analyze their own legal status. An expert meeting is being convened [Manila, Philippines, 5–6 May 2011] which will serve as a venue for consultation and review of technical issues relating to public health law); World Health Organization Europe, 'Guidance on Public Health Law', http://www.euro.who.int/en/what-we-do/health-topics/Health-systems/public-health-services/activities/guidance-on-public-health-law, accessed 11 October 2012 (public health laws have been developed with WHO support and endorsed by national parliaments in Albania, Kyrgyzstan, and the Republic of Moldova. Work on new public health laws is under way in Armenia and the Former Yugoslav Republic of Macedonia).

[87] G Howse, 'Formulating a Model Public Health Law for the Pacific: Programme Methods' (2009) 123 Public Health 237.

state and local legislation, four model public health law acts have been developed via collaboration among academic institutions, NGOs, and government agencies: The Model State Public Health Act,[88] The Model State Emergency Health Powers Act,[89] the Model State Public Health Privacy Act,[90] and the Uniform Emergency Volunteer Health Practitioners Act.[91] State governments have used the model acts as starting points for reform efforts tailored to meet the needs of the communities they serve.[92]

Although these reform efforts are promising, some environmental health concerns remain in danger of slipping through cracks between sectors. In many countries, programming in areas like vector control or sanitation is often carried out through specialized agencies that lack a broader public health mandate. This is concerning because public health agencies traditionally possess fairly expansive authority to regulate private individuals and commercial entities, whereas more specialized agencies are likely to be underfunded and lack significant regulatory authority. The predominant focus of the global public health sector on the control of highly communicable diseases poses a potential barrier to effective climate adaptation.

---

88 The Turning Point Public Health Statute Modernization Collaborative, 'Model State Public Health Act: A Tool for Assessing Public Health Laws' (2003), http://www.hss.state.ak.us/dph/improving/turningpoint/PDFs/MSPHA web.pdf, accessed 11 October 2012.

89 The Center for Law and the Public's Health at Georgetown and Johns Hopkins Universities, 'The Model State Emergency Health Powers Act' (2001), http://www.publichealthlaw.net/MSEHPA/MSEHPA.pdf, accessed 11 October 2012.

90 Lawrence O Gostin and James G Hodge Jr, 'The Model State Public Health Privacy Act' (1999), http://www.publichealthlaw.net/ModelLaws/MS PHPA.php, accessed 11 October 2012.

91 Uniform Law Commission, 'Uniform Emergency Volunteer Health Practitioners Act' (2006), http://www.publichealthlaw.net/ModelLaws/UEVHPA.php, accessed 11 October 2012.

92 The Center for Law and the Public's Health at Georgetown and Johns Hopkins Universities, 'The Turning Point Model State Public Health Act: State Legislative Update Table as of August 15, 2007', http://www.publichealthlaw.net/ Resources/ResourcesPDFs/MSPHA%20LegisTrack.pdf, accessed 11 October 2012.

4 HEALTH ADAPTATION IN THE UNITED NATIONS SYSTEM

Scientists, scholars, and policymakers have only recently begun to focus significant attention on the complex nexus between climate change and human health. But in a relatively short period of time, there's been an explosion of research and scholarship on the ways in which consideration of the health dimension of climate change might be used to shape both mitigation and adaptation regimes. Of course, much of this activity has been within NGOs and government agencies at the country and local level, but an examination of the two intergovernmental bodies with primary responsibility for climate change and health provides a useful lens for assessing the current state of focus on issues at the intersection of climate change and health. A description of recently developed programs that have been created explicitly for the purpose of health adaptation is also a useful starting point for study of the issue.

4.1 Health Adaptation in the UNFCCC

The text of the UNFCCC does not clearly establish a role for the health sector. 'Health' is specifically mentioned twice in the UNFCCC text. Health impacts are acknowledged in Article 1(1)[93] and Article 4(1)(f) includes a commitment to assess mitigation and adaptation measures 'with a view to minimizing adverse effects on … public health … of projects or measures undertaken by them to mitigate or adapt to climate change'.[94] Other provisions, such as general commitments to research,[95] technology transfer,[96] education, training, and public awareness,[97] and to take climate change considerations into account in 'relevant social … policies and actions'[98] obviously implicate the health sector. But those implications are, for the most part, not made explicit in the text.

---

93 Article 1(1) defines 'adverse effects of climate change' as 'changes in the physical environment or biota resulting from climate change which have significant deleterious effects on the composition, resilience or productivity of natural and managed ecosystems or on the operation of socio-economic systems or on *human health* and welfare' (emphasis added).

94 Article 4(1)(f).

95 Article 4(1)(g).

96 Article 9(1).

97 Article 4(1)(i).

98 Article 4(1)(f).

Through most of the 1990s, a time when the UNFCCC annual conferences of parties (COPs) were focused almost exclusively on establishing a mitigation regime, health impacts were only occasionally referenced in COP reports as reasons for adopting mitigation measures.[99] By 1999, the COPs began to devote some attention to adaptation measures and health impacts. Although sections of the UNFCCC text committing states to adaptation do not explicitly discuss health adaptation as a concern,[100] the annual COP reports began to recognize health as one among several key priorities for adaptation.[101] When the 2000 COP established the Adaptation Fund to finance concrete adaptation

---

[99] UNFCCC, 'Report of the Conference of the Parties on its Second Session, held at Geneva from 8 to 19 July 1996' (FCCC/CP/1996/15/Add.1 1996) 72 ('The projected changes in climate will result in significant, often adverse, impacts on many ecological systems and socio-economic sectors, including food supply and water resources, and on human health. In some cases, the impacts are potentially irreversible'); UNFCCC, 'Report of the Conference of the Parties on its Fourth Session, held at Buenos Aires 2 to 14 November 1998: Addendum: Part Two: Action Taken by the Conference of the Parties at its Fourth Session' (FCCC/CP/1998/16/Add.1 1999) 68 (citing the 'considerable loss of life and devastation' caused by Hurricane Mitch in Central America).

[100] In Article 4(1)(e) member states commit to '[c]ooperate in preparing for adaptation to the impacts of climate change; develop and elaborate appropriate and integrated plans for coastal zone management, water resources and agriculture, and for the protection and rehabilitation of areas, particularly in Africa, affected by drought and desertification, as well as floods'. UNFCCC Article 4(1)(e).

[101] UNFCCC, 'Report of the Conference of the Parties on its Fifth Session, held at Bonn from 25 October to 5 November 1999: Addendum: Part Two: Action Taken by the Conference of the Parties at its Fifth Session' (FCCC/CP/1999/6/Add.1 2000) 32–33 (noting 'the specific needs and special situations of the least developed countries where widespread poverty limits adaptive capacity, particularly in relation to the impacts of the adverse effects of climate change on socio-economic conditions, including, *inter alia*, water resources, agriculture and food security, economic activities, coastal zones and health …') emphasis in original; UNFCCC, 'Report of the Conference of the Parties on its Sixth Session, held at the Hague from 13 to 25 November 2000: Addendum: Part Two: Action Taken by the Conference of the Parties at the First Part of its Sixth Session' (FCCC/CP/2000/5/Add.2 2001) 24, 7 ('Encourag[ing] the Third United Nations Conference on the Least Developed Countries, when considering the establishment of debt relief mechanisms, to fully take into account the effects of climate change on productivity in the agricultural and other economic sectors as well as on health' and committing Annex 2 countries to '[i]mproved monitoring of diseases and disease control and prevention for Parties affected by climate change').

activities, health adaptation was not initially included among the concrete adaptation activities to be financed through an innovative 2 percent levy on certified emissions reduction credits issued by the Clean Development Mechanism.[102] The following year, however, health adaptation was explicitly recognized as among the activities to be funded through the Adaptation Fund and the Special Climate Change Fund.[103] In addition to referring to health adaptation generally, the 2001 COP report specifically referred to financing of projects aimed at '[i]mproving the monitoring of diseases and vectors affected by climate change, and related forecasting and early-warning systems, and in this context improving disease control and prevention'.[104] Other activities explicitly recognized as fundable included capacity building for disaster preparedness and management and establishment of centers for rapid response to extreme weather events.[105]

Increasing focus on adaptation in the UNFCCC played a role in raising the profile of health concerns within the realm of climate change law, while also promoting the consideration of climate change concerns within other sectors. The 2002 COP began to emphasize the need for integration of climate change considerations into policies and planning in other sectors, including health.[106] The same year, the COP explicitly included health as one of the sectors in which technology transfer should be strengthened.[107] From 2004 to 2009, however, the UNFCCC COP reports reverted to silence on health impacts as the COP became almost

---

[102] FCCC/CP/2000/5/Add.2, above note 101 at 5 ('The following activities will be included in the category of adaptation activities: avoidance of deforestation, combating land degradation and desertification').

[103] UNFCCC, 'Report of the Conference of the Parties on its Seventh Session, held at Marrakesh from 29 October to 10 November 2001: Addendum: Part Two: Action Taken by the Conference of the Parties at its Seventh Session' (FCCC/CP/2001/13/Add.1 2002) 35–36.

[104] Ibid. at 35.

[105] Ibid. at 36.

[106] UNFCCC, 'Report of the Conference of the Parties on its Eighth Session, held at New Delhi from 23 October to 1 November 2002: Addendum: Part Two: Action Taken by the Conference of the Parties at its Eighth Session' (FCCC/CP/2002/7/Add.1 2003) 3 (noting that 'National sustainable development strategies should integrate more fully climate change objectives in key areas such as water, energy, health, agriculture and biodiversity').

[107] Ibid. at 4 ('Technology transfer should be strengthened, including through concrete projects and capacity-building in all relevant sectors such as energy, transport, industry, health, agriculture, biodiversity, forestry and waste management').

exclusively focused on negotiation of a successor to the Kyoto Protocol on mitigation.

Neglect of health impacts during this time period bled over into the adaptation work of the UNFCCC as well. In developing strategic priorities for funding, the Adaptation Fund Board missed an opportunity to highlight the importance of health impacts. Developing countries that are parties to the Kyoto Protocol and 'particularly vulnerable to the adverse effects of climate change' are eligible for funds. The Board's statement of strategic priorities highlights countries' vulnerabilities to sea-level rise, floods, drought, desertification, and alteration of fragile mountainous ecosystems.[108] Impacts on human health are neglected.[109] of December 2011, the Fund had accumulated USD 243 million, with USD 100 million coming from donation agreements.[110] Between September 2010 and December 2011, the Fund awarded USD 103,698,701.[111] Seventeen projects were approved in areas such as water management, coastal management, agriculture, food security, and rural development.[112] Most of the projects focus on natural disaster risk management – such as Georgia's project to develop climate resilient food and flash flood management practices or Nicaragua's project to reduce the risks from floods and droughts[113] – rather than focusing more directly on 'the pronounced health impact of climate change'.[114]

Although UNFCCC instruments promote 'common but differentiated responsibility' for climate change and impose obligations on developed countries to aid developing countries in their adaptation efforts, these international obligations do little to alter the fact that primary responsibility for addressing the impacts of climate change is still left to the affected states themselves. In spite of this, domestic climate change adaptation

---

108 Adaptation Fund, 'Operational Policies and Guidelines for Parties to Access Resources from the Adaptation Fund', http://adaptation-fund.org/system/files/file/AFB_Operational%20Policies%20and%20Guidelines.pdf, accessed 11 October 2012.

109 William Onzivu, 'Health in Global Climate Change Law: The Long Road to an Effective Legal Regime Protecting both Public Health and the Climate' (2010) 4 Climate and Carbon Law Review 364, 371.

110 Adaptation Fund, 'Report of the Seventeenth Meeting of the Adaptation Fund' (AFB/B.17/6 2012) 26 para. 118.

111 UNFCCC, 'Adaptation Fund Project Data 2012', http://unfccc.int/cooperation_and_support/financial_mechanism/adaptation_fund/items/6668.php, accessed 11 October 2012.

112 Ibid.

113 'Operational Policies and Guidelines', above note 108.

114 Onzivu, above note 109 at 131.

efforts are woefully inadequate to rise to the challenge of health adaptation. Developing country parties to the Kyoto Protocol were obligated to develop National Adaptation Programmes of Action (NAPAs) 'to identify priority activities that respond to their urgent and immediate needs to adapt to climate change'.[115] In 2010, the WHO performed a valuable assessment of the extent to which health was covered in the 41 NAPAs that had been submitted to the UNFCCC at that point.[116] The study found that although nearly all of the 41 NAPAs reviewed included health as one of the sectors that will be affected by climate change, fewer than one-quarter of the plans provided a comprehensive health vulnerability assessment. Similarly, although nearly three-quarters of the plans included health interventions among the actions they proposed, fewer than 20 percent of reviewed NAPAs proposed health interventions adequate to address anticipated needs. Of all priority actions identified in the reviewed NAPAs, only 11 percent focused on health adaptation and health adaptation projects represented only 3 percent of the total estimated cost associated with identified priority actions. These health-focused priority actions included interventions aimed at malaria and vector control, access to drinking water and sanitation, and (to a lesser extent) nutrition and health systems strengthening.[117]

The 2010 COP in Cancun marked an encouraging turning point for the integration of health into the work of the UNFCCC. The Cancun COP 'invited' all countries (not only those that are obligated to produce NAPAs under the Kyoto Protocol) to plan, prioritize, and implement 'adaptation actions, including projects and programmes'.[118] In an accompanying footnote, the report noted that project and program areas include 'areas of water resources; *health*; agriculture and food security; infrastructure; socio-economic activity; terrestrial, freshwater and marine ecosystems; and coastal zones'.[119]

---

[115] See Chapter 2, section 2.2.

[116] Lucien Manga et al., 'Overview of Health Considerations within National Adaptation Programmes of Action for Climate Change in Least Developed Countries and Small Island States' (WHO 2010).

[117] Ibid.

[118] UNFCCC, 'Report of the Conference of the Parties on its Sixteenth Session, held in Cancun from 29 November to 10 December 2010: Addendum: Part Two Action Taken by the Conference of the Parties at its Sixteenth Session' (FCCC/CP/2010/7/Add.1 2011) 4 para. 14(a).

[119] Ibid. (emphasis added).

4.2 Health Adaptation in the World Health Organization

During a time when the UNFCCC backed off of its initial efforts to integrate health concerns into its work, the effort to put a human face on climate change was taken up by NGOs and international governmental organizations outside of the UNFCCC, most notably the WHO.[120] Although the WHO has traditionally 'shied away from the "high politics" of' the 'much larger body of international law that powerfully affects global health in areas ranging from food safety, arms control, and the environment, to trade and human rights',[121] the WHO has been involved in the work of the UNFCCC in a variety of ways. In 2008, the World Health Assembly issued a strong mandate to the WHO to become more engaged in issues at the intersection of climate change and health via World Health Assembly Resolution 61.19.[122] The WHO began to shape four main roles for itself at the intersection of climate change and human health: (1) documenting health impacts to lend urgency to the call to action on mitigation; (2) identifying priorities for research; (3) outlining health adaptation strategies and promoting best practices; and (4) promoting the integration of health concerns into climate change policymaking in other sectors. These roles are highlighted in the WHO's 2009 Work Plan on Climate Change and Health.[123]

The WHO leaders have noted that '[c]onsideration of the health impact of climate change can help political leaders move with appropriate urgency' on mitigation measures.[124] The WHO and its regional offices have published a series of reports on the consequences of climate change for human health dating back to 1990.[125] The WHO has also played an

---

[120] CARE, 'The Human Face of Climate Change', http://www.care.org/campaigns/human-face-of-climate-change/index.asp, accessed 11 October 2012.

[121] Lawrence O Gostin, *Public Health Law and Ethics: A Reader, Revised and Updated* (2nd edn, University of California Press 2010) 277.

[122] WHA61.19 (World Health Assembly 61 2008).

[123] World Health Organization, 'What WHO is doing for Climate and Health: WHO Workplan on Climate Change and Health', http://www.who.int/globalchange/health_policy/who_workplan, accessed 11 October 2012.

[124] Margaret Chan, 'The Impact of Climate Change on Human Health' (statement, WHO 7 April 2008), http://www.who.int/mediacentre/news/statements/2008/s05/en/index.html, accessed 11 October 2012.

[125] World Health Organization, *Potential Health Effects of Climate Change* (WHO/PEP/90/10 1990); A J McMichael et al. (eds), *Climate Change and Human Health* (WHO 1996); R Sari Kovats et al., *Climate Change and Human Health: Impact and Adaptation* (WHO/SDE/OEH/00.4, WHO 2000); A J McMichael et al., *Climate Change and Human Health: Risks and Responses* (WHO

important role in the Intergovernmental Panel on Climate Change (IPCC) impact assessment process.[126] Notably, in 2008, the WHO devoted its annual World Health Day to spotlighting the health impacts of climate change.[127] The timing coincided with the build-up to the 2009 COP in Copenhagen, at which a successor mitigation protocol was under negotiation.

The WHO secretariat is primarily focused on techinical expertise and assistance to member states. The secretariat is thus particularly well suited to identifying priorities for scientific research at the intersection of climate and human health. In response to the 2008 World Health Assembly resolution, the WHO convened a global consultation of public health researchers and practitioners to develop recommendations for priority research areas. The resulting recommendations highlight the fact that much is already known about the disease burdens associated with climate related threats like vector- and water-borne disease, air pollution, and natural disasters. They point to an opportunity for climate and health research to 'contribute to the wider goal of improving overall health and health equity' by '[m]aking use of the vast body of existing research knowledge and capacity that is available in related fields, such as management of environmental health risks, or control of infectious diseases'.[128] The WHO-convened experts did note a significant gap in research regarding 'the health consequences of decisions at all levels, from mitigation of climate change, to management of climate risks and vulnerability factors, to treatment of diseases that are affected by climate'.[129]

The research agenda promoted by the WHO has also focused on developing an evidence base for best practices in health adaptation. In addition to facilitating concrete adaptation programs, the WHO is in a unique position to guide these concrete actions through the articulation of an overarching strategy. In 2007, an article in the Bulletin of the World Health Organization challenged the health sector to develop a comprehensive strategy for health adaptation:

2003); World Health Organization, *Protecting Health from Climate Change: Global Research Priorities* (WHO 2009).

[126] Bettina Menne (WHO Regional Office for Europe/Germany) was one of two coordinating lead authors of the chapter on Human Health in the IPCC's Fourth Assessment Report; Confalonieri, above note 15.

[127] Chan, above note 124.

[128] WHO 2009, above note 125 at 4.

[129] Ibid.

> Threats to health are often cited to justify actions to mitigate or adapt to climate change … For all the attention that these issues receive, however, a comprehensive strategy to support a public health response is conspicuously lacking. This strategy is urgently needed, both because the health community has a duty to counter emerging threats, and because increased attention to climate change offers opportunities to focus on the most disadvantaged populations' current needs.[130]

The challenge of developing a comprehensive strategy has not yet been fully met, but the WHO's efforts have been impressive for an organization whose mandates generally limit it to country-driven processes. The public health adaptation to climate change project launched by the WHO and the United Nations Development Programme (UNDP), with funding from the Special Climate Change Fund, provides a model of cooperation across sectors in a process that is, first and foremost, driven by country-specific needs.[131] Regional bodies have taken on a particularly important role in coordinating country-level adaptation efforts. Most notably, the WHO's African Regional Office has advanced a new Framework for Health Adaptation to Climate Change in Africa, which provides 'a scientific and evidence-based coordinated response to the climate change adaptation needs of African countries in order to support the commitments and priorities of African governments'.[132]

Although agriculture and water management sectors have massive importance to health, the health law and policy community has not been significantly involved in discussion of adaptation measures for food and water security thus far. Better integration of the health community into adaptation planning for water management is particularly crucial in light of evolving understanding of the complex relationships between mitigation, adaptation, and health. Many mitigation and adaptation measures in the water and agricultural sectors can have co-benefits for health, but others may come with negative consequences. For example, water

---

[130] Diarmid Campbell-Lendrum et al., 'Global Climate Change: Implications for International Public Health Policy' (2007) 85 Bulletin of the World Health Organization 235.

[131] World Health Organization, 'Climate Change and Human Health: Climate Change Adaptation to Protect Human Health', http://www.who.int/globalchange/projects/adaptation/en/index.html, accessed 11 October 2012; World Health Organization, (2011) 1&2 Health and Climate Adaptation Bulletin; World Health Organization.

[132] World Health Organization Regional Office for Africa, 'Framework for Public Health Adaptation to Climate Change in the African Region: Report to the Secretariat' (AFR/RC61/10, 2011) ii.

management methods have the potential to exacerbate infectious disease threats by increasing the availability of vector breeding grounds.

For instance, in the face of drought, some have turned to rainwater collection.[133] However, rainwater collectors, especially defective ones, can provide a breeding ground for disease-carrying mosquitoes. In Ethiopia, two studies have linked nearby rainwater harvesting ponds with increased malaria rates.[134] In India, defective rainwater collectors were linked to dengue-carrying mosquitoes.[135] The Dengue Control Program in Cambodia is developing insecticide-impregnated jar lids to keep water sources from providing a breeding ground for mosquitoes.[136]

There has been considerable focus on the development of new agricultural technologies to meet the growing demand for food production in the face of ecological changes that are harmful to crops.[137] The suggestion that genetically modified crops (such as drought or flood resistant strains of staples like rice and wheat) will play an important role in agricultural adaptation has created controversy over potential health and environmental risks associated with genetically modified organisms (GMOs).[138] The health sector is likely to play a role in evaluating these possible risks and weighing them against the potential benefits for fighting famine. The WHO's work on identifying research priorities at the intersection of climate change and health has emphasized the need for more data to

---

133 United Nations Development Programme, 'Rain Water Harvesting Improves Lives of Tanzanian Farmers', http://www.undp.org/content/undp/en/home/ourwork/environmentandenergy/successstories/water-harvesting-improves-lives-tanzanian-farmers.html, accessed 11 October 2012.

134 Daniel Kassahun Waktola, 'Malaria and Pond-based rainwater Harvesting Linkages in the Fringes of Central Highland Ethiopia' (2008) 8(956) Rural and Remote Health; Mekonnen Yohannes and Mitiku Haile, 'The Potential of in situ Rain Water Harvesting for Water Resources Conservation on Malaria Transmission in Tigray, Northern Ethiopia' (2010) 2 Momona Ethiopian Journal of Science 49.

135 T Mariappan et al., 'Defective Rainwater Harvesting Structure and Dengue Vector Productivity Compared with Peridomestic Habitats in a Coastal Town in Southern India' (2008) 45 Journal of Medical Entomology 148.

136 Cambodia National Malaria Center, 'Programs', http://www.cnm.gov.kh/index.php?action=ID4, accessed 11 October 2012.

137 David Tilman et al., 'Beneficial Biofuels – the Food, Energy, Environment Trilemma' (2009) 325 Science 270.

138 Rebecca M Bratspies, 'Biotechnology, Sustainability, & Trust' (2009) 18 Kansas Journal Law and Public Policy 273.

guide 'health promoting mitigation and adaptation decisions in other sectors'.[139]

Through its work plan on health and climate change, the WHO is working to become a focal point for integration across sectors. The WHO and its regional bodies are promoting the integration of health concerns into climate change planning by asserting their own presence in key international and regional negotiations while also encouraging a variety of health sector stakeholders to demand a place at the table in national and local decision-making processes. More recent COPs have witnessed a revival of the health sector's role, largely attributable to the efforts of the WHO and its regional subsidiary bodies. The health sector enjoyed stronger representation at the 2011 COP in Durban than in any previous year. Nearly one dozen side events touched on aspects of health and climate change,[140] including a Climate and Health Summit led by the NGO community, which culminated in the Durban Declaration on Climate and Health.[141] In spite of this high level of health sector activity at the 2011 COP, however, the Durban COP report once again omitted any explicit mention of health.[142]

5 CHALLENGES AND OPPORTUNITIES

Climate change adaptation efforts have largely neglected the enormous impact of climate change on human health. The fact that climate change will act primarily as an intensifier and re-distributor of existing threats to health is no justification for complacency. Although scientific understanding and technical tools for addressing environmental threats to health are reasonably well developed, existing global public health law frameworks tend to privilege very different types of threats. There are three main challenges that must be addressed by global public health law in the context of climate adaptation. First, health adaptation efforts must

---

139 WHO 2009, above note 125 at 9.

140 World Health Organization, 'Health Issues Gain Traction at UN Climate Conference' (2011), http://www.who.int/globalchange/mediacentre/events/2011/durban_conference_update/en/index.html, accessed 11 October 2012.

141 Durban Declaration on Climate and Health' (Durban, 4 December 2011), http://www.climateandhealthcare.org/wp-content/uploads/2011/12/Durban-Declaration-on-Climate-and-Health-Final.pdf, accessed 13 June 2012.

142 UNFCCC, 'Report of the Conference of the Parties on its Seventeenth Session, held in Durban from 23 November to 11 December 2011: Addendum: Part Two: Action Taken by the Conference of the Parties at its Seventeenth Session' (FCCC/CP/2011/9, 2012).

transition from threat-specific programs aimed at vector control, disaster preparedness, or heat wave planning toward a more upstream approach that emphasizes development and reinforcement of basic health and sanitation infrastructure. Current health adaptation planning neglects the climate related threats that are most concerning from a disease burden standpoint and that are also the most neglected by existing frameworks. These threats (associated with climate related stress on water security and sanitation) demand investment in basic infrastructure to promote resilience. Second, adaptation-motivated reform efforts must emphasize integration across sectors. Global health considerations need to be more fully integrated into climate change legal frameworks and climate change considerations need to be more fully integrated into the core of global public health law frameworks. Beyond health-environment integration, health adaptation will also require communication and cooperation across multiple law and policy sectors that shape human exposure to environmental determinants of health: land use, water management, agriculture, transportation, and others. Finally, new global public health law frameworks must be adopted and existing frameworks adjusted to emphasize the role of law in ensuring equitable access to health care and healthy living conditions. Climate adaptation will be far better served by equity-based regimes than security-based ones. Fortunately, in each of these areas, there are promising developments that might serve as models for adaptation-oriented law reform.

5.1 Infrastructure

Adapting to the impacts of climate change on human health will require improved access to health care and increased investment in public health and sanitation infrastructure. If current trends persist, considerable resources might be directed toward the development of new drugs, vaccines and other technologies to address the intensification of particular health threats in the context of climate change. High-profile funders have put hundreds of millions of dollars into the development of a malaria vaccine, for example.[143] Although these efforts have great potential to save and improve lives, health adaptation to climate change must go far beyond the development of new drugs, vaccines, or technologies. Despite the rapid growth of an impressive array of interventions to protect health, disparities in health outcomes have continued to

---

[143] PATH Malaria Vaccine Initiative, 'New Funding to Further Malaria Vaccine Development' (25 September 2008), http://www.path.org/news/an080925-mvi-award.php, accessed 11 October 2012.

widen.[144] 'The power of existing interventions is not matched by the power of health systems to deliver them to those in greatest need, in a comprehensive way, and on an adequate scale.'[145]

An increase in resources for health systems strengthening (an effort that is already underway as part of the UN system's approach to the MDG and the Healthy People 2010 initiative in the US[146]) is an obvious objective for health adaptation to climate change, but doing more with available resources is equally important.[147] Greater spending alone does not necessarily result in better health outcomes.[148] Given what we know about the particular vulnerability of poor countries and poor communities within wealthier countries to environmental health impacts, it is clear that affordability and access issues will be crucial to the health sector's response. Training of the health care workforce on the particular risks associated with extreme heat events, natural disasters, and food- water- and vector-borne disease will also be very important as the geographic range of these threats shifts as a consequence of environmental changes.[149]

Public health infrastructure for vector control and injury prevention, surveillance and early-warning systems (for infectious disease, chronic disease, and toxic exposures), and health communication and education (to arm the public with information about protective health behaviors and available health services) will be crucial to climate change adaptation. These interventions tend to be extremely cost-effective[150] and they tend

---

144 World Health Organization, *Everybody's Business: Strengthening Health Systems to Improve Health Outcomes* (WHO 2007) iii.

145 Ibid.

146 Healthy People, 'Access to Health Services', http://healthypeople.gov/2020/topicsobjectives2020/overview.aspx?topicid=1, accessed 11 October 2012.

147 Ibid.

148 Comparisons of per-capita spending on health with average life expectancy show that higher spending on health does not by itself lead to better health outcomes. UC Atlas of Global Inequality, 'Health Care Spending', http://ucatlas.ucsc.edu/spend.php, accessed 11 October 2012.

149 California Natural Resources Agency, '2009 California Climate Adaptation Strategy: A Report to the Governor of the State of California in Response to Executive Order S-13–2008' (2009) 151, http://resources.ca.gov/climate_adaptation/docs/Statewide_Adaptation_Strategy.pdf, accessed 11 October 2012.

150 Guy Hutton and Laurence Haller, *Evaluation of the Costs and Benefits of Water and Sanitation Improvements at the Global Level* (WHO/SDE/WSH/04.04, WHO 2004) 3; Joshua O Yukich et al., 'Cost and Cost-Effectiveness of Vector Control in Eritrea Using Insecticide-treated Bed Nets' (2009) 8 Malaria Journal 51.

to promote health equity by improving the living conditions of vulnerable populations.[151] Despite these benefits, however, spending on public health is negligible compared to health care spending.[152] As in the health care sector, the public health system must also pay careful attention to planning for groups that are particularly vulnerable to one or more health impacts of climate change (such as the elderly, the poor, urban residents, children, and the mentally disabled).[153]

The single most important focus of health adaptation should be on increasing access to improved sanitation. The WHO reports regularly on progress toward the water and sanitation MDG. Its assessments are consistently dire with regard to improved sanitation. Although progress toward halving the number of people without safe and adequate access to drinking water is on-track, access to improved sanitation has been much slower to improve and is in danger of regressing.[154] The reasons for neglect of sanitation infrastructure are complex. Part of the problem is that the relatively downstream focus of most health adaptation planning efforts and the disease-specific focus of many existing global health programs promotes neglect of upstream interventions like sanitation.[155] Charity-based programs focused on HIV/AIDS, malaria and tuberculosis,

---

[151] Prüss-Üstün and Corvalán, above note 4 at 6.

[152] Trust for America's Health, 'Shortchanging America's Health 2008: A State-by-State Look at how Federal Public Health Dollars are Spent' (2008) 3–4, http://healthyamericans.org/reports/shortchanging08/Shortchanging08.pdf, accessed 11 October 2012.

[153] California Department of Public Health, 'Public Health Impacts of Climate Change in California: Community Vulnerability Assessments and Adaptation Strategies Report No. 1: Heat-Related Illness and Mortality Information for the Public Health Network in California' (2007) 38, http://www.ehib.org/papers/Heat_Vulnerability_2007.pdf, accessed 11 October 2012 (highlighting planning for populations vulnerable to heat-related illnesses and proposing short-term and long-term adaptation strategies for reducing these populations' vulnerability through planning for transportation to cooling centers, education about associated health risks, and reducing social isolation).

[154] UNICEF and World Health Organization, 'Progress on Drinking Water and Sanitation: 2012 Update' (2012) 2, http://whqlibdoc.who.int/publications/2012/9789280646320_eng_full_text.pdf, accessed 11 October 2012.

[155] 'The residents who live [in a village by a river], according to the parable, began noticing increasing numbers of drowning people caught in the river's swift current and so went to work inventing ever more elaborate technologies to resuscitate them. So preoccupied were these heroic villagers with rescue and treatment that they never thought to look upstream to see who was pushing the victims in.' Sandra Steingraber, *Living Downstream: An Ecologist Looks at Cancer and the Environment* (Addison-Wesley 1997) xvi.

for example, emphasize more downstream, disease-specific responses that rely on medical countermeasures like vaccines, antibiotics, and antivirals. The funding criteria used by these programs are driven by demand for measurable results within a relatively short timeframe (generally of the order of five years).[156] It is quite possible to measure the success of a program in terms of how many people have been put on a life-saving drug regimen. It is far more difficult to identify individual people whose lives have been saved by investment in improved sanitation.

5.2 Integration

Health adaptation will require greater integration of health and environmental governance at the international, national, and sub-national levels as well as integration of health concerns into the policymaking of other non-health sectors engaged in adaptation planning.[157] Climate change adaptation planners must pay more attention to health impacts and, at the same time, global public health policymakers must take climate change impacts into account in their other activities. Although the WHO's work on integrating health considerations into climate negotiations has been impressive, there is still much to be done. As the findings of the Framework on Public Health Adaptation to Climate Change in Africa point out:

> [M]anagement of the public health impacts of climate change is still not adequately reflected in the negotiations of the UNFCCC, both at national and international levels. Ministries of environment that represent most governments in these negotiations receive minimal and inadequate technical inputs from ministries of health. Therefore, they are not appropriately informed of the specific needs of public health.[158]

Even as the WHO is working to enhance integration of health considerations into climate change governance, it is missing important opportunities to integrate climate change considerations into its core work in areas like public health emergency preparedness. Climate change experts

---

[156] The United States President's Emergency Plan for AIDS Relief, 'Annex V: Health System Strengthening Priority-Setting', http://www.pepfar.gov/guidance/framework/120741.htm, accessed 11 October 2012.

[157] William Onzivu, 'International Environmental Law, the Public's Health, and Domestic Environmental Governance in Developing Countries' (2006) 21 American University International Law Review 597, 660.

[158] WHO Regional Office for Africa, above note 132 at 3.

working within the WHO have demanded that '[r]esearch on climate change and health must be placed more firmly within the overall context of improving global health, and health equity, rather than being considered as a stand-alone issue'.[159] But the rest of the WHO does not seem to be hearing that message. A 2009 WHO statement on the role of health impacts in climate change mitigation and adaptation is particularly interesting in this regard.[160] Noting that 'the health and humanitarian dimensions of climate change are among the ultimate justifications for taking action on climate change', the WHO stated that both mitigation of and adaptation to climate change 'are important for the protection of health of vulnerable populations from both sudden and slow onset climate-related emergencies'.[161] Although the WHO's statement seemingly places climate adaptation within a context of public health emergencies, this has not yet led to integration of climate adaptation into the WHO's central efforts in the area of public health emergency preparedness and response.

Beyond the issue of health-environment integration, an effective response to climate related health threats will require integration across many other areas of law and policy. This chapter's approach is informed by the vision of public health law as the use of law as a tool for protecting and improving the public's health. Thus, the legal tools that might be useful for health adaptation to climate change are not restricted to those typically defined as 'health laws'. Land use, disaster preparedness, water management, energy use, and other areas of law covered in other chapters of this volume provide useful tools for health adaptation as well as serving other climate change adaptation purposes.

Lack of integration across sectors also contributes to the neglect of basic health and sanitation infrastructure. Investment in sanitation infrastructure, for example, often slips through the gaps between public health, development, and water management sectors. Public health specialists may view investment in basic infrastructure as a development or water management issue. But the primary goals of those sectors are not health-related and so the enormous health benefits, which make investment in sanitation infrastructure an imminently rational use of resources, tend not to come to the fore in water management discussions. The

---

[159] WHO 2009, above note 125 at ii.

[160] World Health Organization, 'Protecting the Health of Vulnerable People from the Humanitarian Consequences of Climate Change and Climate Related Disasters' (2009), http://www.preventionweb.net/files/11036_0471.pdf, accessed 11 October 2012.

[161] Ibid. at 1.

development sector provides a somewhat better model for emphasizing integration as a priority for climate change adaptation. The UNDP has made 'work[ing] with countries to strengthen the governance and improv[e] the management of water accompanied by investment in infrastructure projects, such as wastewater treatment facilities and structures for water storage and flood control' a priority in its climate change adaptation agenda.[162] UNDP's water security work also highlights climate change and sanitation concerns.[163] UNDP's 2010 climate adaptation toolkit, however, mentions health consistently, but not sanitation.[164]

Integration is an enormous challenge presented by the problem of global climate change, but it is also a significant opportunity. As policymakers at every level of government increasingly turn their attention to the problems of climate change adaptation, a number of promising examples are emerging of innovative approaches to integrating health, environmental, and other agencies and programs to address common concerns. In 2010, for example, the WHO and the UNDP launched a 'global project on public health adaptation to climate change', which includes a series of pilot programs designed to 'increase adaptive capacity of health systems institutions, including field practitioners, to respond to climate-sensitive health risks'.[165] This program includes projects in Barbados, Bhutan, China, Fiji, Jordan, Kenya, and Uzbekistan.[166] The WHO–UNDP program is funded by the Global

---

[162] United Nations Development Programme, 'Adaptation to Climate Change', http://www.undp.org/content/undp/en/home/ourwork/environmentandenergy/focus_areas/water_and_ocean_governance/adaptation-climate-change/, para. 4 accessed 11 October 2012.

[163] United Nations Development Programme, 'The Community Water Initiative: Fostering Water Security and Climate Change Mitigation and Adaptation' (2010), http://www.undp.org/content/dam/aplaws/publication/en/publications/environment-energy/www-ee-library/water-governance/cwi-fostering-water-security-and-climate-change-mitigation-and-adaptation/CWI_-_%20Local_Adaptation_to_Climate_Change_Knutson_March_18–1.pdf, accessed 11 October 2012.

[164] United Nations Development Programme, 'A Toolkit for Designing Climate Change Adaptation Initiatives' (2010), http://www.undp.org/content/dam/aplaws/publication/en/publications/environment-energy/www-ee-library/environmental-finance/low-emission-climate-resilient-development/designing-adaptation-initiatives-toolkit/Toolkit%20FINAL%20(new%20cover).pdf, accessed 11 October 2012.

[165] WHO, 'Climate Change Adaptation to Protect Human Health', above note 131.

[166] Ibid. Profiles for Barbados ('To strengthen national adaptive capacity to address health issues related to climate change attributable to water scarcity'), Bhutan ('To strengthen national capacity to identify and prevent adverse climate

Environmental Facility Special Climate Change Fund (about USD 4.5 million) and donations from 2010–2014.[167] WHO Europe also has a seven-country initiative that seeks to 'protect health from climate change through addressing adaptation, strengthening health systems and building institutional capacity'.[168] This project involves Albania, Kazakhstan, the Russian Federation, Tajikistan, the Former Yugoslav Republic of Macedonia, and Uzbekistan.[169] These programs are far from perfect, but they represent a focus on infrastructure investment and integration across sectors that is promising.

The Framework for Health Adaptation to Climate Change in Africa and the Joint Statement on Climate Change and Health of the African Ministers of Health and Environment provide an especially useful model for international law reform at the intersection of health and climate change. The obligations recognized in the Joint Statement go significantly beyond what is required by UNFCCC instruments. Member states have committed to:

---

change related health outcomes' such as glacial lake outburst floods and an increase in range and incidence of vector and water borne diseases), China ('To strengthen the national capacity to respond to the increased health risks due to heat waves'), Fiji ('To increase the adaptive capacity of the health sector to respond to climate sensitive risks' such as severe storms and droughts and climate sensitive diseases), Jordan, Kenya ('To strengthen national focus and adaptive capacity to prevent epidemic highland malaria'), and Uzbekistan (focused on 'adapt[ive] capacity of [the] health care system … to cope with climate induced disease') are available from http://www.who.int/globalchange/projects/adaptation/en/index.html, accessed 11 October 2012.

[167] WHO, 'Climate Change Adaptation to Protect Human Health', above note 131.

[168] World Health Organization Regional Office for Europe, 'Protecting Health from Climate Change: A Seven-Country Initiative in the Eastern Part of the WHO European Region', http://www.euro.who.int/en/what-we-do/health-topics/environment-and-health/Climate-change/publications/2009/protecting-health-from-climate-change-leaflets/protecting-health-from-climate-change-a-seven-country-initiative-in-the-eastern-part-of-the-who-european-region, accessed 11 October 2012.

[169] Ibid.; World Health Organization Regional Office for Europe, 'The Project in Each of the Seven Countries', http://www.euro.who.int/en/what-we-do/health-topics/environment-and-health/Climate-change/publications/2009/protecting-health-from-climate-change-leaflets/protecting-health-from-climate-change-a-seven-country-initiative-in-the-eastern-part-of-the-who-european-region/the-project-in-each-of-the-seven-countries, accessed 11 October 2012.

> Give special attention to the management of the adverse effects of climate change on health with a view to: (i) [m]aking evidence-based, climate-informed and sound decisions to implement a set of preventive actions to reduce population vulnerability and mitigate the additional burden posed by the major climate-sensitive diseases according to their respective epidemiological circumstances; (ii) [a]nticipating, preparing for and responding to the health consequences of extreme weather events, particularly by strengthening the functioning of health systems; [and d]eveloping applied research in order to better understand population vulnerability and the health implications of climate change at the local level.[170]

The statement further obligates member states to '[d]evelop an essential public health package to enhance the climate change resilience status of all our countries by 2014'.[171] The 2011 Framework elaborated on the 'essential public health package' to be developed by countries, noting that it includes 'assessment of the risks posed by climate variability and change to public health and health systems, surveillance, delivery of preventive and curative interventions including preparedness for and response to the public health consequences of extreme weather events and research'.[172] These instruments are still heavily focused on downstream threat-specific interventions rather than infrastructure investment, and implementation and enforcement problems limit their concrete impact, but they do provide a model for how inter-sectoral tools might be developed, at least 'on the books'.

5.3 Equity

Another reason that the basic health and sanitation infrastructure crucial to health adaptation is neglected, even in proportion to its current disease burden, is that current global public health frameworks have been shaped by the interests of resource rich countries. Contrary to aspirational statements that health equity and social justice should be the normative foundations of global health law,[173] the driving force of international health law has, in reality, been a concern for national interests that may

---

170 African Ministers of Health and Environment, 'Joint Statement on Climate Change and Health' para. 15(b), http://www.afro.who.int/index.php?option=com_docman&task=doc_download&gid=6068, accessed 11 October 2012.

171 Ibid. para. 15(e).

172 WHO Regional Office for Africa , above note 132 at para. 19.

173 Jennifer Prah Ruger, 'Normative Foundations of Global Health Law' (2008) 96 The Georgetown Law Journal 423, 424–425; Gostin and Taylor, above note 3.

be affected by health threats arising outside domestic borders.[174] The focus of international health law on the cross-border spread of infectious disease has gone hand in hand with the influence of 'health security' as the dominant norm of international health law.[175] That pathogens 'know no borders' is a common refrain issued by policymakers in demand of international cooperation with respect to health.[176] The foundation of international cooperation with regard to health on mutual vulnerability to threats that are beyond the reach of any individual state, regardless of its resources, is not consistent with environmental health data. The disease burden associated with water scarcity and vector-borne diseases varies enormously depending on a state's resources. Viewing these problems as climate change impacts reveals them as health threats that are, at least in considerable part, imposed by one group of states upon another. Environmental degradation greatly exacerbates health disparities among nations and among communities within nations. A meaningful response to the health threats associated with global environmental change will thus require that global health law more fully realize its commitment to health equity and social justice.[177]

The WHO's focus on 'global public health security' has not been without controversy. William Aldis's critical analysis of health security as a public health concept points to a disconnect between how policymakers in resource rich countries understand the term 'health security' and how the term is understood by health workers and policymakers in the

---

[174] Fidler, above note 62; William Aldis, 'Health Security as a Public Health Concept: A Critical Analysis' (2008) 23 Health Policy and Planning 369; Obijiofor Aginam, 'Bio-terrorism, Human Security and Public Health: Can International Law Bring them Together in an Age of Globalization?' (2005) 24 Medicine and Law 455.

[175] World Health Organization, *World Health Report 2007: A Safer Future: Global Public Health Security in the 21st Century* (WHO 2007) vii.

[176] Laurie Garrett, 'The Challenge of Global Health' (2007) 86 Foreign Affairs 14; Jeffrey Goldberg, 'Microbes on the Move' (*The New York Times Magazine*, 10 October 1999) 21, 22; Barbara Sibbald, 'New Program will Target 45 Million TB Patients in Next Decade' (2001) 164 Canadian Medical Association J 1741 ('"Microbes know no borders," says Ernest Loevinsohn, director general of CIDA's Food Aid Centre. "We have to work overseas to protect Canadian health as well"').

[177] At the heart of the turn from 'international health law' to 'global health law' was a concern about the 'entrenched power structures' reinforced in international health law. Lawrence O Gostin, 'World Health Law: Toward a New Conception of Global Health Governance for the 21st Century' (2005) 5 Yale Journal Health Policy Law & Ethics 413.

developing world. Policymakers representing the interests of wealthy populations generally frame health security as a matter of 'protection of their populations especially against external threats, for example terrorism or pandemics'.[178] This emphasis on health security as a concept related to *national* security is in tension with the very different understanding of health security as a concept related to *human* security – with an emphasis on human rights – among those concerned with the interests of the developing world.[179] An essential question for determining the usefulness of existing global public health law frameworks as tools for health adaptation to climate change is whether the tension between these two conceptions of health security is irresolvable. Is it possible to reconcile wealthy populations' efforts to protect themselves from threats that are perceived to be external with poor populations' efforts to ensure equitable access to health care and healthy conditions?

Recent proposals for global health law reforms, such as Lawrence Gostin's proposed Framework Convention on Global Health,[180] as well as fully realized developments like the Framework Convention on Tobacco Control (FCTC) adopted in 2003, suggest that global health law may already be moving in new directions that emphasize the importance of upstream determinants of health and their effect on health equity.[181] The FCTC addresses a wide range of issues related to advertising, marketing, and consumption of tobacco products, as well as related research and information sharing. Tobacco control is obviously far less relevant to the health impacts of climate change than infectious disease control or measures relating to public health emergencies. However, the FCTC provides an example of international cooperation on a health issue where the transnational aspect of the threat is its *source* (the tobacco industry and its practices)[182] rather than its potential *spread*. The impetus behind the FCTC is not that tobacco related illness will be transmitted across

---

[178] Aldis, above note 174 at 369.

[179] Ibid. at 370.

[180] Lawrence O Gostin, 'Towards a Framework Convention on Global Health: The Joint Action and Learning Initiative (JALI)' (2010), http://www.law.georgetown.edu/oneillinstitute/documents/2010_Towards%20a%20Framework%20Convention%20on%20Global%20Health.pdf, accessed 11 October 2012.

[181] World Health Organization, 'Framework Convention on Tobacco Control' (2005), http://whqlibdoc.who.int/publications/2003/9241591013.pdf, accessed 11 October 2012.

[182] Ruger, above note 173 at 436; Allyn L Taylor and Douglas W Bettcher, 'WHO Framework Convention on Tobacco Control: A Global "Good" for Public Health' (2000) 78 Bulletin of the World Health Organization 920, 923–925 (discussing the globalization of the tobacco epidemic).

borders the way infectious disease spreads, but rather that member states benefit from banding together to address a common threat. The FCTC could arguably be cast as a precedent for the type of international cooperation with respect to health that will be required in the face of climate change.

Lawrence Gostin's groundbreaking proposal for a new Framework Convention on Global Health[183] (FCGH) is particularly timely in light of the growing health burden of environmental degradation and is extremely relevant to health adaptation to climate change. Gostin argues:

> What is truly needed, and what richer countries instinctively (although not always adequately) do for their own citizens, is to meet … basic survival needs. By focusing on the major determinants of health, the international community could dramatically improve prospects for good health.[184]

Gostin defines 'basic survival needs' to include many of the issues that will be crucial to addressing the health impacts of climate change, including sanitation, vector control, clean air and water, and functioning health systems.[185] In many ways, the FCGH proposal builds on what has been achieved in the FCTC. Additionally, it borrows the framework convention approach to treaty making.[186] But it also furthers extends the argument for international cooperation to address the social, economic and environmental determinants, rather than focusing on specific disease threats. As Gostin argues:

> The determinants of health (pathogens, air, food, water, even lifestyle choices) do not originate solely within national borders. Health threats inexorably spread to neighboring countries, regions, and even continents. It is for this reason that safeguarding the world's population requires cooperation and global governance.[187]

This emerging shift from defending (mostly wealthier) populations against the transboundary spread of communicable diseases to defending (mostly poorer) populations against the health impacts of globalization will be crucial to climate change adaptation.

---

183 Gostin, above note 180.

184 Ibid. at 334.

185 Ibid.

186 Scott Barrett, *Environment and Statecraft: The Strategy of Environmental Treatymaking* (Oxford University Press, 2003) 133–164.

187 Gostin, above note 2 at 229.

6 CONCLUSION

Legal obligations regarding climate change adaptation are uniquely positioned to reorient global public health law in ways that promote cost-effective, equity-promoting upstream interventions, which are being neglected by current frameworks. If mutual vulnerability continues to be the dominant justification for international cooperation with respect to health, the appropriate priorities for health adaptation to climate change will remain neglected, and untold numbers will continue to suffer. The UNFCCC principle of common but differentiated responsibility is far more appropriate as a guiding norm for health adaptation. Law reform aimed at climate adaptation has the potential to promote a responsibility-based justification for prioritizing global public health as a supplement to the security-based justifications that currently dominate the relevant legal frameworks. Whether this potential is realized, however, remains to be seen.

8. Climate change adaptation and agricultural and forestry law

Robert W Adler

1 INTRODUCTION

Every person needs sufficient food, clothing, shelter, and some form of heating and cooking fuel to live a decent, dignified life. Indeed, under several sources of international law, even if currently unenforced, everyone has (or should have) a human right to those basic life support goods and services.[1] With the global population expected to grow to 9 billion by 2050,[2] and expected increases in per capita food consumption due to rising affluence and changing food preferences,[3] the United Nations (UN) estimates that agricultural production must increase by 50 to 70 percent by 2050 to meet demand.[4] A growing population will also generate increased demand for energy, fiber and other materials. Moreover, about a fifth of the world's population, and up to 65 percent in developing countries, rely directly on agriculture and forestry as their source of livelihood.[5] As a result, there is a pressing need for individual nations and the global community to design and implement strategies to adapt to the

---

1 Olivier De Schutter, Report submitted by the Special Rapporteur on the Right to Food, Human Rights Council Sixteenth Session, A/HRC/16/49. Applicability of international human rights law to climate change is beyond the scope of this chapter, but is discussed in Chapter 4.

2 United Nations, Department of Economic and Social Affairs, Population Division, *2011 World Population Prospects* (UN, 2010 Rev), http://esa.un.org/unpd/wpp/index.htm, accessed 16 October 2012.

3 Bruce Campbell et al., *Agriculture and Climate Change: A Scoping Report* (Meridian Institute 2011) 1.

4 De Schutter, above note 1 at 4; United Nations Environment Programme, Food and Ecological Security, Identifying Synergy and Tradeoffs (UNEP Policy Series No. 4, June 2011).

5 Campbell et al., above note 3 at 1. Agriculture represents between 20–60 percent of GDP in most developing countries, and an estimated 2.5 billion households are involved in agriculture worldwide. United Nations Conference on Trade and Development, *Assuring Food Security in Developing Countries Under the Challenges of Climate Change: Key Trade and Development Issues of a Fundamental Transformation of Agriculture* (UNCTAD 2011) 2.

potentially significant impacts of climate change on agriculture and forestry.[6]

This chapter will first summarize the most significant predicted impacts of climate change on agriculture and forestry, characterize the nature and extent of the uncertainty surrounding those predictions, and summarize some of the major changes in farming, ranching and forestry technologies and other practices needed to address those impacts. It will then analyze the challenges inherent in supporting and facilitating those practice changes, and describe legal and institutional adaptation strategies that might help to address those challenges. No single strategy will achieve an 'optimal' balance between adaptation and mitigation, between agriculture and forestry, and between short- and long-term adaptation approaches. Instead, laws, policies and decision making processes should focus on making the best possible decisions at any given time and for any particular region, with participation by all affected stakeholders at varying levels of decisions (international, regional, national and local), while maintaining enough flexibility to modify those decisions over time, as the effects of climate change move from less to more certain, and as political, economic and physical conditions evolve.

2 CLIMATE CHANGE IMPACTS TO AGRICULTURE AND FORESTRY

2.1 Changes in Growing Conditions

Climate change will affect agriculture and forestry in a number of ways, although considerable uncertainty surrounds current predictions.[7] There is variability across the global circulation models used to project the amount of impact that greenhouse gases (GHGs) will have on global average temperatures (GATs) over time,[8] and uncertainty about the timing and magnitude of GHG reductions that are likely due to various

---

6 William R Cline, *Global Warming and Agriculture, Impact Estimates by Country* (Center for Global Development 2007) 1; S Mark Howden et al., 'Adapting Agriculture to Climate Change' (2007) 104 PNAS 19691.

7 United Nations Environment Programme, *Climate Change Science Compendium 2009* (Catherine P McMullen & Jason Jabbour eds 2009) 46–47.

8 National Research Council, *Advancing the Science of Climate Change* (National Academies Press 2010) 225–227; IPCC, *Special Report on Emissions Scenarios* (IPCC 2001) Ch 5.

mitigation strategies and economic conditions.[9] Moreover, it is even more challenging to predict the relationship between increasing GATs and other climatic conditions, such as the timing, form and amount of precipitation, and to downscale climate projections from global to regional, national or local conditions.[10] Finally, as detailed below, experts continue to debate the expected relationship between those changes and agricultural productivity, as well as the manner in which domestic and global markets will respond to those changes.

Changes in climate affect the fundamental conditions under which plants are cultivated and livestock are raised. Food and forage crops can be sensitive to temperature in several ways. Growth, seed and fruit setting and production timing and success, and other aspects of plant physiology can be affected by changes in maximum daily temperatures, minimum nightly temperatures, and length or time of onset or termination of the growing season.[11] Higher air temperatures can increase evapotranspiration (ET) and therefore crop water demand.[12] In some regions temperature changes will alter plant growth and therefore crop yield (either upward or downward), but where crops are already being cultivated in areas near their threshold maximum temperatures, further increases can render those crops, or specific crop varieties, no longer viable in the region.[13]

Climate change will also affect precipitation patterns and therefore water supplies on which crops and forests depend around the globe.[14] Global precipitation will rise with higher GATs due to increasing

---

[9] United Nations Environment Programme, *The Emissions Gap Report, Are the Copenhagen Accord Pledges Sufficient to Limit Global Warming to 2°C or 1.5°C? A Preliminary Assessment* (UNEP 2010).

[10] UNEP, above note 7 at 7. Efforts proceed, however, to downscale global models to predict regional climate change effects, for example: United States Geological Survey, Scientific Investigations Report 2011–5077, *Integrated Watershed-Scale Response to Climate Change for Selected Basins Across the United States* (USGS 2012).

[11] M L Parry et al. (eds), *Climate Change 2007: Impacts, Adaptation and Vulnerability. Contribution of Working Group II to the Fourth Assessment Report of the Intergovernmental Panel on Climate Change* (Cambridge University Press 2007) 2--3; US Global Change Research Program, *Global Climate Change Impacts in the United States* (Cambridge University Press 2009) 72.

[12] US Global Change Research Program, above note 11 at 72.

[13] M J Salinger et al., 'Reducing Vulnerability of Agriculture and Forestry to Climate Variability and Change: Workshop Summary and Recommendations' (2005) 70 Climatic Change 341.

[14] Parry et al., above note 11 at 3.

evaporation from oceans and other large water bodies, and therefore increased moisture in the atmosphere.[15] The distribution of precipitation will also change, resulting generally in higher precipitation in the northern latitudes and lower precipitation in the mid-latitudes,[16] with an expected expansion of areas subject to drought and desertification.[17] For forests and agriculture that depend on rainfall rather than artificial irrigation, changes in seasonality or reliability of precipitation could be significant. For agriculture or orchards that rely on artificial irrigation, changes in precipitation amount, timing or type (rain versus snow) can affect storage in dams and distribution networks, and therefore water supply for irrigation and other uses. Warmer temperatures and drier conditions will also lead to an increased incidence and intensity of wildfires in arid and semi-arid forests and grasslands.[18]

Less directly, but also with potentially significant impact, climate change is likely to shift the ranges of pest and weed species that might adversely affect various crops.[19] For example, northern expansion of kudzu, an aggressive weed that carries the fungal disease soybean rust, is a major threat to US soybean production.[20] Large areas of forests in some regions have already been damaged widely due to bark beetles and other pests, as the natural population check of colder winters diminishes over time.[21] Even if additional pests and weeds can be controlled through chemical or other means, those measures can contaminate water resources and generate other adverse environmental effects, and increase input costs.

Livestock is also likely to be sensitive to variations in temperature, precipitation, and other conditions resulting from climate change. Particularly in areas in which livestock are already at the limits of their heat tolerance, higher temperatures can significantly stunt growth or cause livestock deaths.[22] The same factors that might impair growing conditions for food and fiber crops will also affect feed crops and the abundance, condition and plant mix of forage in natural rangeland, which

---

15 US Global Change Research Program, above note 11 at 41–42.

16 William Easterling, Pramod Aggarwal et al., 'Food, Fibre and Forest Products' in Parry et al., above note 11 at 275, 281.

17 UNEP, above note 7 at 38.

18 Parry et al., above note 11 at 3, 9, 10.

19 Easterling, Aggarwal et al., above note 16 at 283.

20 US Global Change Research Program, above note 11 at 75.

21 Barbara J Bentz et al., 'Climate Change and Bark Beetles in the Western United States and Canada: Direct and Indirect Effects' (2010) 60 Bioscience 602.

22 US Global Change Research Program, above note 11 at 78.

could either impair livestock health through reduced nutrition, reduce herd sizes or increase feed costs. Moreover, livestock may also be vulnerable to increases and changes in insects and other pests,[23] or require higher expenditures to keep them healthy in the face of those changes.[24]

The overall estimated impacts of climate change on agriculture and forestry also must account for extreme weather events in addition to changes in 'average' conditions.[25] Droughts,[26] floods,[27] high winds, intense storms, hurricanes and tornadoes already have become more frequent as GATs rise,[28] and individual extreme events can have severe impacts on agricultural production – sometimes wiping out entire crops.[29] They can also pose serious hazards to agricultural workers and buildings and other infrastructure in agricultural communities.

One key uncertainty regarding the relationship between climate change and agricultural production is the potentially offsetting effect of carbon fertilization. The highest quantity GHG responsible for climate change, atmospheric carbon dioxide (CO_2),[30] is also a key plant nutrient because photosynthetic plants use sunlight to convert CO_2 and water to stored energy. It is difficult to predict, however, for different plant species and under different conditions, how much the positive effects of higher CO_2 levels will offset any negative impacts from other climate change impacts. Controlled laboratory experiments document improved plant growth under higher CO_2 concentrations, but in field experiments either less significant improvements occurred, were limited by other environmental factors, or required additional irrigation or other inputs to obtain

---

23 Easterling, Aggarwal et al., above note 16 at 283.

24 US Global Change Research Program, above note 11 at 75–76.

25 Parry et al., above note 11 at 5; US Global Change Research Program, above note 11 at 44.

26 Aiguo Dai, 'Drought Under Global Warming: A Review' (2010) 2 WIREs Climate Change 45; G Wang, 'Agricultural Drought in a Future Climate: Results from 15 Global Climate Models Participating in the IPCC 4th Assessment' (2005) 25 Climate Dynamics 739.

27 P C D Milly et al., 'Increasing Risk of Great Floods in a Changing Climate' (2002) 415 Nature 514.

28 UNEP, above note 7 at 2–3.

29 US Global Change Research Program, above note 11 at 74–75.

30 Although CO_2 is emitted into the atmosphere in higher quantities than other GHGs, other GHGs have higher warming potential. UNFCCC, Report of the Conference of the Parties on its Seventeenth Session, FCCC/CP/2011/9/Add.2, Annex III.

those results.[31] Moreover, higher CO_2 concentrations, especially when coupled with higher temperatures, sometimes benefit weeds more than crops.[32]

2.2 Impacts on Domestic and Global Economies, Markets and Cultures

The economic and social significance of the effects of climate change on agriculture and forestry depend not only on the physical impacts described above, but also on the manner in which existing legal and other institutions (absent intentional adaptation efforts) respond to those physical effects. For example, production increases in some regions might compensate for declines in others, but only if price, income, distribution and trade regimes adequately support the necessary import and export markets.[33] Even in the current global economy more than enough food is produced to feed the entire world population,[34] but price, trade, distribution and other barriers, along with significant income disparity within and among nations, impede the complete flow of products to all who need them. Due to inefficiencies in the food production and distribution chain, at best only two-thirds of all food grown is actually used to meet human needs.[35] The use of grains to feed livestock results in significant additional efficiency losses.[36]

Several models have been used to predict both regional and global effects of climate change on agricultural productivity. As with underlying climate change models, however, there is variability among models in

---

[31] Easterling, Aggarwal et al., above note 16 at 282.

[32] US Global Change Research Program, above note 11 at 75.

[33] Gerald C Nelson et al., *Food Security, Farming, and Climate Change to 2050, Scenarios, Results, Policy Options* (International Food Policy Research Institute 2010) xvii–xx.

[34] For all 174 crops tracked by the UN, global production increased by 28% between 1985 and 2005. Jonathan A Foley et al., 'Solutions for a Cultivated Planet' (2011), 478 Nature 337.

[35] United Nations, FAO, *Global Food Losses and Food Waste* (FAO 2011). Some estimates are considerably lower, considering both pre- and post-harvest losses. UNCTAD, above note 5 at 20.

[36] Only 62% of crop production is used directly for human food, compared to 35% for animal feed, although that percentage varies from 40% for human food in North America and Europe to over 80% in Africa and Asia. Foley et al., above note 34 at 337.

both methodology and result.[37] Generally, models predict that agricultural productivity will decline – in some cases quite significantly – in tropical and subtropical regions where crops are already at the limits of their heat tolerance, or where reduced water supply and other factors will limit productivity.[38] Depending on the assumptions used, in terms of overall global production those declines might be offset in the short-term in temperate zones, where crops are not at the temperature limits and where growing seasons will increase. The models also predict that areas suitable for certain crops will expand northward, increasing productive acreage in some parts of the world. However, as temperatures continue to rise, productivity will decline in those regions as well.[39] As a result, even divergent models agree that global food production will decline relative to current levels by the second half of the twenty-first century, and that food prices will rise globally.[40] As a result, the UN estimates that 600 million more people could be at risk of hunger by 2080 due to climate change.[41]

Existing efforts to predict the effects of climate change on agriculture, however, should be interpreted with caution. First, as noted above, the magnitude of the offsetting effect of carbon fertilization on plant growth is uncertain. Some models use considerably more optimistic assumptions about those effects than others, or than have been borne out in field experiments conducted to date,[42] and reliance on carbon fertilization may assume unrealistically the unlimited availability of additional irrigation water needed to take advantage of higher CO_2 levels.[43] Second, the models typically employ mid-range assumptions, that is, are based on average temperature increases and changes in precipitation. A considerable amount of damage to crops and trees can occur due to extreme events, such as droughts, floods, torrential rains, hurricanes or tornadoes.[44] Third, recent studies suggest that warming is proceeding faster than projected by earlier models, and that the ultimate magnitude of

---

37 Cline, above note 6 at 7–21.

38 Easterling, Aggarwal et al., above note 16 at 275.

39 Cline above note 6 at 2, 95–98; Parry et al., above note 11 at 6.

40 Nelson at al., above note 33 at xvii–xviii.

41 United Nations Development Programme, *Human Development Report 2007/2008. Fighting Climate Change: Human Solidarity in a Divided World* (UNDP 2007).

42 Cline, above note 6 at 23–26.

43 Ibid. at 26–32. Earlier models did not account for project changes in precipitation patterns. Easterling, Aggarwal et al., above note 16 at 280.

44 Easterling, Aggarwal et al., above note 16 at 283.

change will be higher as well.[45] Therefore, adaptation planning for agriculture and forestry should proceed with sufficient flexibility to address a potentially wide range of effects.

Regionally, the predicted impacts of climate change on agriculture differ dramatically. Unfortunately, the largest impacts will occur in countries and entire regions that are most vulnerable because of undiversified economies, shortages of capital to invest in needed improvements in infrastructure, poor access to technological improvements, and insufficient earnings from other industries.[46] The most significant declines in food production are expected in Africa and in south central Asia, and to a lesser degree in Latin America. For example, by 2020 an additional 60 to 90 million hectares in sub-Saharan Africa are projected to become more arid, and yields from rain-fed agriculture could decline by as much as 50 percent.[47] Crops could also decline by up to 30 percent in central and south Asia by the mid-twenty-first century, resulting in continued high risks of hunger.[48]

Available models are even less developed to predict the effects of climate change on grazing systems and economies around the world, due to uncertainties in the effects of CO_2 fertilization, precipitation changes, pests and other factors.[49] However, grazing in arid and semi-arid regions, where a large number of pastoralist communities derive their livelihoods, will be particularly vulnerable to changes in temperature and the amount and seasonal distribution of precipitation, as well as extreme drought events. In recent years, drought has already had devastating effects on pastoralist communities in portions of Africa and other regions.[50] Moreover, although CO_2 fertilization might enhance productivity of row crops

---

45 UNEP, above note 7 at 8.

46 Jodie Keane et al., *Climate Change and Developing Country Agriculture: An Overview of Expected Impacts, Adaptation and Mitigation Challenges, and Funding Requirements* (International Centre for Trade and Sustainable Development 2009).

47 Parry et al., above note 11 at Ch 9.

48 Ibid. at 22.

49 Britta Tietjen and Florian Jeltsch, 'Semi-arid Grazing Systems and Climate Change: A Survey of Present Modeling Potential and Future Needs' (2007) 44 J Applied Ecology 425.

50 Robert W Adler, 'Drought, Sustainability and the Law' (2010) 2 Sustainability 2176; James Forole Jarso, 'Africa and the Climate Change Agenda: Hurdles and Prospects in Sustaining the Outcomes of the Seventh African Development Forum' (2011) Sustainable Development Law & Policy 38; Richard Munang and Johnson N Nkem, 'Using Small-Scale Adaptation Actions to

or trees, it can actually reduce the forage value of grazing lands by changing plant mixes and reducing the nutritive value of forage.[51]

Similarly, although models generally project increases in global timber productivity due to CO_2 fertilization and expansion of forests toward the poles,[52] those models vary considerably across methods, do not account for uncertainties about whether CO_2 fertilization experiments on young trees under controlled conditions will translate to field conditions, or for adverse impacts due to pests, weeds, fire, changes in soil moisture and other factors.[53] Any increases in timber production may also be offset by rising demand for biofuels. In addition, as is the case for row crops, the regional distribution of both natural and plantation forests is likely to change.[54] Especially in the tropics, vulnerable nations that rely on forest products, such as seeds, nuts, berries, hunting habitat, resins, and plants associated with forest ecosystems, may face disproportionate impacts.

Climate change will also cause significant sociological impacts to communities that have engaged in farming, grazing and forestry as a mainstay of their local economies for generations. Even if net global production of food and other agricultural products remains satisfactory, and even if those products can successfully be redistributed through markets or more aggressive intervention strategies, longstanding communities could wither or be required to shift to different sources of livelihood, with accompanying social and economic disturbances or dislocation.[55] Some examples of climate-induced migration have already occurred in recent years, for example, as severe drought has prompted mass migrations of pastoralist and other populations in portions of Africa. Similar impacts are likely due to the effects of deforestation and forest fragmentation on existing communities, which are projected to affect some 1.2 billion impoverished forest-dependent people.[56] Those impacts may also have implications for global security, as migrations of

---

Address the Food Crisis in the Horn of Africa: Going Beyond Food Aid and Cash Transfers' (2011) 3 Sustainability 1510.

[51] US Global Change Research Program, above note 11 at 77–78.

[52] Easterling, Aggarwal et al., above note 16 at 289.

[53] Andrei P Kirilenko and Roger A Sedjo, 'Climate Change Impacts on Forestry' (2007) 104 PNAS 19697.

[54] Ibid. at 19701.

[55] Koko Warner et al., *In Search of Shelter, Mapping the Effects of Climate Change on Human Migration and Displacement* (CARE 2009).

[56] Easterling, Aggarwal et al., above note 16 at 291.

destitute populations generate increased potential for conflict within and among nations.[57]

2.3 Practice Changes to Address Impacts to Agriculture and Forestry

Although adaptation planning for agriculture and forestry remains nascent, and the exact nature of adaptation will vary significantly among different physical, political and economic contexts, a number of works have been published laying out possible adaptation strategies.[58] Physical or technological strategies to adapt to the effects of climate change on agriculture and forestry fall within a spectrum of approaches ranging from least to most intrusive on existing economies and local cultures. However, it is unreasonable to assume that adaptation methods will suffice to avoid significant adverse effects to human populations and economies, and adaptation success will vary across regions facing different climatic or other conditions.[59] Therefore, adaptation will have to include relief strategies to compensate those agricultural regions hit hardest by climate change, to help them to transition to new livelihoods, or even to move to new locations.

First, agricultural practices can be modified to continue to harvest the same crops and trees or to raise the same livestock, or to use existing or remaining production more efficiently. That might be accomplished by changing planting schedules or using new seed or plant varieties, including genetically modified organisms (GMOs) that either boost yields or are more heat tolerant or drought tolerant. It also might involve revised or modernized planting, cultivation, or other agricultural practices; or improved water supplies and irrigation systems or substitution of irrigated for rain-fed agriculture. New pest and weed control methods might be needed as unwanted species spread to new ranges, which may cause regions to face difficult choices between chemical and other control

---

[57] Ben Wisner et al., *Climate Change and Human Security* (Radix 2007), http://www.radixonline.org/cchs.html, accessed 16 October 2012.

[58] See, e.g., Nelson et al., above note 33; Jon Padgram, *Agricultural Development Under a Changing Climate, Opportunities and Challenges for Adaptation* (World Bank 2009); FAO, *Climate-Smart Agriculture, Managing Ecosystems for Sustainable Livelihoods*; *Climate, Agriculture and Food Security, A Strategy for Change* (Consultative Group on International Agricultural Research 2009); Michael H Glantz, René Gommes and Selvaraju Ramasamy, *Coping with a Changing Climate: Considerations for Adaptation and Mitigation in Agriculture* (FAO 2009); US Global Change Research Program, above note 11.

[59] Easterling, Aggarwal et al., above note 16 at 295.

methods, with accompanying environmental and human health risk consequences. Improved weather prediction and information dissemination systems, including advance warning systems for drought, flooding and other extreme events can help agriculturalists to plan appropriate planting and harvesting times and other practices under more variable climatic conditions. Changes in diet can direct a higher percentage of production to human consumption, or to reduce food losses in the field or during storage and processing.[60] All of those methods will require technology transfer and education in new agricultural methods, and in many cases significant capital investments or increased operating costs.

Second, where traditional or other existing crops can no longer be raised under new conditions, due to limits of heat tolerance or water tolerance at either end of the spectrum (too wet or too dry), or other changes, agricultural regions may shift to new crops altogether. Those more dramatic changes can preserve the basic nature of agricultural communities, but may require even more significant transfers of technology, education and training, and other support mechanisms to facilitate the transition to new crops with very different cultivation, harvesting, storage and handling, marketing, or other requirements. Those changes might cause cultural disruption if local or regional diets have been based on existing crop mixes. And if the distribution of major crop production shifts globally, redistribution will have to occur through trade and enhanced or modified storage and distribution networks.

In the most extreme conditions, some areas will no longer be suitable for agriculture, while other regions that have been too cold may become suitable for those economies. Such dramatic changes in agricultural suitability zones are more likely as climate change progresses, which will allow more time for adaptation strategies to be planned and implemented. Adaptation in those circumstances will also be of a very different nature than those designed to modify agricultural practices or to transition to different crops, and may require a combination of financial relief, relocation assistance, retraining for new livelihoods, and efforts to prevent adverse impacts on existing residents where relocations occur.

---

[60] De Schutter, above note 1 at 4, citing United Nations Environment Programme, *The Environmental Food Crisis – The Environment's Role in Averting Future Food Crises*, 2009.

3 POLICY AND INSTITUTIONAL CHALLENGES IN DESIGNING LEGAL SYSTEMS FOR ADAPTATION

In addition to the physical and technological challenges discussed above, the agriculture and forestry sectors face significant policy tradeoffs and institutional issues that will affect, and at times impede, successful adaptation policies. Legal strategies designed to facilitate and support adaptation should be designed with the following general policy considerations in mind.

3.1 Adaptation versus Mitigation

The agriculture and forestry sectors – and the governments and other institutions that support them – bear a core responsibility to feed, clothe, shelter, and to a lesser extent provide energy to a growing world population in a changing climate. Therefore, to prevent widespread suffering in some regions, local or regional declines in production of food and other critical materials either must be prevented or diminished through adaptation strategies, or offset through successful and equitable redistribution of those goods through regional and global trade.[61] On the other hand, agriculture and forestry contribute approximately 30 percent of all anthropogenic GHG emissions globally,[62] and serve as major carbon sinks,[63] making them a significant target for mitigation strategies.[64] That suggests potentially difficult tradeoffs between climate change adaptation and mitigation for agriculture and forestry. Efforts to reduce GHG emissions from agriculture and forestry cannot compromise productivity levels without interfering with the goal of maintaining or increasing stocks of food and other critical goods to support the growing human population. Thus, most current projections assume that it will not

---

[61] Easterling, Aggarwal et al., above note 16 at 281.

[62] Continued deforestation contributes an estimated 17% of all GHG emissions, and forested areas vulnerable to climate change could release billions of additional tons of GHGs by the end of the twenty-first century. UNEP, above note 7 at 49; FAO, above note 58 at 4. Agriculture contributes an additional 13.5% of global emissions. US Global Change Research Program, above note 11 at 71. These estimates, however, are subject to considerable uncertainty. Campbell et al., above note 3 at 3.

[63] Even small increases in Amazon deforestation could eliminate its status as a net carbon sink, with a climate change impact equal to the equivalent of all of the fossil fuel emissions of Western Europe. UNEP, above note 7 at 40, 49–50.

[64] Campbell et al., above note 3 at 5.

be possible to maintain needed levels of agricultural production while simultaneously *reducing* agriculture's contribution to climate change. At most, they suggest that productivity can be maintained or improved without simultaneously increasing GHG emissions from agriculture.[65]

For example, in areas of the world facing declining precipitation, it may be possible to maintain food production through water transport, pumping, storage and irrigation, which will increase energy use. Depending on what fuels are used to produce that energy, and whether gravity-based water transport is possible topographically, enhancing water supplies for agriculture could increase overall GHG emissions.[66] Similarly, maintaining or increasing agricultural productivity might require increased mechanization of farming or grazing practices, for example, to take advantage of more efficient planting and harvesting methods or to move livestock more efficiently to prevent overgrazing of rangeland, and increased use of agricultural inputs such as fertilizer and pesticides. In addition to generating higher production costs, those adaptations bear significant energy costs to obtain raw materials, manufacture, transport and apply those products.

Where increased trade is used to offset rather than prevent regional losses in production of food and other key materials, increased energy use and therefore increased contribution to climate change is also inevitable. Large amounts of energy will be needed to store and package those commodities prior to consumption, particularly where refrigeration or advance processing is required. Moreover, transportation of large volumes of commodity crops and other food and agricultural products, including very large volume materials such as timber, require significant fuel costs and related GHG emissions.

On the other hand, some adaptation strategies, such as those based on principles of agroecology, can both increase food production and access to local populations and reduce climate change impacts by reducing reliance on energy-intensive inputs and mechanization, and by increasing carbon storage in soils.[67] They can also improve resilience to extreme weather events by increasing soil moisture and retention, stabilizing soils against erosion, and reducing the effects on new or increased pests and

---

[65] Campbell et al., above note 3 at 5, 10.

[66] Electric Power Research Inst., Water & Sustainability (Vol. 4): US Electricity Consumption for Water Supply & Treatment – The Next Half Century (2002).

[67] De Schutter, above note 1 at 12–14.

weeds by increasing biodiversity.[68] Similarly, sustainable forest management can contribute to both mitigation and adaptation by promoting optimal stocking and use of forest products for durable goods or as substitutes for fossil fuels, thus preventing additional GHG emissions.[69] Thus, where possible, adaptation efforts should attempt to maximize opportunities to achieve mitigation goals simultaneously.

3.2 Conflicts between Agriculture, Forestry and Energy

Strategy development, legal regimes and institutional arrangements governing climate change adaptation will also have to consider tradeoffs between agriculture, forestry and energy. For example, agriculture has played a key role in mitigation to date through the production of crops used to manufacture biofuels.[70] However, that practice reduces stocks of those commodities for food, and increases global demand and therefore commodity prices. Although food prices generally have been low in recent years, that trend is likely to reverse over the coming decades.[71] Thus, either markets, legal institutions or market incentives imposed through legal institutions will play roles in dictating or guiding the proportion of agricultural production devoted to energy production relative to food supply. Higher allocations for food to adapt to regional or global production declines might impede the use of biofuels as a mitigation strategy, while higher allocations for fuel will make it more challenging to provide enough food to growing populations.

Likewise, there is an inherent tradeoff between the adaptation and mitigation implications of using land for agriculture and forestry, respectively. As noted above, healthy forests are among the Earth's most important carbon sinks. One strategy to adapt to declines in agricultural production, however, is to convert forests to crop agriculture or livestock use, most of which is likely to occur in sub-Saharan Africa and Latin America.[72] Although potentially effective as a strategy to adapt to climate-induced losses in agricultural production, that practice will further reduce the ability of the world's forests to sequester carbon.

---

68 Ibid.

69 UNEP, above note 7 at 49.

70 Mary Jane Angelo, 'Corn, Carbon and Conservation: Rethinking U.S. Agricultural Policy in a Changing Global Environment' (2010) 17 George Mason Law Review 593.

71 International Food Policy Research Institute, *Climate Change, Impact on Agriculture and Costs of Adaptation* (IFPRI 2009) at 6–7.

72 Easterling, Aggarwal et al., above note 16 at 280–281.

International and domestic legal regimes governing climate change (discussed below) will have to address the tradeoff between deforestation as a strategy for agricultural adaptation and maintenance or expansion of forests as a mitigation strategy. Markets, legal institutions or market incentives imposed through legal institutions will play a role in dictating or guiding the proportion of currently forested land that will be cleared for agricultural production, and the proportion of land that is retained or reforested for purposes of carbon sequestration. However, those tradeoffs might be avoided in part if food production is increased through improved efficiency per acre rather than expanding acreage, or if existing production is stored, transported and marketed with less waste.

3.3 Equity and Distributive Issues

Resolution of the conflict between efficiency and equity is particularly important for basic human life support products and has important implications for foreign aid policy and government investment policy. Efforts to offset production declines in developing countries through modernization, technology transfer and investment in irrigation and other new systems will require significant investments of capital as well as higher ongoing operational costs, and therefore must take into account significant existing disparities in global distribution of both monetary capital and knowledge. That suggests the need for changes in international aid policies and their linkage to climate change law and policy, as well as the relationship between international intellectual property law and other aspects of technology transfer.

In theory, redistribution of global food production can be addressed efficiently through international trade.[73] Countries with increased production under changing climate scenarios will sell their surpluses to countries with declining agricultural yields. Even if net global food production declines, there should be increased incentives to improve efficiency of storage, distribution and marketing. Several barriers, however, suggest that solutions to these distributive problems are not necessarily so easy. First, developing nations are not likely to have the income or balance of trade necessary to support large increases in agricultural imports, unless they can successfully develop other economic sectors to generate hard currency and exportable goods of their own.[74] That is likely to increase food insecurity in some regions. For example, by 2080 the proportion of

[73] Easterling, Aggarwal et al., above note 16 at 297.
[74] Cline, above note 6 at 32–33.

the world's undernourished people in sub-Saharan Africa could increase from 24 to 40–50 percent, and under extreme scenarios, as high as 75 percent.[75] Second, there is considerable disagreement about the value of government investment in agricultural economies to boost production and to support existing livelihoods relative to the impact of those subsidies in distorting the agricultural economy. Third, trade barriers such as product duties or tariffs or non-tariff trade barriers, particularly if they have a protectionist impact, can impede the global flow of agricultural goods. Thus, an adaptation strategy for agricultural and forest products will need to consider elements of both international aid and international trade law and policy.

3.4 Balancing Short-term Relief and Long-term Risk

Adapting to the effects of climate change on agriculture and forestry will entail tradeoffs between short-term relief and compensation to farmers, ranchers and pastoralists affected by episodic climate-related disasters (such as droughts, floods, extreme storms or heat waves), and long-term strategies to reduce vulnerability or to adapt to relatively more permanent regional shifts in climate. For example, it may be appropriate to provide subsidized insurance, relief payments, or other compensation in the face of extreme but episodic weather events. Excess reliance on that form of public assistance, however, can encourage landowners to continue risky ventures in an increasingly unfriendly changing climate, rather than adopting new practices to reduce those risks, or even facilitating the transfer of production to regions that can better support those crops.[76]

Another example is the increased use of groundwater to replace declining surface water supplies due to reduced or changed patterns of precipitation. If pumped judiciously to allow aquifer recharge, increased groundwater use can help agricultural and livestock raising communities accustomed to surface water supplies adapt to declining precipitation. If precipitation becomes more variable rather than consistently lower, conjunctive management of surface and groundwater can also be an effective adaptation strategy, with excess surface water collected and stored in aquifers during wet years and then 'harvested' during dry years. However, if groundwater is 'mined' significantly beyond natural recharge

---

[75] Easterling, Aggarwal et al., above note 16 at 298.

[76] Robert W Adler, 'Balancing Compassion and Risk in Climate Adaptation: U.S. Water, Drought, and Agricultural Law' (2012) 64 Florida Law Review 201.

rates, which will decline in areas with decreasing precipitation, shortages will be exacerbated in the long-term.[77]

3.5 Uncertainty and the Balance between Legal Stability and Flexibility

Legal adaptation regimes for agriculture and forestry will have to address all of the above tradeoffs in the face of significant uncertainty and variability in climate change rather than a 'new but stable' climate, either globally or regionally. First, climate change will evolve over a long period of time until (hopefully) GHG emissions and atmospheric levels stabilize or begin to decline. Thus, GATs are likely to continue to rise steadily, causing a range of global and regional climatic responses over time in terms of impacts on regional temperatures, precipitation patterns, storm events and other factors.[78] Second, because of the difficulty in scaling global climate models to increasingly localized regions, the exact impact of climate change on the factors most important to agriculture and forestry in specific regions will only become apparent over time, and even then are likely to remain volatile.

Therefore, legal regimes designed to support climate change adaptation for agriculture and forestry will need to have sufficient stability to encourage adoption of and investment in new practices and equipment. Some of the adaptation strategies discussed above will have very long timeframes for returns on investment, for example, construction of new water supply infrastructure and irrigation systems, or new transportation networks (roads, railroads, shipping facilities) to redistribute agricultural products. However, those systems also must contain sufficient flexibility to support major changes in climate conditions over time, particularly where those changes are unpredictable or predictable only within broad ranges. Indeed, given the many unknowns inherent in climate science, we need to anticipate the possibility of 'surprises' that might have significant impacts on agriculture and forestry, such as unanticipated tipping points

---

[77] Colin Polsky and David W Cash, 'Drought, Climate Change and Vulnerability: The Role of Science and Technology in a Multi-Scale, Multi-Stressor World' in Donald A Wilhite (ed.), *Drought and Water Crises, Science, Technology and Management Issues* (CRC 2005) 215, 225–227.

[78] V Ramanathan and Y Feng, 'On Avoiding Dangerous Anthropogenic Interference with the Climate System' (2008) 105 Proceedings of the National Academy of Sciences 14245–14246; Susan Solomon et al., 'Irreversible Climate Change Due to Carbon Dioxide Emissions' (2009) 106 Proceedings of the National Academy of Sciences 1704.

that trigger more sudden changes in climate, or in related factors such as the rapid spread of pests.

4 LEGAL AND INSTITUTIONAL STRATEGIES FOR ADAPTATION

Legal and institutional approaches to adaptation in agriculture and forestry should be designed in light of the major considerations addressed above. They need to account for the full range of diverse, sometimes significant, and variable physical impacts that climate change will have (although for purposes of domestic policy individual nations obviously should consider only those impacts expected to occur in their regions). They should accommodate a similarly diverse range of 'on the ground' practice changes useful in limiting or offsetting climate change impacts on productivity. And they should consider how to overcome applicable legal and institutional barriers and how to balance competing goals in ways that generate multiple 'wins' or that minimize impacts on other objectives.

Adaptation measures will incorporate legal strategies at both the international and domestic levels. The following is a summary of the most important areas of law that should be developed or further implemented at both the international and domestic levels. Details and specific choices will have to vary to account for diverse physical, economic, political, cultural and other differences, and will have to be sufficiently flexible to respond to changing and uncertain conditions, and to take advantage of knowledge acquired over time.

4.1 International Law and Policy

4.1.1 International climate change agreements

The most fundamental source of international law governing climate change adaptation for agriculture and forestry is the body of multilateral agreements negotiated under the auspices of the United Nations Framework Convention on Climate Change (UNFCCC).[79] The ultimate objective of UNFCCC is 'stabilization of greenhouse gas concentrations in the atmosphere at a level that would prevent dangerous anthropogenic interference with the climate system'; and for that stabilization to occur

---

[79] United Nations Framework Convention on Climate Change (New York, 9 May 1992), (1992) 31 ILM 849, entered into force 21 March 1994.

'within a time frame sufficient to allow ecosystems to adapt naturally to climate change, *to ensure that food production is not threatened*, and to enable economic development to proceed in a sustainable manner.'[80] Thus, maintenance of food production is an essential corollary to the Convention's ultimate objective, implicitly at levels sufficient to support the global population.

From a mitigation perspective, UNFCCC, either implicitly or directly, acknowledges and commits the parties to address the significant role that agriculture and forestry play in contributing to GHG emissions[81] and in serving as carbon sinks and reservoirs.[82] However, UNFCCC also commits the parties to cooperate in planning for adaptation to climate change impacts, including impacts to 'water resources and agriculture', and for protection and rehabilitation of areas affected by drought, desertification and floods,[83] and calls on nations to develop National Adaptation Programmes of Action (NAPAs) appropriate to their circumstances. It also generally commits developed country parties to assist developing country parties that are particularly vulnerable to climate change effects, including impacts to forests and agricultural lands, in meeting adaptation costs.[84]

Finally, UNFCCC addresses two collateral issues that are important to agricultural and forestry adaptation efforts, and that are addressed further below: trade implications and technology transfer. Regarding the former, it declares that measures taken regarding climate change 'should not constitute a means of arbitrary or unjustifiable discrimination or a disguised restriction on international trade'.[85] With respect to the latter, it commits developed country parties to 'take all practicable steps to promote, facilitate and finance, as appropriate, the transfer of, or access to, environmentally sound technologies and know-how' to enable other parties to implement the Convention.[86]

---

[80] Art. 2 (emphasis added).

[81] Art. 4(1)(a), (b) (committing parties to prepare inventories and national control programs to address anthropogenic emissions of all GHGs not controlled by the Montreal Protocol).

[82] Art. 4(1)(d) (committing parties to conserve and enhance GHG sinks and reservoirs, including 'biomass' and 'forests').

[83] Art. 4(1)(e).

[84] Art. 4(4) and (8). See further Chapter 2, section 2.2.

[85] Art. 3(5).

[86] Art. 4(5).

The Kyoto Protocol,[87] adopted during the third Conference of the Parties to UNFCCC in December 1997, was designed primarily to focus on mitigation. In particular, the Protocol translated the general consensus established in UNFCCC to stabilize global emissions of GHGs into firm commitments by individual countries to reduce GHG emissions by specific amounts.[88] Notably, emissions from and removals by land use and forestry activity, including agricultural soils, were included in the activities to be included in the Protocol's accounting mechanisms, so long as they were measurable as verifiable changes in carbon stocks.[89] Moreover, the Protocol specifically envisioned the use of trading mechanisms in which countries could acquire emissions reductions from other parties to meet their reduction commitments, including through enhancement of anthropogenic removals in any economic sector.[90]

Thus, *in theory*, reforestation or changes in agricultural tilling that demonstrably increase carbon sinks, which could simultaneously advance adaptation strategies, might be sold as a credit in the Protocol scheme. Similarly, countries not included in Annex I to UNFCCC may participate in the Clean Development Mechanism (CDM) in which Annex I parties use certified emission reduction credits purchased or otherwise obtained from CDM projects to meet commitments under the Protocol. However, in the first reporting period under the Kyoto Protocol, only four countries elected to calculate and incorporate cropland and grazing contributions in their reporting systems,[91] and the initial rules adopted for CDM did not allow for use of agricultural carbon sequestration due to significant uncertainties in accounting and verification.

Despite this predominant focus on mitigation, the Kyoto Protocol contains a requirement for parties to formulate, implement and update programs to 'facilitate adequate adaptation to climate change',[92] but with no specific requirements or commitments. To a large degree, it makes sense that international agreements on climate change focus heavily on mitigation requirements while being generally hortatory with respect to adaptation. GHG emissions from all nations contribute to global harm,

---

[87] Kyoto Protocol to the United Nations Framework on Climate Change, UN Doc FCCC/CP/1997/7/Add. 1 (Kyoto, 10 December 1997), (1998) 37 ILM 22, entered into force 16 February 2005.

[88] Art. 3 and Annex B.

[89] Art. 3(3),(4),(7) and Annex A (identifying sources of agricultural GHG emissions).

[90] Art. 3(10), Art. 6.

[91] Campbell et al., above note 3 at 85.

[92] Art. 10(b).

and in fact, at least in the short-term, those countries most responsible for emissions will bear the least amount of harm, and vice versa. On the other hand, countries have an inherent incentive to focus on adaptation in proportion to the expected amount of harm to their people, economies and ecosystems. Moreover, to the extent that the trading and CDM mechanisms allow countries to sell credits for actions that simultaneously help with adaptation in agriculture and forestry and increase carbon sinks, the Protocol can be used to promote those twin goals.

However, by not addressing adaptation specifically in more detail, the Kyoto Protocol failed to move forward significantly with the commitment in UNFCCC for developed nations to provide assistance to developing nations to adapt to the expected impacts of climate change. This omission was particularly important for countries in Africa, South Asia, Latin America, and other regions in which impacts to agriculture and forestry are expected to be particularly significant, and for countries in those regions whose vulnerability to those effects is high due to resource limitations, less diversified economies and other factors.[93]

Subsequent agreements and decisions made pursuant to UNFCCC increased the focus on adaptation planning, funding and implementation, but continued to emphasize the links between adaptation and mitigation in agriculture and forestry. However, they also underscored the difficulties of achieving satisfactory progress in those areas. For example, the Bali Action Plan adopted during COP13[94] included directives to address deforestation and forest degradation and enhancement of forest stocks in developing countries; enhanced action on adaptation including vulnerability assessments for drought, desertification and floods as well as associated risk management and disaster reduction strategies; enhanced action on technology transfer; and enhanced action on provision of financial resources.[95] Those measures, however, remained essentially voluntary.

Somewhat more concrete steps were taken at COP16 in Mexico in 2010,[96] at least to the extent of creating more formal institutional arrangements within which to accomplish the various programs agreed to in UNFCCC. The Cancun Adaptation Framework is designed to facilitate

---

93 W Neil Adger and P Mick Kelly, 'Social Vulnerability to Climate Change and the Architecture of Entitlements' (1999) 4 Mitigation and Adaptation Strategies for Global Change 253.

94 Report of COP13, 15 December 2007, Part Two, FCCC/CP/2007/6/Add.1.

95 Bali Action Plan, UNFCCC Decision 1/CP.13.

96 Report of COP16, 10 December 2010, Part Two, FCCC/CP/2010/7/Add.1.

development of national adaptation plans and strategies; national adaptation programs of action; and impact, vulnerability and adaptation assessments, with an Adaptation Committee to promote implementation via technical support, information sharing and other measures.[97] More pointedly, the Cancun Agreements include creation of a Green Climate Fund to effectuate commitments by developed countries to provide USD 30 billion for the period 2010 to 2012 and USD 100 billion annually by 2020 to address developing countries' adaptation needs, with a focus on most vulnerable countries.[98] Likewise, COP16 agreed to create a Technology Mechanism to develop and disseminate environmentally sound technologies and know-how to facilitate implementation of adaptation and mitigation activities, along with the development and enhancement of existing knowledge and technologies in developing countries.[99] Each of those institutional arrangements has significant applicability to adaptation measures for agriculture and forestry.

The two aspects of the UNFCCC regime that have been most relevant to agriculture and forestry adaptation thus far have been the CDM and the program for reducing emissions from deforestation and forest degradation (REDD), or its more recent iteration adding forest conservation, sustainable forest management, and enhancement of carbon stocks in forests (REDD+). Both programs embrace a payment for ecosystem services (PES) approach to climate change mitigation that provides opportunities to address mitigation and adaptation in complementary ways. However, both pose significant problems faced by most PES programs in assuring that projects produce real and lasting environmental benefits, due to issues of additionality, leakage, permanence, and measurement and verification.[100]

'Additionality' refers to the requirement that a project obtain benefits that would not have occurred absent project implementation, such as investments in improved irrigation efficiency that would have occurred anyway due to market forces. 'Leakage' is the loss of benefits achieved by a project because of changes or increases in activities elsewhere to

---

97 The Cancun Agreements, UNFCCC Decision 1/CP.16, Pt. II. See also Chapter 2, section 2.4.

98 The Cancun Agreements, UNFCCC Decision 1/CP.16, Pt. IV.A (Finance).

99 The Cancun Agreements, UNFCCC Decision 1/CP.16, Pt. IV.B (Technology development and transfer).

100 Paulo A Lopes, 'Is REDD Accounting Myopic? Why Reducing Emissions from Deforestation and Forest Degradation Programs Should Recognize Other Ecosystems and Services Beyond CO_2 Sequestration' (2011) 11 Sustainable Development Law & Policy 25.

meet the same demand, such as logging a different area of forest than the one that is preserved. 'Permanence' means the requirement that benefits remain in place after the initial project period, for example, assurance that a restored forest is not logged at a later time. 'Measurement and verification' is the ability to quantify or reasonably estimate actual project benefits, such as the amount of carbon storage obtained by improving soil conditions or by planting cover crops, and the ability to verify that project actions are in fact implemented, especially because single management changes can have multiple and unexpected GHG impacts depending on soil and climate conditions. The rules governing implementation of the CDM and REDD programs attempt to address each of those issues in order to assure that benefits are achieved, and to provide investors sufficient certainty to encourage project investment. Likewise, the manner in which payments are structured will affect the allocation of risk between sellers and buyers of project credits.[101]

The CDM established in the Kyoto Protocol has always been available conceptually for agricultural projects that can generate GHG reduction credits, but the percentage of credits that could be used through land use, land use changes and forestry changes was capped at 1 percent for the 2008–2012 commitment period. Especially for countries not party to the Protocol, especially the United States, similar efforts can be undertaken through voluntary markets, which can be governed by third party standards rather than the rules adopted for the CDM.[102]

Some agricultural mitigation projects could cut against adaptation goals, for example, where agricultural crops are used as biofuels rather than to increase food production, especially in nations with food security problems. Other kinds of agricultural mitigation projects could enhance adaptation efforts concurrently with mitigation goals by improving soil fertility. Examples include conversion to higher-yielding crop varieties, planting cover crops between perennial tree and other crops, and crop rotation with nitrogen-fixing legumes.[103] A recent study for the World Bank, however, found that 'most projects have to do with agriculture as a source of bio-energy, and that few projects approved by the CDM Board tap the mitigation potential associated with changing how agricultural

---

[101] Donald F Larson, Ariel Dinar and J Aapris Frisbie, *Agriculture and the Clean Development Mechanism*, World Bank Development Research Group, Policy Research Working paper 5621, April 2011.

[102] Katherine Hamilton, Milo Sjardin, Molly Peters-Stanley and Thomas Marcello, *Building Bridges: State of the Voluntary Carbon Markets 2010* (Ecosystem Marketplace and Bloomberg New Energy Finance 2010).

[103] Larson et al., above note 101 at 3.

lands are used', reflecting more than 80 percent of projects and mitigation benefits.[104] In part, this is because soil sequestration of carbon is excluded from the CDM due to perceived problems with measurement and verification. Thus, there is significant room to increase the percentage of CDM projects that would simultaneously benefit mitigation and adaptation. Given that developed countries have committed significant funding to assist more vulnerable nations in adaptation efforts, one potential incentive to maximize those dual benefits would be to credit a percentage of an investment of either public or private capital in a CDM project with concurrent adaptation benefits toward the investing nation's commitment of adaptation funding.

The range of forestry-related projects subject to the Kyoto regime has evolved over time. Article 3(3) of the Kyoto Protocol limited forestry activities that would be counted in the GHG accounting system to afforestation, reforestation and deforestation. At subsequent negotiations, activities such as re-vegetation and improved forest management were added to recognized measures by individual countries, and in the Bali Action Plan adopted at COP13, and at COP16 in Cancun, the parties agreed to work toward establishment of the improved accounting mechanisms and rules necessary to account properly for emissions reductions from improved forestry practices, for potential inclusion in a post-Kyoto climate regime. The result has been a cooperative UN program to develop and promote a comprehensive REDD regime that includes rules and decision processes sufficient to satisfy the accountability requirements discussed above (additionality, leakage, permanence, and measurement and verifiability) while still allowing countries to maximize the sustainability and climate change adaptation attributes of REDD projects.[105] Especially given the longer timeframes applicable to forestry relative to row crop agriculture or grazing, payment structures that allocate the risk of project failure properly are even more important to the functionality of the REDD program.

Moreover, as is true for agriculture, REDD projects designed primarily to serve mitigation goals could either run counter to adaptation goals or generate synergistic mitigation and adaptation efforts. Efforts to prevent deforestation might impede efforts to increase food productivity to offset losses in other areas or lower production per acre by increasing the amount of land under cultivation for row crops or for grazing. However,

---

[104] Ibid. at 2, 8.

[105] UN Collaborative Programme on Reducing Emissions from Deforestation and Forest Degradation in Developing Countries (UN-REDD, FAO, UNDP, UNEP Framework Document, June 2008).

reforestation or improved forest management can support adaptation efforts by improving watershed health and thus enhancing water supplies and reliability, preventing soil erosion, increasing biodiversity, and potentially maintaining or preserving other sources of food produced by forests. The ultimate rules and procedures adopted in the REDD process and any subsequent multilateral agreements will have to address those tradeoffs.

It should be noted that a number of other multilateral agreements and other international regimes have attempted to address the problem of deforestation, including the 1985 Tropical Forestry Action Plan,[106] the 1992 Convention on Biological Diversity,[107] the International Tropical Timber Organization,[108] private initiatives such as the Forest Stewardship Council,[109] or voluntary agreements promoted by the UN Forum on Forests.[110] Commentators have decried the limited degree of success those systems have generated in slowing rates of deforestation.[111] However, any system of climate change mitigation and adaptation for forestry considered in the REDD or subsequent processes will have to account for, and be consistent with, those existing legal requirements.

4.1.2 International trade law and policy

International trade law and policy will increasingly need to accommodate shifting patterns of trade in agricultural products as production increases in some regions and declines in others.[112] Reductions in agricultural productivity in one nation can be offset by increased production in others only if trade law allows the free flow of goods between those countries, which can be facilitated through reduced tariffs, subsidies or other trade barriers. Indeed, early studies predicting that climate change would not

---

[106] UN FAO, Tropical Forestry Action Plan (1985).

[107] Convention on Biological Diversity (Rio de Janeiro 5 June 1992), 31 ILM 818, entered into force 29 December 1993.

[108] See http://www.itto.int, accessed 16 October 2012.

[109] See http://www.fsc.org, accessed 16 October 2012.

[110] See http://www.un.org/esa/forests, accessed 16 October 2012. Non-Legally Binding Instrument on Sustainable Management of All Types of Forests, UNFF 2007 A/C.2/62/1.5 (22 October 2007).

[111] William Boyd, 'Ways of Seeing Environmental Law: How Deforestation Became an Object of Climate Governance' (2010) 37 Ecology Law Quarterly 844.

[112] Ulrich Hoffmann, *Assuring Food Security in Developing Countries Under the Challenges of Climate Change: Key Trade and Development Issues of a Fundamental Transformation of Agriculture* (United Nations Conference on Trade and Development, Discussion Paper No. 201, February 2011) 4.

have significant adverse effects on food security (which have now been questioned on multiple grounds) relied in part on assumptions about a relatively free global trading system, which have not occurred consistently in recent years.[113]

To date, the relationship between international trade and climate change has focused largely on mitigation rather than adaptation.[114] UNFCCC includes a general aspiration that measures to combat climate change should not be used as 'a means of arbitrary or unjustifiable discrimination or a disguised restriction on international trade',[115] and the Kyoto Protocol similarly urges parties to implement the Protocol in ways that 'minimize effects on international trade'.[116] Yet no more specific agreement has translated those aspirations into specific measures. To some degree this is a chicken and egg problem, as World Trade Organization (WTO) officials and negotiators prefer to wait for a comprehensive agreement to replace the expiring Kyoto Protocol before addressing the relationship between the climate change regime and trade law.[117]

One major, longstanding market distortion occurs due to the high levels of subsidies given to agriculture by developed countries, which in turn stimulates exports to developing nations. The average level of support provided to farmers in developed countries between 2003 and 2005 was 30 percent of the gross value of farm receipts, which cost developing countries an estimated USD 17 billion per year.[118] Subsidies at these levels mask the true cost of production, making it easy for producers in developing nations to export surplus crops to countries with lower production. Continued subsidies at those levels will exacerbate the costs to developing nations to offset predicted declines in agricultural production due to climate change. Moreover, to the extent that imports from developed nations still undercut small local farmers, an even higher

---

113 Gerald Nelson et al., *The Role of International Trade in Climate Change Adaptation* (International Food and Agricultural Trade Policy Council and International Centre for Trade and Sustainable Development, Issue Brief No. 4, 2009).

114 Gary Clyde Hufbauer and Jisun Kim, *The WTO and Climate Change: Challenges and Options* (Center for Trade and Economic Integration, Conference Draft 2009).

115 Art. 3(5).

116 Art. 2(3).

117 Ad Hoc Working Group on Trade and Climate Change, *From Collision to Vision: Climate Change and World Trade, A Discussion Paper* (World Economic Forum, November 2010).

118 Hoffmann, above note 112 at 28.

percentage of a nation's food supply will come from external sources, thus increasing vulnerability to food shortages.

Developing countries are also significant net importers of agricultural inputs such as fertilizers and other agricultural chemicals.[119] Shifts to organic fertilizers and mixed production systems that reduce the need for pesticides and other chemicals will reduce input costs while improving agricultural resilience. Yet some countries continue to subsidize inputs such as fertilizer and energy use, which are likely to reduce resilience to climate change, relative to public investments in research and development (R&D), water infrastructure and other public goods that will enhance adaptive capacity.[120]

On the other hand, it may be useful or necessary for countries to adopt subsidies or other incentives to encourage or enable farmers to invest in agricultural adaptation methods. Given the expiration of the WTO exemption for subsidies designed to promote environmental practices,[121] it will be challenging for countries to do so in ways that do not risk violation of WTO or other trade rules. Efforts to promote adaptation to the agricultural effects of climate change might be considered protectionist and therefore to violate WTO rules, as have some efforts to mitigate climate change.[122] Absent a renewal or reformulation of an environmental exemption, one possible solution is to promote adaptation through devices such as land tax exemptions or access to low-cost credit rather than direct subsidies.[123]

Similarly, tariffs or other trade barriers might impede the transfer of sustainable agricultural technologies.[124] Particularly where food security or production of other essential life support goods and services are at stake, WTO negotiations should strive to strike a balance between legitimate efforts to promote climate change adaptation and misuse of those incentives to acquire an unfair competitive trade advantage. The 2001 Doha Declaration launched a negotiating effort to reduce or eliminate tariffs and non-tariff trade barriers to environmental goods

---

[119] Hoffmann, above note 112 at 23.

[120] Ibid.

[121] World Economic Forum, above note 117 at 14.

[122] World Economic Forum, above note 117 at 9–11; Hufbauer and Kim, above note 114 at 5.

[123] Hoffmann, above note 112 at 24 (citing Herren et al. 2011).

[124] World Economic Forum, above note 117 at 13.

and services,[125] but little progress has been made in achieving an accord.[126]

A second key factor is the ability of countries facing increased demand for food due to declining production to pay for the imports necessary to offset those losses. Even without climate change, models predict a 40–72 percent increase in the price of key agricultural commodities (rice, maize, soybeans, wheat) by 2050 due to population and income growth globally; but those price increases will be significantly larger due to climate change.[127] The combination of these price increases and changes in supply and demand due to the production impacts of climate change are expected to impose significant economic burdens on developing countries, especially in Africa, south and east Asia, the Pacific, and to a lesser degree Latin America and the Caribbean. For the developing world as a whole, economic models predict increases in the value of net food imports rising as much as 361 percent under the drier climate change scenarios.[128] Trade will only be able to offset food shortages partially under those circumstances, leading to a 20 percent increase in child malnutrition (about 25 million additional malnourished children).[129]

4.1.3 International intellectual property rights

Similarly, to adapt to the impacts of climate change on agriculture and forestry the most vulnerable nations will need sufficient access to new or modified agricultural technologies and practices, the intellectual property rights to which may lie in individuals or businesses in developed nations. Most investment in biotechnology research is in the private sector, with concentrated intellectual property rights (IPRs) in relatively few large companies.[130] On the one hand, there will be little incentive for private investment in the development of more sustainable agricultural methods and products absent appropriate protection of IPRs.[131] Small farmers,

---

125 World Trade Organization, Ministerial Declaration (Doha, 20 November 2001) at 31(iii), WT/MIN(01)DEC/1, http://www.wto.org/english/thewto_e/minist_e/min01_e/mindecl_e.htm, accessed 16 October 2012.

126 'Dead Man Talking, Ten Years of Trade Talks Have Sharpened Divisions Not Smoothed Them' (*The Economist* 28 April 2011); Hufbauer and Kim, above note 114.

127 Nelson et al., above note 113 at 8–9.

128 Ibid. at 12–14.

129 Ibid. at 14–15.

130 Hoffmann, above note 112 at 21. As of 2004, for example, over 40% of US biotechnology patents were held by five firms, and only 24% were in the public sector. Ibid. at 28, Table 4.

131 World Economic Forum, above note 117 at 13.

however, particularly in developing countries with low incomes and an extremely high percentage of smallholders, are less likely to be able to afford investments necessary to use innovations such as drought-resistant, heat tolerant, pest-resistant or higher yield crops, especially if prices are high or supplies are limited due to the lack of competition ensured (for prescribed periods of time) by patents and other IPRs.

In general, IPRs are governed at the international level by the WTO Agreement on Trade-Related Aspects of Intellectual Property Rights (TRIPS),[132] although TRIPS itself incorporates several pre-existing international accords on IPRs.[133] The objective of TRIPS recognizes the basic tradeoff described above in the context of climate change and agriculture: The protection and enforcement of IPRs should contribute to the promotion of technological innovation and to the transfer and dissemination of technology, to the mutual advantage of producers and users of technological knowledge and in a manner conducive to social and economic welfare, and to a balance of rights and obligations.[134]

To accomplish this objective, TRIPS requires a member to afford the same IPR protection to the nationals of other members as it does for its own nationals ('national treatment')[135] and as it does for the nationals of any other nation ('most-favored-nation treatment'),[136] and to adopt and implement fair and expeditious procedures to enforce those rights.[137] Moreover, although TRIPS specifically excludes from patentability 'plants and animals other than micro-organisms, and essentially biological processes for the production of plants or animals', and inventions deemed necessary 'to protect human, animal or plant life or health or to

---

[132] Formally, the TRIPS Agreement is Annex 1C to the Agreement Establishing the World Trade Organization, adopted during the Uruguay Round negotiations and signed at the Marrakesh Ministerial Meeting in April 1994.

[133] Art. 2. Existing agreements of the World Intellectual Property Organization (WIPO) incorporated into TRIPS include the Paris Convention for the Protection of Industrial Property, the Berne Convention for the Protection of Literary and Artistic Works, the International Convention for the Protection of Performers, Producers of Phonograms and Broadcasting Organizations (the Rome Convention) and the Treaty on Intellectual Property in Respect of Integrated Circuits.

[134] Art. 7.

[135] Art. 3.

[136] Art. 4.

[137] Art. 41.

avoid serious prejudice to the environment', it does require patents or other equivalent protections for plant varieties.[138]

At the same time, TRIPS requires that developed countries 'shall provide incentives to enterprises and institutions in their territories for the purpose of encouraging technology transfer to least-developed country Members',[139] which logically includes technologies and practices useful in adapting agriculture and forestry to climate change. Despite use of the mandatory term 'shall', this provision is far less specific than the provisions requiring protection of IPRs, making it difficult to predict how useful it will be in encouraging the timely transfer of agricultural and forestry adaptation products and technologies. However, TRIPS allows exceptions where a patent owner abuses its rights by refusing to provide a technology on 'reasonable terms and conditions', under conditions that properly respect the IPRs, and waives even those requirements 'in the case of a national emergency or other circumstances of extreme urgency'.[140] Moreover, TRIPS members may 'adopt measures necessary to protect public health and nutrition, and to promote the public interest in sectors of vital importance to their socio-economic and technological development' where otherwise consistent with the Agreement.[141] The fact that various UNFCCC provisions regarding IPRs call for additional measures to ensure technology transfer in the climate change context underscores the ambiguity in whether the various sources of flexibility in TRIPS will suffice to allow adequate and affordable transfer and use of agricultural and forestry innovations to help adapt to changing climatic conditions.

To the extent that TRIPS and other aspects of international intellectual property (IP) law do not foster sufficient development and transfer of agricultural and forestry adaptation technologies, one solution is to increase public investment in those technologies, with requirements to retain any resulting IPRs in the public domain. In recent decades, public investment in agricultural R&D has declined or remained stagnant in most developing nations,[142] in part because of food surpluses and resulting low commodity prices. Some evolving agricultural practices, however, such as agroecology, are not even amenable to patents or other IPRs, and therefore are more likely to be supported by public than by

---

[138] Art. 27. New plant varieties may also be covered by the conventions adopted by the International Union for the Protection of New Varieties of Plants.

[139] Art. 66(2).

[140] Art. 31(b).

[141] Art. 8(1).

[142] Hoffmann, above note 112 at 25, Table 2.

private investment. Another possible solution is to target financial assistance to end users who might most productively use new privately developed and owned biotechnologies so that they become more affordable, or to use available public aid funding to purchase licensing rights for use by farmers who might not otherwise be able to afford them.

4.2 Domestic Laws and Policies

An even wider range of domestic laws and policies are relevant to climate change adaptation in the agricultural and forestry sectors. It is not possible to address all of them here, or to address any of them comprehensively, particularly given that legal systems or contexts for these legal issues vary so widely among countries. The following is a very brief summary of the most important areas of law nations should evaluate to provide the appropriate support for adaptation efforts.

4.2.1 Agricultural laws and subsidy policies

Given the importance of the agriculture and forestry sector to individual livelihoods and to provide fundamental life support goods and services, most nations have an incentive to adopt laws and policies designed to support and promote production of food and fiber products and, more recently, to encourage production of biofuels.[143] The nature and detail of those laws, however, may be instrumental in either supporting or impeding climate change adaptation efforts.

The United States, for example, has a long history dating to the 1930s of national agricultural laws designed to support agricultural markets in light of frequent shifts in weather and market conditions and the inherent risks those conditions pose to agriculturalists.[144] Over various periods of time, those policies have included price supports and deficiency payments when prices fell below prescribed levels, subsidized crop insurance, drought and other disaster relief policies, subsidized inputs (such as feed and seed), low-interest loans, and conservation incentives to protect erodible or otherwise sensitive soils, wetlands, and other components of agricultural and natural ecosystems. In recent years, US farm policy has shifted from payments tied to production and surplus levels to payments based on historical acreage planted for designated commodity crops.[145]

[143] Angelo, above note 70.

[144] Adler, above note 76; US Department of Agriculture, Economic Research Service, Agriculture Information Bulletin No. 485: History of Agricultural Price-Support and Adjustment Programs, 1933–84 (1984).

[145] Adler, above note 76 at 260–261.

Although designed to reduce incentives for overproduction and to allow farmers to fallow lands when economic or other conditions are not suitable for planting, and to ensure that the agricultural sector receives sufficiently stable income to remain healthy during adverse economic or physical conditions, those policies are relatively decoupled from the incentives that may be needed under a changing climate. Moreover, because they reduce the risk of poor planting decisions, in the long run they could encourage farmers to continue risky decisions as conditions become warmer and, in some places, considerably drier.

The kinds of future agricultural laws and policies chosen by individual nations to facilitate adaptation to climate change will vary considerably depending on whether growing conditions become more or less favorable, and will reflect a balance between maintaining or increasing production of agricultural products. In regions in which growing conditions improve or expand, policies might be designed to encourage surplus production to support export markets to less fortunate regions, in ways that promote fair prices and reduced trade barriers. Those policies, however, should avoid overproduction, especially at the expense of long-term soil fertility or other attributes of sustainable agriculture. In regions in which conditions will deteriorate due to increasing aridity, higher temperatures, or increased pests or weeds, legal incentives should incentivize more sustainable agricultural methods designed, for example, to protect and make the most efficient use of water supplies, enhance and maintain soil conditions, and take advantage of mixed agricultural systems, including mixed cropping and livestock systems, that enhance local and national food security.[146]

Two related issues involving public investment in adaptation measures for agriculture and forestry are the total amount of public investment and the ways in which those investments are targeted. Existing investments might reap better returns if spent on the most effective strategies, while even large increases in public spending might be ineffective if devoted to the wrong ends. In response to the food price shocks of the late 1990s, foreign direct investment in agriculture skyrocketed from an average of USD 600 million per year in the 1990s to USD 3 billion annually from 2005–2007.[147] Public investment, on the other hand either remained level or declined during that period, and existing public investments subsidize private input goods (such as chemical fertilizers and pesticides and other

---

146 Hoffmann, above note 112 at 14–18.

147 United Nations Conference on Trade and Development, *World Investment Report 2009. Transnational Corporations, Agricultural Production and Development* (UNCTAD 2009).

capital-intensive farming methods) rather than public infrastructure, research and education, or sustainable farming methods.[148] New or increased public investments in agriculture, of course, must comply with applicable international trade rules discussed above, while supporting adaptations needed in light of climate change.

The manner in which public funds are invested will also affect the benefits obtained. For example, studies indicate that public subsidies for farmers to purchase private goods, such as pesticides and fertilizers, do not yield as significant increases in productivity as do investments in public goods, such as educational programs, local roads, and research and development in sustainable agriculture.[149] Payments for environmental services, reflected in the international climate change legal and financial regime through CDM and REDD, among others, can also be used at the domestic level to stimulate adaptation measures.[150]

4.2.2 Land tenure and land use law

According to one analysis, between 60–70 percent of farms in the world are run by people with no contractual or other legally recognized land rights.[151] Moreover, international investors have been increasingly purchasing or entering into long-term leases in developing countries, in some cases displacing or generating tenure conflicts with traditional residents with customary or other land use 'rights' that lack formal legal recognition.[152] Yet both mitigation and adaptation strategies related to climate change benefit from, if not require, attention to domestic laws and policies regarding secure land tenure.

First, it is difficult to stimulate either capital or labor-intensive investments to generate long-term improvements in land sustainability by those who lack long-term property rights or other forms of land tenure. Some strategies, such as planting drought-resistant or heat-resistant annual crops may be economically viable even with short-term or unsecure land tenure. Longer-term investments, however, such as planting trees or new varieties of perennial crops, building new irrigation systems, erosion control methods, and diversification of agricultural systems (integrated

148 Hoffmann, above note 112 at 23–27.

149 De Schutter, above note 1 at 16–17.

150 Hoffmann, above note 112 at 27.

151 Hoffmann, above note 112 at 24 (citing Parsons 2010).

152 Lorenzo Cotula, Sonja Vermeulen, Rebeca Leonard and James Keeley, *Land Grab or Development Opportunity? Agricultural Investment and International Land Deals in Africa* (FAO, International Fund for Agricultural Development and International Institute for Environment and Development 2009).

cropping and livestock systems, etc.) have longer timeframes for returns on investment, making secure land rights of some form essential.[153] Similarly, PES programs such as REDD, CDM, and similar domestic efforts are likely to benefit those with secure land tenure, and measures adopted under those programs that require fallowing of land for carbon sequestration purposes might displace existing users who have relied on those resources.[154]

Although property rights systems vary greatly around the world, and while international law is not likely to dictate domestic land rights and other property law regimes, the Food and Agriculture Organization (FAO) has adopted a set of voluntary guidelines regarding land tenure, with some focus on climate change and a strong focus on assuring adequate food security for all people.[155] The guidelines establish principles for recognition, governance and protection of legitimate tenure rights and rights of access to public lands and resources, including those based on traditional or customary systems. However, they also recognize and accommodate the need to promote external investment as a strategy for improving food security, and propose means of balancing them against the needs of smallholders and other existing users.

Agricultural adaptation strategies will need to balance competing land uses and livelihood patterns, in addition to the conflicts between agricultural land and forests discussed above. For example, many parts of the world are already urbanizing rapidly, and that trend is expected to continue in the forthcoming decades. Loss of farmland to development at the urban fringe will reduce food production close to population centers, which will complicate transport and increase energy costs of providing food, especially fresh produce, to areas of high demand. Migration of existing agricultural workers to cities might result in insufficient numbers of workers to support changing farm economies, or require more mechanization, with accompanying capital and energy costs. Expansion of cultivated land to meet rising demand for agricultural products will also compete with other land uses, including ecosystem preservation and restoration uses such as wetland protection and restoration. That competition will require attention through land use control and management legal systems, and both the stability and transferability of property rights,

---

[153] CGiAR Systemwide Program on Collective Action and Property Rights, Policy Brief No. 7 (February 2010).

[154] Ibid.

[155] FAO, Voluntary Guidelines on the Responsible Governance of Tenure of Land, Fisheries and Forests in the Context of National Food Security (FAO 2012).

properly balanced against public needs, will play a key role in those shifting land uses.

4.2.3 Water rights and environmental quality

Water law is addressed more comprehensively in Chapter 9, but is addressed briefly here because it is so fundamental to agriculture and forestry. Agriculture and forestry adaptation strategies will, in many cases, involve arid regions that rely on water collection, storage and distribution for municipal and industrial as well as agricultural uses. Dwindling supplies will generate legal and other conflicts over water allocation, which will have to be resolved through existing, modified or new systems of water rights. The certainty of those legal regimes will affect the ability to invest in water development and distribution systems, and the degree to which those rights are marketable. Although it may be critical to allocate enough water to support transitions from rain-fed to irrigated agriculture, in arid regions it will be equally important to provide sufficient water for basic municipal drinking water and other human uses, and to support other water-intensive economies, such as manufacturing, or in-stream economic uses such as recreation and tourism. At the same time, significant environmental harm could occur if aquatic ecosystems are dried up through excessive water withdrawals. Thus, legal regimes will need to resolve conflicts between these competing uses of water, in ways that encourage efficient use and demand-side as well as supply-side responses.

The adaptation law and policy regime will also need to consider tradeoffs between water and energy needs, and consequently between related adaptation and mitigation strategies. Some forms of energy generation designed to reduce GHG emissions, such as hydroelectric or nuclear power, require more water to be left in rivers to power turbines or to cool the power plants, respectively. That water will not be as readily available for agricultural uses, unless power generation occurs upstream of irrigation needs. Systems of integrated land and water use planning may be needed to coordinate multiple uses of water.

Adaptation strategies for agriculture and forestry also might jeopardize other environmental protection needs. Increased agricultural production to feed, clothe and shelter a growing world population, even without climate change, inevitably will increase the already significant human health and environmental impacts of agriculture, including impacts on ecosystems and ecosystem services, particularly if additional measures are not taken to mitigate those effects. Adaptation strategies discussed above could significantly increase those impacts. Expansion of row crop agriculture will risk increased soil erosion, water pollution and other environmental effects. Increased mechanization, especially if not employed properly, can increase erosion and localized air pollution. Increased agricultural inputs, including fertilizers and pesticides, can pollute both surface and ground water and generate health risks to farm workers and water users. Water transfers and irrigation systems can deplete aquatic ecosystems and cause soil salinization, reducing long-term agricultural viability. Either those impacts need to be addressed as a fundamental part of agricultural adaptation planning, or existing, revised or new environmental laws and policies will have to be integrated into those used to support adaptation.

Similarly, existing legal regimes are struggling to deal with the increased use of GMOs even absent climate change. GMOs could significantly enhance the ability of farmers to adapt crops to a changing climate by increasing drought tolerance, heat tolerance, etc. However, considerable dispute persists regarding the potential risks that GMOs may pose to human health and the environment, and climate adaptation law will have to consider how to balance those risks and benefits. Likewise, GMOs present particular challenges for intellectual property law, which can be assessed as part of the overall challenges (addressed above) associated with technology transfer as a strategy for climate change adaptation.

5 CONCLUSION

A large range of international and domestic legal systems will be involved in efforts to facilitate climate change adaptation in the agriculture and forestry sectors. No single set of laws and policies is presumptively 'correct' in this area, for several related reasons. First, future conditions for farms, forests and grazing will vary dramatically around the globe, meaning that solutions appropriate in some regions will not work in others. Second, the world will not see a simple shift from one existing climate 'condition' to another; rather, the climate will change sequentially and, at least in some respects, unpredictably. Third, climate change adaptation for agriculture and forestry must address a series of potentially conflicting objectives, such as mitigation versus adaptation and maintaining production versus maintaining or enhancing carbon sinks. As a result, legal systems designed to facilitate adaptation in agricultural and forestry communities and economies should be designed to be flexible and responsive to changing physical, social, economic and political circumstances.

9. Climate change adaptation and water law

Jonathan Verschuuren

1 INTRODUCTION

This chapter deals with adaptation in water (management) law, focusing primarily on floods and droughts in fresh water systems. I will explore whether current water law is sufficiently equipped to meet the challenges of climate change. In other words, through a smart use of existing water laws, can the necessary adaptation measures in the field of fresh water management be carried out, or do we need an adaptation of the law? After a brief overview of the impacts of climate change on fresh water resources in section 2, these questions will be answered through an analysis of the basic elements of international and domestic water law: the right to water, the principle of reasonable and equitable use, and the river basin approach (sections 3, 4 and 5 respectively). Under these basic elements, most water adaptation issues, such as floods and droughts, stormwater management and drinking water supply will be dealt with. Before drawing some conclusions and deciding whether the law needs to be changed (and if so, how) in section 7, the relationship between mitigation and adaptation in the field of water law is briefly discussed in section 6.

This chapter does not address coastal adaptation and coastal flooding, as this is dealt with in Chapter 10 on marine and coastal adaptation. This chapter also does not specifically deal with water as a habitat for biodiversity resources, as that is discussed in Chapter 11 on biodiversity. Finally, it should be noted that water is an important issue in some of the other chapters as well, particularly Chapter 8 on agriculture and Chapter 12 on planning law.

2 IMPACTS OF CLIMATE CHANGE ON FRESH WATER RESOURCES

The impacts of climate change on freshwater systems are so vast that they are almost impossible to grasp (see Box 9.1).

BOX 9.1

The Intergovernmental Panel on Climate Change, in its 2007 Fourth Assessment Report, summarizes the impacts as follows:[1]

> More than one-sixth of the world's population live in glacier- or snowmelt-fed river basins and will be affected by the seasonal shift in streamflow, an increase in the ratio of winter to annual flows, and possibly the reduction in low flows caused by decreased glacier extent or snow water storage. Sea-level rise will extend areas of salinisation of groundwater and estuaries, resulting in a decrease in freshwater availability for humans and ecosystems in coastal areas. Increased precipitation intensity and variability is projected to increase the risks of flooding and drought in many areas ...
>
> Many of the semi-arid and arid areas (e.g., Mediterranean basin, western USA, southern Africa, and north-eastern Brazil) will suffer a decrease in water resources due to climate change. Efforts to offset declining surface water availability due to increasing precipitation variability will be hampered by the fact that groundwater recharge will decrease considerably in some already water-stressed regions, where vulnerability is often exacerbated by the rapid increase in population and water demand.
>
> Higher water temperatures, increased precipitation intensity, and longer periods of low flows exacerbate many forms of water pollution [such as sediments, nutrients, dissolved organic carbon, pathogens, pesticides, salt, and thermal pollution], with impacts on ecosystems, human health, water system reliability and operating costs ...

Note:

1 M L Parry et al. (eds), Climate Change 2007: Impacts, Adaptation and Vulnerability: Contribution of Working Group II to the Fourth Assessment Report of the Intergovernmental Panel on Climate Change (Cambridge University Press 2007) 175.

According to the IPCC, current water management practices are very likely to be inadequate to reduce the negative impacts of climate change on water supply reliability, flood risk, health, energy, and aquatic ecosystems. Improved incorporation of current climate variability into water-related management would make adaptation to future climate change easier.[1] Across the globe, authorities involved in water management are discussing what such incorporation means in practice.

The first and foremost theme discussed at the World Water Forum – the world's most important international meeting on water issues, with more than 25,000 participants including many heads of state and ministers – during its 5th session held in Istanbul in 2009, was climate change adaptation. In the Ministerial Declaration adopted at that conference, it was stated that transnational, national, and sub-national plans and programmes have to be developed and implemented 'to anticipate and address the possible impacts of global changes'.[2] The Declaration especially mentioned assessments of varying hydrological conditions, extreme water events and the shape and functionality of existing infrastructure as being essential, and it stated that investment efforts were needed to establish necessary infrastructure and to increase storage and drainage capacity.[3] The Ministerial Declaration adopted during the 5th World Water Forum then continued with a section on extreme water events, like floods and droughts, stating that we have to proceed from crisis management to disaster preparedness and prevention of human-induced disasters, as well as risk management, by developing early warning systems, implementing structural and non-structural measures, both for water resources and access to water and sanitation, and building capacity at all levels; further stating that necessary post-disaster mitigation and rehabilitation measures for affected people and hydrological systems should be taken as well.[4]

In specific river basins, the consequences of climate change have already been witnessed. A 2007 assessment of the transboundary rivers, lakes and groundwaters in the UN Economic Commission for Europe (UNECE) region, i.e., the countries of Europe, plus Canada, the United States, Israel, and several countries in Central Asia (Kazakhstan, Kyrgyzstan, Tajikistan, Turkmenistan, and Uzbekistan), found the effects of climate change to be visible in almost all the analysed river basins. According to the assessment:

---

1 M L Parry et al. (eds), *Climate Change 2007: Impacts, Adaptation and Vulnerability: Contribution of Working Group II to the Fourth Assessment Report of the Intergovernmental Panel on Climate Change* (Cambridge University Press 2007) 175.

2 Note that the word 'climate change' is avoided here, showing the sensitivity of the topic in an international diplomatic context in 2009.

3 Istanbul Ministerial Declaration, 22 March 2009, at 9, http://www.worldwatercouncil.org, accessed 17 October 2012.

4 Ibid. at 10.

> Most basins experience an impact of climate change on water quantity (e.g. decreasing water resources availability and extreme hydrological events, including severe floods and long-lasting droughts). With a reduction in precipitation of up to 30% over the last decade, water resources availability, for example, is decreasing in river basins in the discharge area of the Mediterranean Sea. The effects of climate change on the ecological regime of rivers are also becoming visible in transboundary basins in Central Asia, where the rise in air temperatures leads to significant melting of glaciers, resulting in noteworthy changes of the rivers' hydrological and ecological regimes. Thus, climate change adaptation measures in water management and water-depended activities and services (e.g. agriculture, forestry, water supply, hydropower generation) are needed in the entire UNECE region.[5]

The same region saw an increase in the average number of annual disastrous weather and climate-related events by about 65 per cent between 1998 and 2007,[6] many of which impacted water and wastewater services systems. The most costly climate change impacts in western, northern and eastern Europe are projected to be river floods.[7]

3 THE RIGHT TO WATER

Perhaps the most basic element of water law is the right to water. After the existence of the right to water as an individual human right had been disputed in the international arena for many years,[8] the legal recognition of such a right is now a fact. In 2010, the UN General Assembly, as well

---

[5] UNECE, Our Waters: Joining Hands across Borders: First Assessment of Transboundary Rivers, Lakes and Groundwaters (New York and Geneva: UNECE 2007) 29.

[6] European Environment Agency, *Impacts of Europe's Changing Climate, 2008 Indicator-based Assessment* (EEA 2008) 169.

[7] European Environment Agency, *Climate Change, Impacts and Vulnerability in Europe 2012: An Indicator-based Report* (EEA 2012) 18.

[8] Much of the debate during the 4th World Water Forum in Mexico in 2006, for example, dealt with the right to water. The Ministerial Declaration adopted at the conference, however, does not even mention it because no agreement on any formulation could be reached; P Martinez Austria and P van Hofwegen (eds), *Synthesis of the 4th World Water Forum* (Copilco El Bajo: Comisión Nacional de Agua 2006) 90. Progress was made during the 5th Forum in 2009, but the Ministerial Declaration shows reluctance to simply acknowledge the existing of the right to water: 'We acknowledge the discussions within the UN system regarding human rights and access to safe drinking water and sanitation. We recognize that access to safe drinking water and sanitation is a basic human need.' Above note 3 at 15.

as the UN Human Rights Council, explicitly acknowledged the existence of the right to water and sanitation as a basic human right as part of the right to health as well as the right to life.[9] The right to water has been specifically mentioned in several binding human rights documents, such as the Convention on the Rights of the Child 1989 and the Convention on the Elimination of all Forms of Discrimination against Women 1979.[10] The most prominent recognition of the right to water, however, took place under the UN International Covenant on Economic, Social and Cultural Rights 1966 (ICESCR). Article 12(1) ICESCR encompasses everyone's right to the enjoyment of the highest attainable standards of physical and mental health. To achieve the benefits of this right, states should improve all aspects of environmental and industrial hygiene, and provide for the healthy development of children.[11] In 2000, the Committee on Economic, Social and Cultural Rights (the Committee) adopted a General Comment on this human right, stating that Article 12 not only deals with health care, but also with all other factors that determine the enjoyment of good health, such as access to safe drinking water, personal hygiene, sufficient safe food and shelter.[12] In 2002, the Committee adopted General Comment No. 15, which deals entirely with the right to water and which is considered to be the most influential document on this right. In this document, the right to water has been defined as follows:[13]

> The right to drinking water entitles everyone to safe, sufficient, affordable and accessible drinking water that is adequate for daily individual requirements (drinking, household sanitation, food preparation and hygiene).

National constitutions sometimes explicitly acknowledge the right to water as well, especially in Africa. The constitutions of Gambia, Uganda,

---

[9] UN General Assembly Resolution 64/292 of 7 August 2010, UN A/RES/64/292, and Human Rights Council Decision of 24 September 2010, UN A/HRC/15/L.14.

[10] An overview of all texts that implicitly or explicitly refer to the right to water in human rights conventions is available at http://www.righttowater.org.uk, accessed 17 October 2012. For even more texts that refer to the right to water, for instance, those of the International Labour Organisation, see B Toebes, *The Right to Health as a Human Right in International Law* (Intersentia 1999).

[11] ICESCR Art. 12(2)(a) and (b).

[12] UN Committee on Economic, Social and Cultural Rights, General Comment No. 14, UN Doc. E/C.12/2000/4 (2000).

[13] UN Committee on Economic, Social and Cultural Rights, General Comment No. 15, UN Doc. E/C.12/2002/11 (2002).

Zambia, South Africa, and Ethiopia recognize the right to water. For example, Section 27 of the Bill of Rights of the South African Constitution reads:

> 1. Everyone has the right to have access to (a) health care services, including reproductive health care; (b) sufficient food and water; and (c) social security, including, if they are unable to support themselves and their dependants, appropriate social assistance.
> 2. The state must take reasonable legislative and other measures, within its available resources, to achieve the progressive realisation of each of these rights.

Subsection 2 of Section 27 makes clear that the right to water is a social right, so that government intervention is needed to achieve it.[14]

The availability of clean drinking water and water for personal hygiene can be threatened by climate change in various ways. There may be a water shortage during prolonged heat spells. Safe drinking water supply may be hampered by pollution caused by overloaded sewer systems; polluted stormwater runoff; damaged infrastructure, such as corrosion and destruction of pipes and underground water transportation systems after flooding; saline water intrusion in aquifers used for drinking water (due to sea level rise); and possibly also by an increase in water-borne diseases. Similar issues may occur in bathing waters. The water treatment facilities, as well as other relevant water management authorities, have to monitor these potential impacts closely and have to be able to address all these impacts so as to ensure adequate water supply. The World Health Organization (WHO) advocates the adoption of Water Safety Plans through which risk assessment and risk management is applied throughout the entire chain of water supply, from catchment to consumer ('source to tap approach'). In their guidelines on water supply and sanitation in extreme weather events, adopted by both the WHO and the UNECE in 2010, the WHO identifies elements of such Water Safety Plans that are considered important to make the water supply resilient against extreme events, such as:[15] creation of interdisciplinary teams of experts ('the WSP team'); carrying out a risk assessment; development

---

[14] This does not mean that this right cannot be enforced by individual citizens. The South African Constitutional Court has, on various occasions, judged that a certain government policy was contrary to certain socio-economic rights; see extensively Danie Brand and Christof Heyns (eds), *Socio-Economic Rights in South Africa* (Pretoria University Law Press 2005) .

[15] L Sinisi and R Aertgeerts (eds), *Guidance on Water Supply and Sanitation in Extreme Weather Events* (Copenhagen: WHO 2010) 52–59.

and implementation of an improvement plan; and periodic overall review by the team. The risk assessment and consequent improvements not only should be aimed at preventing damage to the water supply system, but also at regaining water supply systems after a drought or a flood.[16]

Ideally, such a comprehensive water safety approach needs to be integrated in the existing river basin management system, so that both the Water Safety Plan and the wider river basin plan work together to make the water supply system resilient. Flood prevention and management measures (see section 5.2) and drought adaptation measures (see section 5.3) have to be aligned with water safety measures, such as: measures preventing polluted water enter drinking water aquifers by, for instance, diversions to other areas; raising boundary walls at intake sites; or installing (other) physical flood barriers around a specific site that is crucial for drinking water supply. Emergency drills and exercises throughout the basin should be conducted; recovery plans should ensure that drinking water supply is safe for consumption or appropriate advice has been given to consumers.[17] Regulation is also needed on the disinfection of wells, boreholes, and domestic distribution systems by adding chlorine as part of the recovery of a water supply system after a drought or a flood. Pollution of the natural environment and the creation of health risks because of disinfection processes need to be prevented.[18] Measures like these should be part of a bigger flood management plan. In countries where water supply services have been privatized, additional legal instruments need to be in place to compel private water suppliers to deliver basic services even if climate change makes this more costly, for instance, by regulating the tariff and supply standard process.[19]

Another important reason why water safety needs to be integrated into river basin management systems is the fact that in situations of water scarcity, the fulfilment of the right to water for individual people and households may conflict with other water uses. Other human uses of rivers and aquifers are: irrigation for agriculture, the intake of water for industrial processes or for cooling processes in power generation, the discharge of waste water and cooling water, shipping, fishing, and recreation. Obviously, non-human, natural ecosystem uses are equally essential. The WHO was one of the first international organizations to

---

[16] Ibid. at 75–77.

[17] Ibid. at 77.

[18] Ibid. at 76.

[19] M G Quesada, *Water Sanitation Services in Europe: Do Legal Frameworks Provide for Good Governance*, (Centre for Water Law and Science University of Dundee 2011) 279.

notice that the right to water should be balanced in an integrated catchment policy with all other water needs, such as irrigation, power generation, and nature conservation.[20] The UNECE Protocol on Water and Health,[21] a protocol to the 1992 Convention on the Protection and Use of Transboundary Watercourses and International Lakes,[22] takes the same approach by stating, in Article 6(1), that the Parties shall pursue the aims of access to drinking water for everyone and provision of sanitation for everyone within a framework of integrated water management systems aimed at sustainable use of water resources, ambient water quality that does not endanger human health, and protection of water ecosystems. This issue will be addressed in more detail in section 5.

4 PRINCIPLE OF REASONABLE AND EQUITABLE USE

The principle of reasonable and equitable use is considered to be the most important principle in international freshwater law.[23] It is considered to be a principle of customary law, and it has been laid down in the Berlin Rules on Water Resources,[24] as well as in the 1992 Convention on the Protection and Use of Transboundary Watercourses and International Lakes[25] and the 1997 UN Convention on the Law of the Non-Navigational Uses of International Watercourses.[26]

---

[20] WHO, *Right to Water* (WHO 2003) 18–21.

[21] Protocol on Water and Health, London, 17 June 1999, Doc. MP.WAT/2000/1.

[22] Convention on the Protection and Use of Transboundary Watercourses and International Lakes, Helsinki, 17 March 1992, (1992) 31 ILM 1312.

[23] Patricia Birnie, Alan Boyle and Catherine Redgwell, *International Law & the Environment* (3rd edn, OUP 2009) 541–544; Philippe Sands and Jacqueline Peel, *Principles of International Environmental Law* (3rd edn, Cambridge University Press 2012) 305–307.

[24] ILA, Berlin Rules on Water Resources and Commentary, (2004) 71 International Law Association Conference Reports 334. The Berlin Rules replace the 1966 Helsinki Rules on the Uses of Waters of International Rivers. The latter already firmly recognized the principle of reasonable and equitable use. The Berlin Rules modernized the principle by putting it in the broader integrated river basin approach.

[25] Art. 2(2)(c).

[26] Convention on the Law of the Non-Navigational Uses of International Watercourses, New York, 21 May 1997, (1997) 36 ILM 700, Art. 5.

According to the principle, states may not use water in such a manner as to prevent or otherwise limit other riparian states from making full use of their equitable and reasonable entitlements in relation to that shared water.[27] The principle thus limits the right of upstream states to control the flow of a river, for instance through damming it, while downstream states cannot demand an unimpeded flow either. Hence, the principle is the basis for many bi- or multi-lateral agreements between riparian states around the globe. On the basis of the principle, equilibrium has been reached in most river basins on the share of the water each riparian state is entitled to. To give an example, the four riparian states of the Orange-Senqu river in southern Africa, upon the establishment of the Orange-Senqu River Commission (ORASECOM) in 2000, agreed to 'utilize the resources of the river system in an equitable and reasonable manner with a view to attaining optimal and sustainable utilization thereof, and benefits therefrom, consistent with adequate protection of the river system'.[28]

Increased demand, and decreasing water levels caused by climate change, challenge the idea that there are fixed entitlements.[29] As a consequence, tensions among states rise. A current example is the Nile river basin, where upstream states such as Rwanda, Ethiopia, Uganda, and Tanzania, challenge the share that in the past was attributed to downstream states Egypt and Sudan.[30] Thanks to Chinese investments in infrastructure in, for instance, Ethiopia, the upstream countries gain access to modern hydro technology, which enables them to use bigger portions of Nile water.[31] Under the Nile River Basin Initiative, the Nile riparian states are trying to solve the issue. Integrated river basin planning is considered to be the only instrument that would lead to the establishment of a new legal equilibrium.[32] Transboundary water agreements have to address large variations in water availability, for instance,

---

27 Sands and Peel, above note 23 at 306.

28 Agreement on the Establishment of the Orange-Senqu River Commission, signed in Windhoek on 3 November 2000, Art. 7(2).

29 Dan Tarlock, 'Four Challenges for International Water Law' (2009) 23 Tulane Environmental Law Journal 384.

30 In 2010, the upstream states concluded the Cooperative Framework Agreement, establishing the Nile River Basin Commission; A K Martens, *Impacts of Global Change on the Nile Basin: Options for Hydrological Reform in Egypt and Ethiopia* (IFPRI Discussion Paper 01052, 2011) 7, http://www.ifpri.org/sites/default/files/publications/ifpridp01052.pdf, accessed 17 October 2012.

31 Ibid. at 14.

32 Martens, above note 30 at 20; Tarlock, above note 29 at 404.

by allocating percentages of the overall flow of rivers to countries in the river basin instead of total numbers.[33] Worst case scenarios have to be developed as well. How, for instance, will the river basin countries deal with droughts? Countries can negotiate a temporary change in the original distribution, accompanied by some kind of mechanism that compensates the country that has to redistribute some of its 'own' water to the other state that is impacted by the drought.[34] Finally, periodic review and adjustment of water allocations seem to be important so as to follow the latest impacts of climate change on the river basin.[35]

5 INTEGRATED RIVER BASIN APPROACH

In the area of water management, we have the relative advantage that, almost universally, an integrated river basin, or catchment, approach is applied. Such a holistic approach to all interrelated water issues is particularly relevant to water adaptation measures.[36] International law has paved the way for such an integrated approach, most notably the UNECE Convention on International Watercourses and Transboundary Lakes and the UN Convention on the Law of the Non-navigational Uses of International Watercourses mentioned above. Both conventions require the establishment of a joint body in order to achieve a common

---

[33] UNECE, Guidance on Water and Adaptation to Climate Change, ECE/MP.WAT/30 (2009) 40.

[34] Ibid.

[35] Ibid.

[36] There is a wealth of academic papers in which this is acknowledged; see, for instance: C A Arnold, 'Adaptive Watershed Planning and Climate Change' (2010) 5(2) Environmental and Energy Law and Policy Journal 417; Richard T Kingsford, 'Conservation Management of Rivers and Wetlands under Climate Change – A Synthesis' (2011) 62 Marine and Freshwater Research 217–222; Dan Tarlock, 'Four Challenges for International Water Law' (2009) 23 Tulane Environmental Law Journal 369–408; D N Cassuto and R S R Sampaio, 'Water Law in the United States and Brazil – Climate Change & Two Approaches to Emerging Water Poverty' (2011) 35 William & Mary Environmental Law and Policy Review 371–413; Poh-Ling Tan, 'Adaptation Measures for Water Security in a Changing Climate: Policy, Planning and Law' in Tim Bonyhady, Andrew Macintosh and Jan McDonald (eds), *Adaptation to Climate Change. Law and Policy* (The Federation Press 2010) 135–156; B Cosens and M Williams, Resilience and Water Governance: Adaptive Governance in the Columbia River Basin (SSRN paper 2012), http://ssrn.com/abstract=1942587, accessed 17 October 2012.

management of international watercourses,[37] and both conventions support an ecosystem approach, i.e., an approach under which all consequences of human activities on the entire ecosystem are considered, respecting the integrity of the ecosystem as a whole.[38]

The UNECE Convention most elaborately defines the measures that have to be taken to protect transboundary water systems, such as: (a) prevention and control of pollution; (b) ecologically and rationally sound water management, conservation of water resources and environmental protection; (c) reasonable and equitable use[39] taking into account the transboundary character; (d) conservation, and, where necessary, restoration of ecosystems.[40] Several legal principles, such as the precautionary principle and the polluter pays principle, apply.[41] Under the UNECE Convention, the joint bodies have a wide range of tasks, including elaborating monitoring programmes concerning water quality and water quantity; exchanging information; elaborating emission limits for waste water and joint water-quality objectives; developing concerted action programmes for the reduction of pollution loads; and implementing environmental impact assessments.[42] A large number of guidelines are available for the application of the UNECE Convention, such as the Guidelines on Monitoring and Assessment of Transboundary Rivers. Again, these guidelines stress the need for an integrated approach. The state of the river and related ecosystem should be assessed in an

---

[37] However, the wording of the UN Convention is much weaker. The UNECE Convention requires riparian states to establish a joint body, whereas the UN Convention only requires the states to consider the establishment of a joint body.

[38] J Brunnée and S J Toope, 'Environmental Security and Freshwater Resources: A Case for International Ecosystem Law' (1994) 5 Yearbook of International Law 41, 55.

[39] Using both the principle of equitable use and the principle of prevention of harm has been criticized for their inherent upstream/downstream conflict; some authors advocate a 'needs based' approach rather than a 'rights based' approach; Heather L Beach, Jesse Hamner and J Joseph Hewitt et al., *Transboundary Freshwater Dispute Resolution: Theory, Practice, and Annotated References* (United Nations University Press 2000) 74.

[40] Art. 2(2). Again, from an environmental protection point of view, the UN Convention is much weaker. There, equitable and reasonable utilization is the only principle and ecological factors seem to be only a minor consideration. Cf. Art. 6(1).

[41] Art. 2(3). These principles are absent in the UN Convention.

[42] Art. 9(2). See in more detail on these issues, M Fitzmaurice and Olufemi Elias, *Watercourse Co-operation in Northern Europe: A Model for the Future* (TMC Asser Press 2004).

integrated manner, based on criteria that include water quality and quantity for different human uses as well as flora and fauna.[43] In 2009, the Conference of the Parties (COP) to the UNECE Convention adopted a specific guideline on adaptation, which will be dealt with below.

The basic instrument that is applied to implement a river basin approach is river basin planning. River basin planning is central to water management adaptation.

5.1 Integrated River Basin Planning

River basin planning is an essential instrument for adaptation, because it allows the competent authorities to regularly assess whether the distribution of water among the various users within the river basin is still compatible with the changes that occur because of climate change. Such an assessment will make clear which measures need to be taken to ensure that the changes that occur do not harm the long term use of water by all users (human and non-human). Such measures may have to be aimed at preventing floods and minimizing flood damage, as well as mitigating the impact of droughts, even in one geographic area.

In 2009, the COP to the UNECE Convention adopted the Guidance on Water and Adaptation to Climate Change,[44] prepared by the Task Force on Water and Climate.[45] These elaborate guidelines take an integrated river basin approach as a starting point for adaptation measures and indicate that integrated river basin planning should plan for extreme events. Flood prevention and mitigation policies can make use of a wide range of instruments, such as the designation of water retention areas that are used for controlled flooding (thus releasing the pressure on other, for instance, more densely populated areas); the widening of the river bed (thus enhancing the capacity of the river to discharge larger amounts of water to the ocean); and/or the creation and management of controlled water levels through dams and other technical measures. In cities, stormwater management is an important element of any water policy. Increased precipitation will put a greater pressure on stormwater systems,

---

43 UNECE Task Force on Monitoring & Assessment, *Guidelines on Monitoring and Assessment of Transboundary Rivers* (UNECE 2000) 10.

44 UNECE/MP.WAT/29, 10.

45 UNECE, Guidance on Water and Adaptation to Climate Change, ECE/MP.WAT/30 (2009).

which may lead to pollution if the sewage system was not adapted to deal with an increased amount of stormwater.[46] See, further, section 5.2 below.

Water scarcity and droughts have to be addressed in integrated river basin plans as well. Regulating flow regimes, altering dam operations, adapting irrigation schemes, storing water on agricultural lands for later use, and crop changes to less water intensive crops or to crops that are resistant to salination, are examples of measures that can be integrated into river basin plans. See, further, section 5.3 below.

In Europe, the Water Framework Directive provides a holistic legal framework for all water issues, i.e., water quality and water quantity issues, both of surface and ground waters, as well as coastal waters.[47] It sets the institutional framework through river basin districts, river basin management plans, and programmes of measures. The Water Framework Directive obliges EU Member States to adopt river basin management plans every six years, as well as programmes of measures detailing the specific measures that will be taken to achieve the goals of the plan. Monitoring the physical status of all waters within the river basin is required so that the river basin authorities have a comprehensive overview of the water status. Such constant monitoring will show variations that occur due to climate change, such as changes in the volume and level or rate of flow, and changes in groundwater levels. The six-year planning cycle then allows the authorities to adapt the plan and take additional or different measures. Climate change adaptation, however, is not explicitly mentioned in the Water Framework Directive. Hence, there is no explicit obligation to set the targets in such a way as to make the river basin resilient to future changes. Keessen and Van Rijswick argue that the current text of the Directive introduces too much flexibility and discretion for Member States, for instance, by allowing Member States to invoke exemptions if goals become unattainable, whereas it would be better to force the authorities to plan for climate change.[48]

---

[46] In 2010, the WHO and the UNECE issued *the Guidance on Water Supply and Sanitation in Extreme Weather Events*, above note 15.

[47] Directive 2000/60/EC establishing a framework for the Community action in the field of water policy (2000) OJ L 327.

[48] Andrea M Keessen and Helena F M W van Rijswick, 'Adaptation to Climate Change in European Water Law and Policy' (2012) 8(3) Utrecht Law Review 38, 49.

5.2 Dealing with Floods

The EU Floods Directive provides a good example of what a legislative framework to deal with increased floods as a consequence of climate change looks like.[49] The Floods Directive was explicitly adopted to address increased inland and coastal flooding caused by climate change. The Floods Directive largely uses the institutional framework offered by the Water Framework Directive mentioned above.

The Floods Directive advises EU Member States to integrate flood mitigation measures into river basin management plans and programmes of measures.[50] It introduces three steps that are to be taken to address increasing flood risks. Firstly, a Preliminary Flood Risk Assessment (PFRA) must be carried out for each river basin district, based on past flooding but taking into account long-term developments as a consequence of climate change.[51] The assessment must include maps of areas prone to flooding, as well as an assessment of the potential adverse consequences for human health, the environment, cultural heritage, and economic activity.[52] All assessments had to be completed by December 2011. The second step is to draft Flood Hazard Maps and Flood Risk Maps.[53] Flood Hazard Maps cover potentially flooded areas according to various scenarios (low/high probability, extreme events). For each scenario, Flood Hazard Maps must indicate the flood extent, water depths or water levels, and the flow velocity. Flood Risk Maps show the potential negative consequences for each of the scenarios, i.e., the number of people affected, the types of economic activity affected, the industrial installations affected (because of pollution that may spread as a cause of flooding), and the areas where sediments, debris or pollution might end up. Both sets of maps are due in December 2013. Based on the PFRA and the Maps, the third step involves adoption of Flood Risk Management Plans, aimed at the 'three P's': prevention, protection, preparedness.[54] Plans must include measures leading to a reduction in the negative consequences of flooding and in the likelihood of flooding through such devices as flood conveyance routes, flood water retention areas, and areas of controlled flooding. These measures may be

---

[49] Directive 2007/60/EC on the assessment and management of flood risks, (2007) OJ L 288.

[50] Art. 9.

[51] Art. 4(2).

[52] Art. 4(2)(d).

[53] Art. 6.

[54] Art. 7 and Annex A.

implemented through reforms to land use planning, nature conservation, and reforms to port infrastructure laws and requirements. Measures must not simply shift flood risks upstream or downstream. Flood forecasts and other early warning systems must also be set up.[55] Flood Risk Management Plans are due in December 2015. All of these instruments must be updated every six years.[56] The Floods Directive may become integrated into a wider climate change disaster management policy, which is currently being developed at the EU level.[57]

In the EU Member States, domestic law builds on both the Water Framework Directive and the Floods Directive. The Netherlands, for example, is especially vulnerable to flooding. The country basically is one large estuary of four major international rivers (Rhine, Meuse, Scheldt, and Ems), the basins of which cover large parts of Belgium, France, Germany, and Switzerland. Without the protection of dikes, dunes, and hydraulic structures, such as storm surge barriers, approximately 60 per cent of the country would be flooded regularly.[58] The Netherlands has taken a holistic approach to all water issues, in line with the EU's Water Framework Directive. The Water Act and the accompanying Water Ordinance cover all water law and implement all the relevant EU law.[59] The National Water Plan 2009–2015, with its legal basis in the Water Act, articulates the nation's overarching water policies and incorporates the four river basin management plans as required under the Water Framework Directive.[60] The Plan not only deals with coastal adaptation (see further Chapter 5), but also deals with the increased risk

---

[55] There already exists a European Floods Alert System (EFAS), but this system is only meant to inform national authorities and not the general public. The latter has to be done by the competent national authorities, who may rely on information provided by EFAS. See the EFAS website at http://floods.jrc.ec.europa.eu, accessed 17 October 2012.

[56] Floods Directive Art. 14.

[57] See the Commission staff working paper, Risk Assessment and Mapping Guidelines for Disaster Management, 21 December 2010, SEC (2010) 1626 final, and Commission Communication of 23 February 2009, 'A Community approach on the prevention of natural and man-made disasters', COM(2009) 82 final.

[58] S N Jonkman et al., 'Flood Risk Assessment in the Netherlands: A Case Study for Dike Ring South Holland' (2008) 28(5) Risk Analysis 1358.

[59] Water Act of 29 January 2009, (2009) 107 Bulletin of Acts and Ordinances; Water Ordinance of 30 November 2009, (2009) 548 Bulletin of Acts and Ordinances.

[60] The National Water Plan 2009–2015 is accompanied by a more specific policy document called 'Water safety 2009–2015'. Both documents are available

of inland flooding as a consequence of more intense precipitation across the international river basins mentioned above and increased melt-water from the Alps. The basic programme to deal with inland flooding is the so-called 'Room for the River' programme. This programme goes back to a high water event which occurred in 1995, during which 250,000 people had to be evacuated from their homes and extensive floods were only just prevented.[61] Under the Room for the River programme, water storage areas to be used for controlled flooding are designated in land use plans, and natural floodplains are expanded using a combination of land use controls and compulsory acquisition.[62] These natural floodplains were developed to deal with high water levels in the river and simultaneously to create additional wetlands under nature conservation laws.[63] Most of these policies are implemented under the Water Act, but some are undertaken pursuant to the Spatial Planning Act.[64]

The regulatory instruments of the Spatial Planning Act are used to avoid unwanted land use developments taking place. The national Spatial Planning Key Decision to implement the Room for the River programme, for instance, states that a land reservation applies where land will or can be given to the river by setting the dike further back from it. This means that:

- land needed for measures in the basic package will be safeguarded from developments which might stand in the way of a flood defence installation;
- land where measures are expected to be needed in the long term will be safeguarded against large scale and/or capital-intensive

---

in Dutch (only) from the national governments website at http://www.rijksoverheid.nl/onderwerpen/water-en-veiligheid/, accessed 17 October 2012.

61 (2009–2010) 32 304(3) Parliamentary documents 1–2.

62 The programme's international website contains much information on the programme, including its main implementing spatial plans, http://www.roomfortheriver.nl, accessed 17 October 2012.

63 Primarily the EU's Natura 2000 network, a network of protected areas instituted as a consequence of the EU's Birds and Habitats Directives; see Jonathan Verschuuren, 'Climate Change: Rethinking Restoration in the European Union's Birds and Habitats Directives' (2010) 28(4) Ecological Restoration 431–439. See also Chapter 11.

64 Spatial Planning Act of 20 October 2006 (2006) 566 Bulletin of Acts and Ordinances.

> developments which will seriously impede future river relief measures from being taken.[65]

Provincial and local zoning plans have to be adjusted to accommodate the implementation of the national Spatial Planning Key Decision. In general, it can be concluded that planning, land use and development laws play an essential role in the implementation of any flood policy, especially through the designation of areas that are prone to floods and the regulation of activities within those areas.[66]

The Netherlands Water Act contains safety norms for dikes and embankments, which vary between a 1:250 and 1:10,000 probability that critical water levels might be reached in any given year, depending on the number of people and amount of infrastructure protected by the dike.[67] The specific requirements for dikes and embankments in terms of height and strength are derived from that norm. The Act contains a range of provisions aimed at protecting land against flooding, including:

- procedural provisions on decisions to create or change river defence works;[68]
- the conferral on the Minister of power to take all necessary measures in case of danger.[69] A danger is defined as 'circumstances as a consequence of which water management works are under an immediate and serious threat or can become under such a threat'.[70] The Minister is even allowed to take measures that are against the law, as long as they do not infringe the constitution or international law;[71]
- the obligation to organize practice drills and exercises to deal with dangerous situations;[72]

---

65 Spatial Planning Key Decision Room for the River, Approved Decision of 19 December 2006, 16. This document is available in English at http://www.roomfortheriver.nl, accessed 17 October 2012.

66 See further on the role of planning law, Chapter 12.

67 Art. 2.2 and Annex II.

68 Arts 5.5–5.13.

69 Art. 5.30.

70 Art. 5.28.

71 Art. 5.30.

72 Art. 5.29. Sometimes, international exercises are organized as well. In 2009, the exercise 'EU FloodEx' tested international assistance during a worst credible flood scenario in the North Sea area on the Dutch coast. The exercise showed that in such a case an international response is necessary, but also that there are many shortcomings associated with poor co-operation of the various

- the duty on property owners to allow authorized officers to enter or do works in any place that they deem necessary, and the power of authorized officers to enter a property without the owner's consent;[73]
- the duty on property owners in water storage areas to allow their land and other property to be flooded;[74]
- a prohibition on property owners in a water storage area to build anything that is considered to be an obstacle to water storage;[75]
- the possibility for property owners in water storage areas to claim compensation in respect of loss or damage suffered as a result of flooding or restrictions on land use;[76]
- compulsory acquisition of land where this is necessary for dike and embankment works;[77]
- a prohibition on interfering with river defence works without a permit.[78]

In 2011, three important amendments were made to the Netherlands Water Act in order to complete the regulatory framework for coastal and river basin adaptation.[79] These amendments establish a Delta Programme, a Delta Programme Commissioner to oversee implementation of the Programme,[80] and a Delta Fund. The Delta Programme is a new annual plan with a six-year planning horizon detailing all measures necessary to combat floods and water scarcity as a consequence of climate change. The Delta Fund will provide the resources required to implement the

---

response services involved. For an account of the exercise, see R J J Beerens et al., *EU FloodEx: an Analysis of Testing International Assistance during a Worst Credible Flood Scenario in the North Sea Area* (Wageningen University 2010), http://library.wur.nl/WebQuery/hydrotheek/lang/1949028, accessed 17 October 2012.

73 Arts 5.20–5.24.

74 Art. 5.26.

75 This is regulated through the relevant spatial plans at provincial and municipal level.

76 Arts 7.14–7.15.

77 Art. 5.14.

78 Art. 6.12 Water Ordinance.

79 Delta Act on Water Safety of 1 December 2011, (2011) 636 Bulletin of Acts and Ordinances.

80 The first Delta Commissioner and his staff have been in office since 2010. For more information, see the Delta Commissioner's website at http://www.deltacommissaris.nl.

Delta Programme. The Act stipulates that, as of 2020, EUR 1 billion has to be made available annually under the Fund.

Finally, some attention should be paid here to dam operations and stormwater management. Dam operations play a big role in river basin management. Altered flood risks due to climate change require alteration of dam operations, especially in cases of extreme weather events. The recent investigations into the Queensland Floods of 2011 show that unexpected massive amounts of rainfall may catch dam operators by surprise, even if they work according to the existing dam operation rules.[81] These rules, therefore, have to be adapted so as to provide for rapid decisions to prevent urban inundations. The Queensland investigations show that only a few hours' delay in adjusting dam operations may have catastrophic results.

Stormwater management in urban areas is an important water adaptation issue because increased precipitation, especially during extreme weather events, can exceed the capacity of the existing stormwater system, leading to sewage overload, flooding, and pollution. Although most adaptation measures with regard to stormwater have to be local measures, the river basin management plan does deliver the background for these local measures. Excessive stormwater must be able to reach receiving waters as soon as possible, so the receiving waters should be prepared for that.

Already in 2008, the New York City authorities had adopted a Sustainable Stormwater Management plan in which several measures were proposed to adapt the stormwater system to the changing requirements.[82] These measures ranged from zoning law measures to practical street level measures. Planning considerations had to take into account the impact on the city's stormwater infrastructure (changes in density and use may have a big impact on the amount of stormwater that needs to be handled). Adaptive measures through zoning plans included, for instance, planning sufficient vegetated and otherwise pervious surfaces. Zoning plans may require a certain percentage of private properties to be planted or to remain pervious. Building codes may set standards to prevent basements flooding, and may require green roofs.[83]

---

[81] Queensland Floods Commission of Inquiry, Final Report (QFCI 2012) 526, http://www.floodcommission.qld.gov.au, accessed 17 October 2012.

[82] E Sussman et al., 'Climate Change Adaptation: Fostering Progress through Law and Regulation' (2010) 18 NYU Environmental Law Journal 65.

[83] See further on green buildings, Chapter 13.

5.3 Water Scarcity and Droughts

Changes in the hydrological cycle, such as altered timing and flow of snowmelt, altered precipitation levels, or increased evaporation due to temperature rise, can lead to water scarcity, i.e., a situation in which there are insufficient water resources to satisfy long-term average requirements. Water scarcity also occurs when the demand rises, for instance, as a consequence of an increase in temperatures. Water scarcity must not be confused with droughts. Droughts are short-term deficiencies of water caused by temporary absence of precipitation, although, depending on the normal climatic conditions in the region, for some regions droughts may last for months or even several years. Adaptation strategies for water scarcity aim at long-term measures such as regulating flow regimes and alterations in dam operations. Adaptation policies aimed at dealing with more intensive and more frequent droughts focus more on increasing resilience on the side of the water users, i.e., through water storage on agricultural lands, changes in crop choices, and the implementation of water efficient irrigation technologies.

As stated above, it is mainly through river basin management planning that these measures are to be developed and integrated into the general water management policy for the area covered by the river basin, on the basis of the principle of reasonable and equitable use (see section 4) and taking into account basic requirements following from the human right to water and from laws aimed at protecting ecosystems (see section 3 above). River basin management plans have to plan ahead for both water scarcity and droughts, given the expected regional and local conditions. Aligning competing water uses, however, is an extremely difficult task, which can only be accomplished through a co-operative governance approach, where all relevant stakeholders together try to figure out how the available water is to be reasonably and equitably shared.[84] For transboundary river basins, such a process has to take place both at the international and local levels. Case studies show that a co-operative governance approach involving all relevant stakeholders can be successful.[85] Conflicts of interests can be overcome, paving the way towards long-term integrated and sustainable management of the site, avoiding

[84] Jonathan Verschuuren, 'The Right to Water as a Human Right or a Bird's Right: Does Co-operative Governance Offer a Way Out of a Conflict of Interests and Legal Complexity?' in P Cullet et al. (eds), *Water Governance in Motion: Towards Socially and Environmentally Sustainable Water Laws* (India: Cambridge University Press 2010) 359–387.

[85] Ibid.

legal conflicts within or between the states involved. However, such a stakeholder process is time consuming and slow, and should be carefully led, keeping a close eye on all sensitive positions. Although often successful at first, the process may run into a wall of legal complexity once the carefully reached agreements are consolidated into legal decision-making at all levels of government in all the countries involved. Therefore, it is important that the co-operative governance process continues during the translation of the agreements into legal decisions, and that all relevant government institutions are actively involved in the process and committed to its outcome.[86]

For more specific water legislation, legal instruments should be provided to implement the decisions on water distribution reached in these stakeholder processes. It is obvious that a strong regulatory oversight is needed.[87] A permissive system that applies to any actions that affect water quality or water quantity, be it surface water or ground water, in which the permitting authority has significant discretion is inevitable. Such a system is based upon the assumption that water is a public good.[88] Other legal instruments can and are used to back up the basic command and control instruments. Financial instruments, such as levies and taxes, are especially relevant in this respect. They aim at having the user pay for water uses, both to ensure sustainable use and to recover costs from water services provided by the government. A more flexible system that seems better suited to address fluctuations in water availability as a consequence of climate change, is a system of tradable water use rights, such as those implemented in Spain, the US, and Australia. Under the Australian National Water Market, for example, several types of water rights are tradable within a complex regulatory system involving both generic water law and specific rules for specific river basins, such as the Murray-Darling basin.[89] Land use law or spatial planning law may also be used to protect aquifers and surface waters with an aim to enhance their adaptive capacity. Land use law can, for instance, be applied to protect small-scale water storage areas, which are created in rural areas as part of a wider adaptation plan. Storage of surplus rainfall in wetter periods, for use during drier periods is a typical adaptation measure (for

---

86 Ibid.

87 Cassuto and Sampaio, above note 36 at 412.

88 Ibid.

89 For more information, see the National Water Market's website at http://www.nationalwatermarket.gov.au, accessed 17 October 2012. See also J Rolfe, 'Water Trading and Market Design' in L Crase (ed.), *Water Policy in Australia: The Impact of Change and Uncertainty* (RFF 2008) 202–213.

instance, in parts of Europe where climate change brings more rainfall in winter and more warm and dry periods in summer).[90]

Finally, there are some recent insights into adaptation for inland waterway transport.[91] Research shows that both the private sector and the public sector have to take measures to ensure that inland waterway transport can continue even in times of unusually low water levels due to droughts. The private sector should adapt the barge sizes (smaller barges, less capacity) so as to be able to continue using a waterway even in times of low water levels. Public sector measures are aimed at adapting the infrastructure to periods of decreased water levels, for instance through dredging and reservoir building.

6 POSSIBLE NEGATIVE IMPACT OF MITIGATION MEASURES ON WATER RESOURCES

It should be noted that mitigation measures may have a negative impact on water resources. Carbon capture and storage, for instance, may lead to groundwater degradation in cases of leakages. The same is true for geothermal energy extraction. Large-scale hydro-electric power may have a negative impact on river ecosystems and fisheries. Carbon farming may lead to enhanced contamination of groundwater with nutrients and pesticides and to high water demand, as is the case with large-scale biofuel production.[92] Similarly, adaptation measures in the water sector can have a negative impact on mitigation. Depending on the specific circumstances, reservoirs can emit greenhouse gases; and irrigation and desalinization consume a lot of energy.[93] Hence the need to integrate mitigation and adaptation policies in the field of water management. Again, river basin management plans can facilitate such integration. For instance, these plans can assure that carbon capture and storage sites are planned in areas where they cannot affect groundwater.

---

[90] See further Chapter 8.

[91] E Demirel, *Economic Models of Inland Navigation in the Context of Climate Change* (Free University Amsterdam 2011), http://dare.ubvu.vu.nl/bitstream/1871/19082/1/dissertation.pdf, accessed 17 October 2012.

[92] UNECE, Guidance on Water and Adaptation to Climate Change, ECE/MP.WAT/30 (2009) 22.

[93] Ibid. at 23.

7 CONCLUSION

Probably the most discussions concerning adaptation have been with regard to water management and water law. The enormous impact of climate change on the water cycle and the paramount importance of water for all life on Earth, as well as the fact that many parts of the world already witness more intense floods and droughts, trigger massive governmental attention towards water adaptation law, both at international and domestic levels. Water law has a relative advantage over most other fields of law that face adaptation challenges in that, in most parts of the world, an integrated river basin approach is applied to all water issues. Such an approach offers a good framework for adaptation measures to be integrated into the existing water management system. The holistic approach, the multilevel governance structure within a river basin, and the regularly evaluated and updated plans and programmes, all offer the flexibility that is needed continuously to adapt to changes in the water cycle. This, however, does not mean that adaptation is an easy thing to do. In some parts of the world, the fulfilment of the right to water is further endangered. In international river basins, conflicts between states over water are aggravated and can only be solved when states are willing to forget about historical claims and instead apply the principle of reasonable and equitable use in a flexible and co-operative way. Huge investment is needed in new or upgraded water defence works. Many legal instruments have to be applied to accommodate higher water levels or even occasional floods. This chapter has shown that much of the thinking and designing of adaptive water law has been done and that the first steps are being taken to put it into practice. This process will probably continue for decades to come.

10. Climate change adaptation and marine and coastal law

Tim Stephens

1 INTRODUCTION

Human induced climate change is driving rapid biophysical changes throughout the oceans and seas that cover 71 per cent of the Earth's surface, and which are the primary drivers of the global climate system. The oceans have been critically important in maintaining global climatic stability, absorbing around 40 per cent of carbon dioxide (CO_2) emissions from human activities and around 90 per cent of the heat added to the Earth's system by climate change.[1] The most visible effect of climate change in the marine and coastal environment has been sea level rise, and the transformation of coastal geography poses a profound climate adaptation challenge. Even if atmospheric concentrations of greenhouse gases are stabilised, sea levels will continue to rise for millennia,[2] making coastal adaptation an issue that will demand ongoing attention.[3]

Accompanying sea level rise are other climate change impacts upon coastal and marine areas. These include rapid melting of sea ice (most notably in the Arctic Ocean,[4] where the extent of summer sea ice reached a record low in 2012), saltwater intrusion into freshwater aquifers

---

1 S D Solomon et al. (eds), *Climate Change 2007: the Physical Science Basis: Contribution of Working Group I to the Fourth Assessment Report of the Intergovernmental Panel on Climate Change* (Cambridge University Press 2007) 404 and 392.

2 H Goelzer et al., 'Millennial Total Sea-Level Commitments Projected with the Earth System Model of Intermediate Complexity' (2012) 7 Environmental Research Letters 1.

3 Ben Saul, Steven Sherwood, Jane McAdam, Tim Stephens and James Slezak, *Climate Change and Australia: Warming to the Global Challenge* (Federation Press 2012) 20.

4 Peter Wadhams, 'Arctic Ice Cover, Ice Thickness and Tipping Points' (2012) 41 AMBIO 23.

(affecting water supplies), rising water temperatures (affecting the productivity and distribution of fisheries, causing algae blooms and damaging corals), storms and storm surges (which are exacerbating erosion) and ocean acidification (the changing chemistry of the oceans damaging many marine organisms). All of these impacts can be tempered to some extent by adaptive responses. As has been noted in previous chapters, the Intergovernmental Panel on Climate Change (IPCC) defines adaptation as 'the adjustment of natural or human systems in response to actual or expected climatic stimuli or their effects, which moderates harm or exploits beneficial opportunities'.[5] However, there are limits to the capacity of natural and human systems to adjust to climate change impacts; for instance, there is no way that many marine organisms and ecosystems will be able to acclimatise to ocean acidification beyond certain thresholds, and so it is only by mitigating emissions of CO_2 that the conditions for a healthy marine environment can be sustained.[6]

This chapter surveys the international legal framework relevant to the adjustment of natural and human systems in coastal, near shore and offshore marine environments in response to climate change. National and sub-national legal systems are developing adaptive responses to climate change in coastal and marine areas by adjusting coastal planning regimes, strengthening building codes, and, through 'connectivity conservation', by establishing protected areas and wildlife corridors.[7] However, developments on the international legal plane with respect to coastal and marine adaptation have been modest. The 1992 United Nations Framework Convention on Climate Change (UNFCCC)[8] addresses coastal adaptation to sea level rise to a limited extent, but the international climate regime is largely silent on adaptation in the marine environment. The main framework for adaptive responses in the maritime domain is supplied by the law of the sea, in particular the 1982 United Nations

[5] M L Parry et al. (eds), *Climate Change 2007: Impacts, Adaptation and Vulnerability: Contribution of Working Group II to the Fourth Assessment Report of the Intergovernmental Panel on Climate Change* (Cambridge University Press 2007) 750.

[6] See Richard E Zeebe, 'History of Seawater Carbonate Chemistry, Atmospheric CO_2 and Ocean Acidification' (2012) 40 Annual Review of Earth and Planetary Sciences 141.

[7] See, e.g., Craig R Groves, 'Incorporating Climate Change into Systematic Conservation Planning' (2012) 21 Biodiversity and Conservation 1651.

[8] United Nations Framework Convention on Climate Change (New York, 9 May 1992), (1992) 31 ILM 849, entered into force 21 March 1994.

Convention on the Law of the Sea (LOSC),[9] and there are several complementary regimes of relevance to adaptation in specific sectors and at regional scales.

2 CLIMATE CHANGE IMPACTS ON COASTAL AND MARINE ENVIRONMENTS

The impacts of climate change and connected phenomena on coastal and marine environments are manifold. The IPCC's Fourth Assessment Report summarised the observed physical and biological changes to the oceans to date, and made projections about likely impacts into the future under several emissions scenarios.

IPCC Working Group I, which assesses the physical science basis of climate change, described significant observed changes in global scale temperature and salinity, regional changes in ocean circulation, marine biogeochemical changes, and sea level rise. It concluded that the heat content of the oceans had increased substantially since the beginning of the twentieth century, that thermal expansion of the oceans was the main driver of sea level rise, that cooler waters towards the poles were becoming fresher, while waters nearer the equator were becoming more saline, that ocean circulation is changing, and that the pH of the world's oceans is decreasing (i.e., becoming acidic).[10]

Working Group II, which assesses the impacts of climate change, adaptation options and vulnerability, considered the effects of climate change on marine and coastal ecosystems in its contribution to the Fourth Assessment Report. However, while a whole chapter of Working Group II's report is devoted to 'Coastal systems and low-lying areas', there is no dedicated analysis of climate change impacts on marine areas beyond the coastline. Where marine issues are dealt with, they are considered in the context of specific regions (e.g., the polar regions) and there is next to no treatment of adaptation options for marine environmental management. This contrasts with Working Group II's extensive and influential analysis of coastal adaptation.[11]

Studies since the IPCC's Fourth Assessment Report in 2007 have documented more fully the impact on the biogeochemistry and ecology of the open oceans as a result of greenhouse gases emitted from human

---

[9] United Nations Convention on the Law of the Sea (Montego Bay, 10 December 1982), (1982) 21 ILM 1261, entered into force 16 November 1994.

[10] Solomon et al., above note 1 at 420–421.

[11] Parry et al., above note 5 at 340–344.

activities.[12] In the discussion that follows the main climate change impacts upon coastal and marine areas requiring an adaptive response are highlighted.

2.1 Sea Level Rise

Sea level rise remains of primary concern from a coastal adaptation perspective because of its capacity to cause large-scale damage to ecosystems and to human settlements and infrastructure.[13] Close to one quarter of the world's population lives in a zone within 100 km of the coast,[14] and the coastal population is projected to grow to up to 5.2 billion by the 2080s.[15]

The IPCC projects that sea levels will rise between 0.18 and 0.59 m this century as a result of thermal expansion of the oceans and the melting of ice sheets.[16] However, this projection is conservative and does not fully account for ice sheet dynamics. Given the recent melting of the largest-ever recorded area of the Greenland ice sheet,[17] it is possible that sea level rise could be several times the IPCC estimates.[18] There is also significant regional variation in projected sea level rise, and, hence, some regions may be more seriously affected than others. Some areas, such as low-lying regions, particularly deltas, are already highly vulnerable to changes in sea level and are already experiencing changes such as subsidence and flooding.

---

12 See, among many others, P J Gleckler et al., 'Human-Induced Global Ocean Warming on Multidecadal Timescales' (2012) 2 Nature Climate Change 524; S C Dorney, 'The Growing Human Footprint on Coastal and Open-Ocean Biogeochemistry' (2010) 328 Science 152; and J B C Jackson, 'Ecological Extinction and Evolution in the Brave New Ocean' (2008) 105 Proceedings of the National Academy of Sciences 11458.

13 Mark Stallworthy, 'Sustainability, Coastal Erosion and Climate Change: An Environmental Justice Analysis' (2006) 18 Journal of Environmental Law 357.

14 Parry et al., above note 5 at 319.

15 Ibid. at 317.

16 Ibid. at 45.

17 Suzanne Goldenberg, 'Greenland Ice Sheet Melted at Unprecedented Rate During July' (*The Guardian*, 24 July 2012).

18 Stefan Rahmstorf, 'A New View on Sea Level Rise' (2010) 4 Nature Reports Climate Change 44.

2.2 Warming Waters

Beyond sea level rise and coastal erosion and inundation, climate change will affect coastal and marine ecosystems in many other ways. One of the most obvious changes will be increasing water temperatures and changes in ocean currents. Rising water temperatures are damaging many coral habitats worldwide, by inducing bleaching (a stress reaction in which the symbiotic relationship between the corals and the algae living within them breaks down, and the algae is expelled). Most corals are sensitive to relatively small increases in temperature, and up to 90 per cent of corals worldwide may be lost if temperatures increase beyond 1.5°C.[19]

Increasing temperatures and changing patterns in oceanic circulation are also having a range of impacts upon fish species, affecting the productivity, distribution and yields of many fisheries.[20] One recent study examining the eco-physiology of marine water-breathing organisms has estimated that the average body weight of over 600 species of fish will decline by up to 24 per cent by 2050 under a high emission scenario,[21] posing major challenges for populations reliant on fish for primary animal protein. Warming waters not only affect fisheries, but can also exacerbate oceanic dead zones by promoting hypoxia (low oxygen levels),[22] the growth of parasites and other harmful pathogens,[23] and the transport, settlement and invasion of alien marine species.[24]

For marine wildlife, climate change is fundamentally a habitat issue, with rising temperatures and other ocean conditions making once stable environments unrecognisable to dependent species. The mobility of fish, cetaceans and many other organisms in the marine environment may mean that they enjoy some level of adaptive resilience, in that they may

---

[19] K Frieler, 'Limiting Global Warming to 2°C is Unlikely to Save Most Coral Reefs' (2012) Nature Climate Change (advance online publication doi:10.1038/nclimate1674).

[20] Parry et al., above note 5 at 333. See also U Rashid Sumaila et al., 'Climate Change Impacts on the Biophysics and Economics of World Fisheries' (2011) 1 Nature Climate Change 449.

[21] William W L Cheung et al., 'Shrinking of Fishes Exacerbates Impacts of Global Ocean Changes on Marine Ecosystems' (2012) Nature Climate Change (advance online publication doi:10.1038/nclimate1691).

[22] Cheryl Lyn Dybas, 'Dead Zones Spreading in World Oceans' (2005) 55 BioScience 552.

[23] C Drew Harvell, 'Climate Warming and Disease Risks for Terrestrial and Marine Biota' (2002) 296 Science 2158.

[24] Anna Occhipinti-Ambrogi, 'Global Change and Marine Communities: Alien Species and Climate Change' (2007) 55 Marine Pollution Bulletin 342.

shift their range in response to rising temperatures and other ocean changes. However, even for highly migratory species, such as whales and dolphins, their mobility may in fact be an elusive protection 'in view of known site fidelity for breeding, calving, nursing and feeding grounds'.[25]

2.3 Ocean Acidification[26]

Another increasingly well-documented impact upon the world's oceans, occurring in tandem with climate change, is ocean acidification, which is damaging coral ecosystems and other pH sensitive organisms. Ocean acidification provides a clear example of an impact on the marine environment from greenhouse gas emissions for which there are few adaptation opportunities. Ocean acidification has major economic and food security implications given its potential to decimate commercially valuable fish stocks.[27]

The oceans act as a buffer for changes in atmospheric CO_2, absorbing large quantities of this greenhouse gas, which has been emitted from human activities since the industrial revolution. When CO_2 dissolves in the oceans it reacts with H_2O to form carbonic acid. The oceans are naturally slightly basic, but oceanic pH is declining and the oceans are more acidic now than they have been for around 500,000 years, with flow-on effects for marine organisms, especially those which form shells.

As CO_2 is absorbed in sea water and carbonic acid is formed, hydrogen ions are released, which then combine with carbonate ions to form bicarbonate. This chemical process removes substantial volumes of carbonate ions from the water, which are essential for many marine photosynthetic organisms and animals, such as molluscs, corals, echinoderms, foraminifera and calcareous algae. Oceanic pH has declined by around 0.1 pH units, and if it falls by as much as 0.4 pH units then levels

---

25 Erich Hoyt, *Marine Protected Areas for Whales, Dolphins and Porpoises* (2nd edn, Earthscan 2011) 80.

26 This section draws from Rachel Baird, Meredith Simons and Tim Stephens, 'Ocean Acidification: A Litmus Test for International Law' (2009) 4 Carbon and Climate Law Review 459.

27 See further R Allan and A Bergin, 'Ocean Acidification: An emerging Australian Environmental Security Challenge' (2009) 1 Australian Journal of Maritime and Ocean Affairs 49. See also S R Cooley and S C Doney, 'Anticipating Ocean Acidification's Economic Consequences for Commercial Fisheries' (2009) 4 Environmental Research Letters 024007.

of carbonate ions may slip below those required to sustain coral reef accretion, an outcome that may be reached as early as 2050.[28]

3 COASTAL AND MARINE ADAPTATION: THE CONCEPTUAL AND POLICY FRAMEWORK

3.1 Coastal Adaptation

Given the significant impact that sea level rise will have on many coastlines, it is not surprising that there has been extensive analysis of the options for coastal adaptation.

Much of this knowledge is consolidated in the contribution of Working Group II to the IPCC's Fourth Assessment Report,[29] which devotes a chapter to 'Coastal Systems and Low-Lying Areas'. As the IPCC observes, 'reactive and standalone efforts to reduce climate-related risks to coastal systems are less effective than responses which are part of integrated coastal zone management (ICZM), including long-term national and community planning'.[30] This reinforces the importance of integrated coastal and ocean management generally.[31]

While there is a need for on-going research into the coastal impacts of climate change, especially at localised coastal scales, there is now good agreement on the types of adaptation interventions that are effective in the coastal zone.[32] The IPCC has provided a conceptual framework for coastal adaptation practices, categorising adaptation measures by reference to the goals of protection, accommodation and/or retreat (see Figure 10.1[33]).

---

[28] Ove Hoegh-Guldberg et al., 'Coral Reefs under Rapid Climate Change and Ocean Acidification' (2007) 318 Science 1737.

[29] Parry et al., above note 5.

[30] Ibid. at 340.

[31] See generally Donald R Rothwell, 'Environmental Integration and Coastal and Marine Law' in Rachel Baird and Donald R Rothwell (eds), *Australian Coastal and Marine Law* (Federation Press 2011) 348.

[32] Richard Klein et al., 'Coastal Adaptation to Climate Change: Can the IPCC Technical Guidelines be Applied?' 4 Mitigation and Adaptation Strategies for Global Change 239.

[33] Figure reproduced from R J Nicholls et al., 'Coastal Systems and Low-Lying Areas' in Parry et al., above note 5 at 342 (Figure 6.11). The figure is based on IPCC CZMS, Strategies for Adaptation to Sea-Level Rise, Report of the Coastal Zone Management Subgroup, Response Strategies Working Group of the Intergovernmental Panel on Climate Change (Netherlands Ministry of Transport,

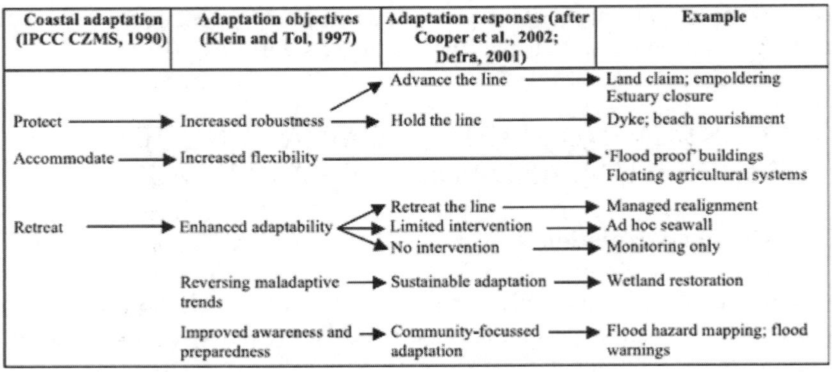

Figure 10.1 Coastal adaptation responses

The central coastal adaptation challenge is therefore less one of knowledge and more one of translating the adaptation imperative into an effective legal and policy framework within a timeframe that will avert the most serious environmental and socio-economic risks. There are various impediments in developing and implementing an appropriate regulatory framework, including institutional complexity and weaknesses, the lengthy functional life of most coastal settlements and infrastructure, and also societal opposition to anticipatory adaptation measures. For instance, the construction of seawalls and other hard defences may cause significant controversy. There may be opposition from some sections of the community because of the impact of construction upon natural ecosystems and the visual environment, yet others may prefer to 'stay and defend', rather than allow coastal geography to shift as the seas rise.[34]

There are several tensions at play in the development of coastal adaptation regimes. These include the need to anticipate change but not

Public Works and Water Management 1990); R J T Klein and R S J Tol, 'Adaptation to Climate Change: Options and Technologies: An Overview Paper', Technical Paper FCCC/TP/1997/3, UNFCCC Secretariat, 1997; N Cooper et al., 'Shoreline Management Plans: A National Review and an Engineering Perspective' (2002) 154(3) Proceedings of the Institution of Civil Engineers: Water and Maritime Engineering 221–228; Department of Environment, Food and Rural Affairs (DEFRA), 'Shoreline Management Plans: A Guide for Coastal Defence Authorities' (DEFRA 2001).

[34] See further Jonathan Verschuuren and Jan McDonald, 'Towards a Legal Framework for Coastal Adaptation: Assessing the First Steps in Europe and Australia' (2012) 1(2) Transnational Environmental Law 355.

overestimate the likely harm, and thereby incur opportunity costs by ruling out necessary development.[35] This is the tension between certainty and uncertainty, and is essentially a temporal problem, as laws and policies need to be developed now in respect of impacts that have not yet fully materialised. It can be addressed by on-going research and coastal assessment, and by iterative systems of adaptive coastal management. Another tension is between global, regional, national and local planning. This is the spatial challenge of coastal adaptation. An overarching approach is needed to ensure consistency in the adaptation framework, but if an umbrella regime is too prescriptive it can prevent authorities from responding to climate change impacts at the local scale. A further tension is between private rights and public goods. This is the societal challenge of coastal adaptation, recognising the interests of private property owners so far as possible without governments becoming insurers of last resort for poor coastal planning. The effective management of these tensions is the central objective of a policy framework for coastal adaptation.

3.2 Marine Adaptation

While coastal adaptation policy has been the subject of extensive analysis, marine adaptation has not attracted the same level of scrutiny, despite the threat that climate change poses to key economic sectors including fisheries, aquaculture and tourism.

Explanations for this policy blind-spot at national and international levels include the dynamic character of the marine environment, the natural variability in oceanic conditions, and the fact that impacts upon the marine environment are less visible and tangible than coastal climate change impacts, such as coastal erosion from sea level rise. However, some climate impacts upon the marine environment are increasingly apparent, such as the bleaching of coral reefs and large-scale algae blooms in enclosed and semi-enclosed seas. These very visible changes have served to highlight the need for adaptive responses to improve the resilience of marine ecosystems in the face of environmental change. The impact of climate change upon fisheries is also increasingly visible because of the flow-on economic impacts to the fishing industry and

---

[35] Andrew Macintosh, 'Coastal Climate Hazards and Urban Planning: How Planning Responses Can Lead to Maladaptation' (2012) Mitigation and Adaptation Strategies for Global Change (published online on 15 July 2012 doi:10.1007/s11027–012.9406–2).

fishing dependent communities. Around 520 million people globally are supported by fishing activities.[36]

We have seen that the main goals of coastal adaptive policy have been identified by the IPCC as protection, accommodation and, where necessary, retreat. This conceptual framework can also be applied, with some modifications, to marine adaptation.[37]

Protective adaptive policies in the marine domain include measures to conserve marine ecosystems, including through the establishment of marine protected areas. However, there is a significant lead-time (several years or even decades) in devising and implementing marine protected areas, meaning that this response is seldom sufficient to deal with rapid environmental change. Other management tools, such as reduced fishery quotas or limits on gear and fishing effort, can normally be implemented more rapidly as there exist well-established regimes for fisheries regulation, and these can often be adjusted by administrative decision rather than requiring legislative action.[38]

In terms of accommodation approaches to adaptation in the marine environment, these may include adjustments to fisheries management areas to reflect the shifting range of a particular fishery, or managing fishing pressure by setting more conservative total allowable catches where the productivity of a fishery is declining. Where protection and accommodation are not sufficient to safeguard marine ecosystems and the services and resources they supply, retreat from existing marine uses and activities may be the only adaptive option.[39] In the fisheries context this could entail the complete closure of fishing access, and the shift of fishing effort to other fisheries.

It is the extent and rate of change that determines the emphasis that will need to be placed upon these three adaptive strategies; if change occurs slowly then marine ecosystems may be cushioned against changing ocean conditions; however, a warming of several degrees within a century (as we are on track to experience) will leave many organisms and ecosystems unable to cope and, as a consequence, the permanent retreat from current marine uses will be inevitable.

---

36 Sumaila, above note 20 at 449.

37 See, e.g., Elizabeth M P Madin et al., 'Socio-Economic and Management Implications of Range-Shifting Species in Marine Systems' (2012) 22 Global Environmental Change 137, 142–143.

38 Ibid.

39 Alistair McIlgorm et al., 'How will Climate Change Alter Fishery Governance? Insights from Seven International Cases Studies' (2010) 34 Marine Policy 170, 175.

There are, therefore, limits to the extent that adaptation is a feasible response to climate change in the marine environment. Where those limits lie is not fully known, and it should not be expected that climatic changes or the ecological responses to them will be linear. As Barnosky et al. report in a recent review of ecosystem responses to climate change, localised ecosystems may shift quickly from one state to another when they are pushed beyond certain critical thresholds, or tipping points.[40] In the marine realm the example given is the phenomenon of vast 'dead zones' in several ocean regions, most notably Chesapeake Bay and the Gulf of Mexico. Barnosky et al. note that there are risks that the global ecosystem as a whole, encompassing both terrestrial and marine systems, may be overwhelmed by human influence, leading to a state shift in Earth's biosphere.[41] Given the risks of such a state change, a key adaptive strategy is improving early warning systems of impending transformations, so that human systems are able to respond in advance of the worst impacts being felt.

Policies of protection, accommodation and retreat in marine adaptive management are likely to involve significant flow-on socio-economic impacts; subsistence fishermen may not have viable alternative fisheries to generate income; food security may be threatened if other nutritional sources are not available; and conflict may be generated from competition over dwindling fish stocks and the displacement of fishermen to new fisheries. This underscores the need for adaptation strategies for fisheries to include not only traditional and contemporary tools of fisheries management (the setting of total allowable catches, quotas, season windows, vessel and gear limits, and marine reserves) but also livelihood diversification options.[42]

The immediate priority in developing marine adaptation options is improved monitoring and assessment of the marine environment. Once knowledge gaps are filled then the availability of appropriate adaptation strategies can be determined. Central questions that need to be addressed include: how the distribution and productivity of marine species and communities are changing and will change, whether certain species can be used as indicators for climate change impacts and whether reducing

---

40 Anthony D Barnosky et al., 'Approaching a State Shift in Earth's Biosphere' (2012) 486 Nature 52.

41 Ibid.

42 Sumaila et al., above note 20 at 6; Food and Agriculture Organization, *The State of World Fisheries and Aquaculture: 2012* (FAO 2012) 119.

other environmental stressors (such as pollution) will enhance ecosystem robustness.[43]

Improved knowledge of impacts, the constraints in adaptive responses, and the potential for rapid and irreversible change in the marine environment, also ought to shift attention back to the primary importance of mitigation policy.

4 COASTAL ADAPTATION: THE INTERNATIONAL LEGAL FRAMEWORK

4.1 Physical Adaptation

There is an expanding body of international law of relevance to adaptation to climate change, although, as Freestone notes, 'adaptation is widely regarded as the poor relation of the international climate change treaty regime' in comparison with the mitigation imperative.[44] The UNFCCC is premised on the capacity of human and natural systems to adapt to some level of gradual climate change. Article 2 of the UNFCCC states that the ultimate objective of the international climate change regime is to achieve the stabilisation of greenhouse gases in the atmosphere at a level that would prevent dangerous anthropogenic interference with the climate system, and this level is to be achieved 'within a time frame sufficient to allow ecosystems to adapt naturally to climate change, to ensure that food production is not threatened and to enable economic development to proceed in a sustainable manner'. This recognises the natural capacity of ecosystems to respond to environmental change, so long as the pace of that change does not exceed certain tolerances.

The UNFCCC specifically addresses coastal adaptation. Article 4(1)(e) requires states to '[c]ooperate in preparing for adaptation to the impacts of climate change' and 'develop and elaborate appropriate and integrated plans for coastal zone management'. Parties are also required to focus on the specific needs and concerns of developing states vulnerable to coastal impacts, including small island countries with low-lying coastal areas (Article 4(8)). To advance this objective, Decision 5/CP.7, adopted at Marrakesh in 2001, founded the Least Developed Countries Fund and

---

43 CSIRO, Marine Climate Impacts and Adaptation, http://www.cmar.csiro.au/climateimpacts, accessed 17 October 2012.

44 David Freestone, 'The International Legal Framework for Adaptation' in Michael B Gerrard and Katrina Fischer Kuh (eds), *The Law of Adaptation to Climate Change: United States and International Perspectives* (ABA 2012) 601.

required it to assist in the development of National Adaptation Programmes of Action (NAPAs).[45] After a hiatus in the development of further adaptation responses under the climate regime, at the 2010 Cancun Conference of the Parties (COP) the Parties adopted the Cancun Adaptation Framework, which for the first time identified adaptation as a priority of equal importance to mitigation.[46] Moreover, the UNFCCC makes specific reference to the importance of parties adopting national adaptation plans that address 'marine ecosystems' and 'coastal zones'. The Least Developed Countries Fund, which is managed by the Global Environment Facility, is dispersing funds for the development and implementation of NAPAs, many of which are directed at protecting coastal regions.[47]

The LOSC provides the global legal framework for all mitigation and adaptation measures that impact upon oceans. The LOSC was concluded in 1982, well before there was widespread international understanding of the threat of climate change (which coalesced much later, in 1990, with the adoption of the IPCC's First Assessment Report). Nonetheless, by virtue of its broad definition of 'pollution of the marine environment', which includes the introduction of 'substances or energy into the marine environment, including estuaries',[48] and its extensive provisions seeking to control pollution, the LOSC clearly has relevance to the mitigation of climate change in order to protect the ocean environment. The LOSC, which now has 162 parties, seeks to prevent marine pollution, requiring all states to take, individually or jointly as appropriate, all measures necessary to prevent, reduce and control pollution of the marine environment from any source.[49]

The LOSC also sets limits on the extent to which coastal states may take physical adaptive steps in responding to climate change in the coastal zone.[50] Coastal states are, for instance, not permitted to implement large-scale coastal engineering works unless there is an assessment of their potential marine environmental effects,[51] and they must adopt laws and regulations to 'prevent, reduce and control pollution of the

---

45 FCCC/CP/2001/13/Add.1. See also Chapter 2, section 2.2.

46 Freestone, above note 44 at 608.

47 Ibid. at 610–612.

48 Art. 1(1)(4).

49 Art. 194(1).

50 See further David Freestone, 'International Law and Sea Level Rise' in Robin Churchill and David Freestone (eds), *International Law and Global Climate Change* (Graham and Trotman 1991).

51 Art. 206.

marine environment from land-based sources',[52] which would include coastal works such as sea walls or land reclamation works. In *Land Reclamation by Singapore in and Around the Straits of Johor*[53] the International Tribunal for the Law of the Sea (ITLOS) invoked several provisions of the LOSC relevant to land reclamation works,[54] and the Tribunal's decision has some relevance to coastal works for climate adaptation purposes.

Implicitly referencing the precautionary principle, ITLOS held that 'prudence and caution' required that Malaysia and Singapore cooperate in assessing and limiting the impact of the land reclamation activities in the Strait in order to protect the marine environment. In a more recent decision, ITLOS, in its Advisory Opinion in *Responsibilities and Obligations of States Sponsoring Persons and Entities with Respect to Activities in the Area*, specifically recognised that there was a 'trend towards making this approach part of customary international law'.[55] This indicates that international law requires coastal states to exercise a high degree of caution when planning coastal adaptation works, and suggests that, in the first instance, states should investigate adaption options that do not involve substantial physical alteration of the coastal environment.

Beyond the general provisions of the LOSC regarding marine environmental protection, there is no agreed international framework addressing the coastal protective measures that states can or should take in response to climate change. It is beyond the scope of this chapter to offer an assessment of the many domestic legal frameworks that are being developed by coastal states worldwide to facilitate coastal adaptive governance. There is considerable variability in the types of national and sub-national legal responses that are being pursued, and these are reflective of a range of social, economic, cultural and political differences in these jurisdictions.[56]

---

52 Art. 207(1).

53 *Land Reclamation by Singapore in and Around the Straits of Johor (Malaysia v Singapore) (Provisional measures)*, ITLOS Case No 12, 8 October 2003.

54 See generally Tommy Koh and Jolene Lin, 'The Land Reclamation Case: Thoughts and Reflections' (2006) 10 Singapore Yearbook of International Law 1.

55 (2011) 50 ILM 455, [135]. See further Duncan French, 'From the Depths: Rich Pickings of Principles of Sustainable Development and General International Law on the Ocean Floor – the Seabed Disputes Chamber's 2011 Advisory Opinion' (2011) 26 International Journal of Marine and Coastal Law 525.

56 See Verschuuren and McDonald, above note 34.

Hence, as a matter of international law states enjoy considerable discretion in determining whether and what measures they will implement in response to coastal threats from rising seas and associated risks. In some regions an effort is being made to fill this gap by providing some general guidance to coastal states. This has tended to emphasise the need for integrated, 'ecosystem' or 'nature' based adaptive responses. For instance the OSPAR Commission, established by the 1992 Convention for the Protection of the Marine Environment of the North-East Atlantic (OSPAR Convention),[57] has urged states in the region of the North Atlantic to integrate adaptation measures into Integrated Coastal Zone Management and marine spatial planning.[58] In relation to the Seas of East Asia, in 2009 ministers from Cambodia, China, Korea, Indonesia, Japan, Laos, Philippines, North Korea, Singapore, Timor-Leste and Vietnam signed the Manila Declaration on Strengthening the Implementation of Integrated Coastal Management for Sustainable Development and Climate Change Adaptation in the Seas of East Asia Region, which recognises the importance of coastal and marine adaptation policies in order to protect water resources, food security and livelihoods, and to safeguard biological diversity.[59]

The parties to the 1971 Convention on Wetlands of International Importance[60] (Ramsar Convention) have also addressed adaptation issues in relation to coastal wetlands such as mangroves, tidal flats and saltmarshes that are important ecosystems and also provide important natural barriers to storm and tidal surges. At COP11 in 2012, the Ramsar parties adopted Resolution XI.14[61] updating and consolidating

---

[57] Convention for the Protection of the Marine Environment of the North-East Atlantic (Paris, 22 September 1992), (1992) 32 ILM 1069, entered into force 25 March 1998.

[58] OSPAR Commission, Assessment of Climate Change Mitigation and Adaptation (OSPAR Commission 2009), http://www.ospar.org, accessed 17 October 2012.

[59] Manila Declaration on Strengthening the Implementation of Integrated Coastal Management for Sustainable Development and Climate Change Adaptation in the Seas of East Asia Region (Manila, 26 November 2009), http://pemsea.org/eascongress/section-support-files/manila_declaration.pdf, accessed 17 October 2012.

[60] The Convention on Wetlands of International Importance, especially as Waterfowl Habitat (Ramsar, 2 February 1971), (1971) 996 UNTS 245, entered into force 21 December 1975.

[61] See http://www.ramsar.org/pdf/cop11/res/cop11-res14-e.pdf, accessed 17 October 2012.

previous resolutions on climate change and wetlands. Resolution X.14 urged parties:

> to maintain or improve the ecological character of wetlands, including their ecosystem services, to enhance the resilience of wetlands as far as possible in the face of climate-driven ecological changes including, where necessary, to promote the restoration of degraded wetlands, and further to promote the ability of wetlands to contribute to nature-based climate change adaptation, particularly the roles of wetlands in regulating water, including reducing risks from water-related disasters, and to sequester and store carbon as important responses for climate change mitigation through the maintenance and enhancement of their ecological functions, and to reduce or halt the release of stored carbon that can result from the degradation and loss of wetlands.

4.2 Legal Adaptation

International law, and specifically the international law of the sea, is relevant not only to physical adaptation measures but also to legal responses to climate change. Foremost among these responses is the continued assertion of territorial sea baselines in the face of rising seas that are altering coastal geographies. A related (but more straightforward) issue is whether maritime boundaries agreed between opposite or adjacent coastal states need to be altered if coastlines shift. The latter is a less complicated question, in that boundary agreements are generally understood to be permanent and dispositive, and not susceptible to termination or withdrawal (except on mutually agreed terms), even if there is a fundamental change of circumstances.[62]

Under the LOSC, the normal baseline which sets the boundary between internal waters and the territorial sea, and from which coastal maritime zones are projected, is the low water mark.[63] Low-tide elevations (or 'drying rocks'), if they are within the 12 nm territorial sea, may be used to generate their own territorial sea.[64] Where the coastline is deeply indented, or where there is a fringe of islands, states may draw straight baselines, including from low-tide elevations even beyond 12 nm having lighthouses or other similar installations upon them.[65] States may only assert an Exclusive Economic Zone (EEZ) and continental shelf

---

62 Vienna Convention on the Law of Treaties (Vienna, 22 May 1969), 1155 UNTS 332, entered into force 27 January 1980, Art. 62(2)(a).

63 LOSC Art. 5.

64 Ibid. Art. 13.

65 Ibid. Art. 7.

from islands that are above water at high tide and which can sustain human habitation or economic life.[66]

The central international legal question posed by sea level rise is whether states may continue to insist upon their existing baselines and basepoints along their coastline, on low-tide elevations and reefs and from islands, or whether these lines and points are ambulatory, automatically receding landwards as territory is inundated. Schofield has observed that 'measures to physically protect the coast from sea-level rise are generally unrealistic, save in exceptional circumstances for critical basepoints, in light of the sheer scale of the challenge' and he notes that '[a]lternative legal options may therefore be pursued'.[67] With the exception of deltas and highly unstable coasts,[68] the LOSC does not expressly countenance the fixing of baselines or basepoints. But likewise it does not provide that baselines move automatically, or must be adjusted by states in response to accretion or avulsion.

It is unlikely that states will simply allow maritime space to be lost without a legal response, and Schofield observes that coastal states may decide to declare unilaterally the continued applicability of their baselines to preserve the extent of maritime estate under coastal state sovereignty and jurisdiction. In doing so, Purcell argues, coastal states will be on firm legal foundations.[69] Following an exhaustive study of the history of the baseline provisions of the LOSC, she rejects the 'ambulatory baseline thesis' and argues that states may, if they wish, adopt the ambulatory approach (as the United States appears to do), or continue to rely on their existing baselines. Purcell contends that '[a]daptation measures aimed at the retention of existing rights to maritime space in their present local and full extent … are largely unnecessary', and that '[t]he current legal regime provides for the establishment of maritime baselines and zonal limits that take geographical circumstances into account but are not in the thrall to the variable coastal and marine environment'.[70]

---

66 Ibid. Art. 121(3).

67 Clive Schofield, 'Sea Level Rise and Options to Secure Maritime Jurisdictional Claims' (2009) 4 Carbon and Climate Law Review 405, 406. See also David Caron, 'When Law Makes Climate Change Worse: Rethinking the Law of Baselines in Light of a Rising Sea Level' (1990) 17 Ecology Law Quarterly 621.

68 LOSC Art. 7(2).

69 Kate Purcell, 'Maritime Jurisdiction in a Changing Climate' in Michael B Gerrard and Katrina Fischer Kuh (eds), *The Law of Adaptation to Climate Change: United States and International Perspectives* (ABA 2012) 731.

70 Ibid. at 737.

5 MARINE ADAPTATION: THE INTERNATIONAL LEGAL FRAMEWORK

The UNFCCC makes limited reference to the marine environment beyond the coastal zone. The threat of sea level rise to coastlines is clearly recognised, with Article 4(8) requiring parties to give full consideration to the effects of climate change on small island countries and countries with low-lying coastal areas. Yet only passing mention is made of marine ecosystems, and only in connection with the environmental services they provide as carbon sinks, rather than the importance of protecting the marine environment per se. The Preamble to the UNFCCC acknowledges the importance of marine ecosystems as sinks and reservoirs of greenhouse gases, and Article 1(d) requires Parties to promote sustainable management, conservation and enhancement of reservoirs of greenhouse gases including those found in marine ecosystems. It is possible to read into Article 4(1)(e), under which Parties are required to 'develop and implement integrated plans for coastal zone management' in adapting to the impacts of climate change, an obligation to develop adaptation plans in near-shore environments to address the effects of climate change. However, the UNFCCC does not expressly extend any such obligation to offshore environments.

In addressing adaptation, the UNFCCC has an almost exclusive terrestrial and coastal focus, and the Kyoto Protocol[71] does not alter this. Moreover the UNFCCC and the Kyoto Protocol deal only with 'climate change', defined in Article 1(2) of the UNFCCC as the change of climate attributed to human activity that alters the composition of the global atmosphere. This does not encompass ocean acidification, which is a separate process from climate change, even though it is caused by a greenhouse gas (CO_2), emissions of which are the primary cause of rising global temperatures. The upshot is that Article 3 of the UNFCCC, which requires states to protect the climate system from adverse effects of climate change, does not include any requirement to prevent ocean acidification. Hence, neither mitigation of, nor adaptation to, ocean acidification is recognised as an objective of the international climate change regime.

---

[71] Kyoto Protocol to the United Nations Framework on Climate Change, UN Doc FCCC/CP/1997/7/Add. 1 (Kyoto, 10 December 1997), (1998) 37 ILM 22, entered into force 16 February 2005.

It is necessary, therefore, to turn to the body of international norms directed specifically to the marine environment to gauge the responsiveness of public international law to the marine adaptation challenge. As noted above, the central regime of relevance is the LOSC, which provides an extensive, globally applicable regulatory framework for all uses of ocean space. The LOSC recognises coastal state sovereignty and jurisdiction over a wide range of maritime zones extending from internal waters, through the territorial sea, the EEZ and the continental shelf. Of particular importance is Part V of the LOSC, which concerns the EEZ, and Part XII, which relates to the protection of the marine environment. These provide coastal states with extensive capacity to protect coastal and marine environments.[72]

Also of relevance is the 1992 Convention on Biological Diversity (CBD),[73] which has engaged with the linkages between biodiversity conservation and marine and coastal adaptation measures. In 2008, the Conference of Parties to the CBD adopted a decision on marine and coastal biodiversity that set out scientific criteria for identifying areas that require protection.[74] And in 2010, the CBD Parties adopted a comprehensive decision on 'Marine and Coastal Biodiversity' that catalogued the impacts that climate change is having on marine and coastal biodiversity and acknowledged the importance of maintaining this biodiversity in adapting to climate change.[75]

5.1 Areas within National Jurisdiction

An assessment of the international legal framework for marine adaptation needs to make reference to the maritime zones in which adaptation measures are proposed. Within the territorial sea, which extends to a limit of 12 nm from the territorial sea baseline, a coastal state has broad ranging capacity to implement rigorous marine environmental protections consistent with its sovereignty over the zone. Hence, coastal states may limit fishing or close fisheries, establish marine protected areas and parks, and impose special protective measures for areas such as reefs. However, there are some limits placed upon coastal states by the innocent passage regime under which foreign flagged vessels are entitled to

---

[72] See further Donald R Rothwell and Tim Stephens, *The International Law of the Sea* (Hart Publishing 2010) 461ff.

[73] Convention on Biological Diversity (Rio de Janeiro 5 June 1992), 31 ILM 818, entered into force 29 December 1993.

[74] Decision IX/20, UN Doc UNEP/CBD/COP/DEC/IX/20 (2008).

[75] Decision X/29, UN Doc UNEP/CBD/COP/DEC/X/29 (2010).

navigate freely through the territorial sea. Under Article 21 of the LOSC, coastal states may adopt laws and regulations relating to innocent passage, including navigational safety and the regulation of maritime traffic, conservation of living resources and preservation of the marine environment, including the prevention of pollution. Such laws and regulations may clearly be directed not only to the protection of the marine environment per se, but also to the objective of climate change adaptation which may be advanced by, *inter alia,* reducing marine pollution.

Beyond 12 nm, and out to a maximum distance of 200 nm, coastal states may claim an EEZ, a zone in which coastal states enjoy exclusive sovereign rights over the living and non-living resources of the water column and seabed. This means that coastal states may implement adaptive measures in relation to resources with little, if any, consideration of the interests of other states. The coastal state is given sole discretion in setting allowable catches for fisheries, taking into account the best scientific evidence, and the responsibility to conserve and manage fisheries so that they are not endangered by over-exploitation.[76] In principle, a coastal state is required to give other states access to its EEZ living resources if it has insufficient capacity to harvest the allowable catch it has set, with particular regard to landlocked, geographically disadvantaged and least developed states.[77] But in practice this right of access to the surplus accorded to foreign states is unenforceable, because coastal state decisions determining the allowable catch, the extent of harvesting capacity and the allocation of surpluses, fall within one of the few exceptions to the compulsory dispute settlement system set out in Part XV of the LOSC.[78] Coastal states, therefore, have largely unfettered rights and jurisdiction over EEZ fisheries and can therefore take the adaptive measures that are considered appropriate for fisheries located within this zone. Indeed, this capacity is implicit in the reference in Article 61(2) to the requirement that coastal states take into account 'the best scientific evidence' available and that fisheries management authorities should take a precautionary approach so as to enable fisheries to adapt to changing conditions.

In relation to non-living resources, coastal states have a clearly unencumbered right to decide whether and under what circumstances mineral resources are exploited within their continental shelf and claimed

---

76 LOSC Art. 61.

77 Ibid. Art. 62.

78 Ibid. Art. 297(3).

EEZ areas. Article 56(3) of the LOSC provides that the rights of coastal states with respect to the seabed and subsoil in the EEZ are made coterminous with those rights under the provisions of the convention addressing the continental shelf (Part VI). In essence it is for the coastal state alone to determine whether to explore or exploit the natural resources of the continental shelf, and no other state may undertake such activities without the express consent of the coastal state.[79] As a consequence, coastal states are fully within their rights to include climate change adaptation considerations in the planning, assessment and approvals processes for proposed oil and gas development of the EEZ. Consistent with a policy of protection or retreat, they may adopt moratoria or permanent prohibitions on offshore oil and gas development.

While in relation to the fisheries and mineral resources of the 200 nm EEZ and continental shelf coastal states have wide if not completely unconstrained discretion in taking adaptive measures, the same cannot be said for measures targeting other elements of the marine environment. In relation to marine pollution from vessels and the risks posed by vessel movements through sensitive marine environments, which remain key concerns for marine environmental protection, coastal states must normally adhere to those international standards adopted under the auspices of the International Maritime Organization (IMO), which are incorporated by reference in the LOSC. However, the IMO regime does allow the designation of certain ecologically vulnerable marine areas as Particularly Sensitive Sea Areas (PSSAs), which in turn allows for the adoption of 'associated protective measures' to prevent, reduce or eliminate the threat posed by international shipping. The Great Barrier Reef in Australia was the first PSSA to be designated by the IMO, in 1990.

PSSAs are normally, but need not always be, included within marine protected areas (MPAs). MPAs are emerging as a critically important area-based tool for marine conservation in the era of climate change. While there are around 120,000 protected areas on land worldwide, covering around 12 per cent of the Earth's land surface,[80] MPAs cover only around 1 per cent of ocean space beyond the territorial sea. There is increasing recognition of the value of MPAs in enhancing the resilience of marine areas in order to adapt to climate change. The 2009 Manado Ocean Declaration, adopted in Indonesia during the World Ocean Conference, emphasised the importance of MPAs, with states resolving 'to

---

79 Ibid. Art. 77.

80 L Coad et al., *State of the World's Protected Areas 2007: An Annual Review of Global Conservation Progress* (UNEP/WCMC 2008) 17.

further establish and effectively manage marine protected areas, including representative resilient networks, in accordance with international law, as reflected in [the LOSC], and on the basis of the best available science, recognizing the importance of their contribution to ecosystem goods and services, and to contribute to the effort to conserve biodiversity, sustainable livelihoods and to adapt to climate change'.[81] Several governments have taken this objective seriously, including Australia, which in 2012 announced the establishment of the world's largest network of marine reserves.[82] This reserve system is in part designed to sustain the health and resilience of the marine environment threatened by climate change.[83]

5.2 Shared Maritime Domains and Areas Beyond National Jurisdiction

A key arena in which adaptation to climate change in the marine environment is being addressed is in relation to fisheries that are shared across multiple coastal state EEZs, that straddle EEZ areas and high seas, or are found exclusively on the high seas. Many such fisheries are subject to management by Regional Fisheries Management Organisations (RFMOs), several of which are beginning to consider adaptive responses.[84] Most wild capture fisheries are already under significant pressure, with the Food and Agriculture Organisation estimating that 90 per cent of fisheries are being harvested beyond their sustainable limits.[85]

---

[81] Manado Ocean Declaration (Manado, 14 May 2009) at 15, http://www.grida.no/marine/news.aspx?id=3691, accessed 17 October 2012.

[82] See http://www.environment.gov.au/coasts/mbp/reserves, accessed 17 October 2012. See also Lorne Kriwoken, Julie Davidson and Michael Lockwood, 'Marine Protected Areas and Transboundary Governance' in Robin Warner and Simon Marsden (eds), *Transboundary Environmental Governance: Inland, Coastal and Marine Perspectives* (Ashgate 2012) 85.

[83] See, e.g., Director of National Parks, Draft South-East Commonwealth Marine Reserves Network Management Plan 2012–22 (2012) 14. See further Lorne Kriwoken, 'Australian Marine Protected Areas: Charting a Course Towards a Representative System' in Warwick Gullett, Clive Schofield and Joanna Vince (eds), *Marine Resources Management* (LexisNexis 2011) 171. For an assessment of the impact of climate change upon Australia's maritime estate see E S Poloczanska, A J Hobday and A J Richardson (eds), *A Marine Climate Change Impacts and Adaptation Report Card for Australia 2009* (NCCARF Publication 2009).

[84] Mark Axelrod, 'Climate Change and Global Fisheries Management: Linking Issues to Protect Ecosystems or to Save Political Interests?' (2011) 11 Global Environmental Politics 64.

[85] FAO, above note 42 at 56.

It was noted earlier in this chapter that well-developed tools of fisheries management may be deployed to control access to fisheries in response to the effects that climate change is having upon their productivity. Where a fishery falls wholly within a coastal state's maritime jurisdiction there is greater likelihood that the fishery management system will be anticipatory of potential change and responsive to observed climatic impacts. This is because nationally managed fisheries are less susceptible to the 'tragedy of the commons' problem besetting multi-jurisdictional and high seas fisheries. The performance of RFMOs varies significantly depending upon their membership and institutional dynamics, and given the failure of many RFMOs to manage the stocks under their jurisdiction in a sustainable manner there are reasons to doubt they will respond effectively to climate change. A case in point is the Commission for the Conservation of Southern Bluefin Tuna, established by the 1993 Convention for the Conservation of Southern Bluefin Tuna,[86] which has not halted the progressive decline in the abundance of this species because of an insistence by key members, most notably Japan, that the Commission adopt an unsustainable Total Allowable Catch.[87]

Climate change is affecting not only the productivity of fisheries but also their distribution, which calls for a reconsideration of management approaches and, in some RFMOs, for a reassessment of an even more fundamental question, namely the legal boundaries of fisheries management areas set by RFMO constituent instruments. An example is the 1980 Convention on the Conservation of Antarctic Marine Living Resources (CCAMLR).[88] CCAMLR is part of the Antarctic Treaty System of which the 1959 Antarctic Treaty[89] is the linchpin. CCAMLR operates within the 'CCAMLR boundary', which is a saw-toothed circumpolar line that roughly approximates the polar front – the oceanic boundary where colder Antarctic waters meet warmer northern waters (often called the 'Antarctic convergence').[90] All marine ecosystems in the Antarctic are vulnerable to climate change, and key food chain species such as krill are

---

[86] Convention for the Conservation of Southern Bluefin Tuna (Canberra, 10 May 1993), 1819 UNTS 359, entered into force 20 May 1994.

[87] 1329 UNTS 48.

[88] Convention on the Conservation of Antarctic Marine Living Resources (Canberra, 1 August 1980), 1329 UNTS 47, entered into force 7 April 1982.

[89] Antarctic Treaty (Washington, 1 December 1959), 402 UNTS 71, entered into force 23 June 1961.

[90] Art. 1(4) CCAMLR.

declining as a result of warming waters.[91] Moreover, there is evidence of a southward shift in the polar ocean front, and Haward and Jabour have noted that if this continues or accelerates then from a management perspective the 30-year-old CCAMLR boundary will no longer embrace all relevant Southern Ocean conservation and fisheries interests.[92]

6 CONCLUSION

The international legal framework for climate change adaptation in coastal and marine environments is in an early stage of development, by comparison with national and sub-national legal responses. In the coastal and in-shore zones, the focus nationally and internationally has been on adaptive policies that can allow communities and ecosystems to adjust to rising sea levels. The imperative for coastal adaptation to climate change is a long-standing one, having been considered since the early 1990s, and is supported by several clear references in the texts underpinning the international climate change regime.

There is also growing understanding of the relevance of the law of the sea to physical and legal measures taken by coastal states to protect their entitlements in littoral areas, whether this is by coastal defence measures and land reclamation, or through technical legal responses to fix baselines despite the shifting sands and rising seas that climate change is bringing to many coastal zones. In relation to physical responses, coastal states may only take those measures that do not further harm the marine environment, in order to meet the obligations under the LOSC to protect and conserve the totality of the marine environment, and in conformity with the precautionary principle which has been recognised by ITLOS as an emergent rule of customary international law. This clearly suggests that coastal states should be giving priority to soft protection measures that build the resilience of ecosystems and habitats, such as mangroves and reefs that offer protection against climate change impacts such as storm surges. In relation to legal responses, whether these be proactive decisions to reaffirm existing baselines, or through studied inactivity by

---

[91] P N Trathan and D Agnew, 'Climate Change and the Antarctic Marine Ecosystem: An Essay on Management Implications' (2010) 22 Antarctic Science 387.

[92] Marcus Haward and Julia Jabour, 'Environmental Change and Governance Challenges in the Southern Ocean' in Tim Stephens and David VanderZwaag (eds), *Polar Oceans Governance in an Era of Environmental Change* (Edward Elgar forthcoming).

declining to reopen existing lines in the sea, the LOSC offers a more resilient system than often thought, with jurisdiction determined primarily by objective physical factors at the time of claim and delineation.[93] This will, of course, offer only a marginal level of comfort to small island states, the very existence of which is threatened, and in relation to which serious consideration needs to be given to legal mechanisms for preserving existing entitlements to offshore marine resources.[94]

It is in areas beyond the coastal zone, in the offshore marine environment, where there is an urgent need for greater attention to climate change adaptation issues. With the exception of some fisheries, there are virtually no steps being taken to safeguard large marine ecosystems that are directly threatened by climate change impacts. This is seen most clearly in relation to ocean acidification, which is a climate change related impact just as serious as global warming of the atmosphere, yet has not merited specific mention in any legally binding international agreement. As in the littoral zone, the first step towards an effective legal and policy framework is understanding the extent of the changes requiring adaptive responses. However, for most marine areas, while there is improving scientific knowledge, there has been little analysis of the flow-on social and economic effects to which human civilisation will need to adapt if it is to weather the coming climate storm.

---

93 Purcell, above note 69 at 760.

94 See Emily Crawford and Rosemary Rayfuse, 'Climate Change and Statehood' in Rosemary Rayfuse and Shirley V Scott (eds), *International Law in the Era of Climate Change* (Edward Elgar 2012) 243; L Yamamoto and M Esteban, 'Vanishing Island States and sovereignty' (2010) 53 Ocean and Coastal Management 1; Emily Follett, 'Sinking Islands, Shrinking Maritime Entitlements: Is a New Rule of International Law Needed to Protect and Promote Offshore Minerals Exploration and Mining?' (2012) 29 Environmental and Planning Law Journal 75.

11. Climate change adaptation and biodiversity law

Arie Trouwborst

1 INTRODUCTION

Climate change appears to be adding significant new challenges to the array of anthropogenic threats already faced by wild flora and fauna, the ecosystems of which they are part, and biological diversity (biodiversity) at large – that is, the variability of life on Earth.[1] The composition of ecosystems is profoundly affected, as species' ranges are shifting in response to altering climatic conditions, and populations struggle to cope with increasingly frequent extreme weather events.

Unsurprisingly, climate change is also posing significant challenges to biodiversity *law* (alternatively referred to as nature conservation law) at national, regional and global levels. It is now evident that in order to meet (inter)national biodiversity targets specific conservation action is required to help species adapt to climate change with minimal losses. This imperative to help nature adapt to climate change was not apparent at the time most of the existing legal instruments aimed at nature conservation came into being. The leading paradigm for most conservation instruments is maintaining or restoring the status quo rather than facilitating change. It is also evident that climate change substantially augments the need for international cooperation in biodiversity conservation.

In line with the central objective of this book, the current chapter is intended to contribute to the identification of the legal challenges involved in the adaptation of biodiversity, and of biodiversity law, to

[1] In Article 2 of the Convention on Biological Diversity (adopted 5 June 1992; in force 29 December 1993; 1760 UNTS 79) (CBD), biological diversity is defined as 'the variability among living organisms from all sources including, inter alia, terrestrial, marine and other aquatic ecosystems and the ecological complexes of which they are part; this includes diversity within species, between species and of ecosystems'.

climate change, and of suitable ways to address these challenges.[2] In doing so, the chapter focuses primarily, albeit not exclusively, on international legal instruments. The chapter is structured as follows. Section 2 concisely introduces the various impacts of climate change on species and ecosystems, as well as the biodiversity adaptation measures commonly recommended in the scientific literature. The third section provides an overall analysis of the role of biodiversity law in this regard, including by distilling and combining key findings from previous studies. This is complemented in the fourth section by an in-depth analysis of one type of action which is universally regarded as crucial to a successful adaptation of biodiversity to climate change, namely the maintenance and restoration of adequate connectivity in order to allow organisms to disperse naturally through landscapes in pursuit of suitable climatic conditions. The fifth section contains concluding observations.

2 CLIMATE CHANGE AND BIODIVERSITY CONSERVATION

It is currently well established that climate change is exercising significant and increasing effects on biological diversity across the globe, and that action is urgently required if species and ecosystems are to adapt to the changing climate with as few losses as possible.[3] Plant and animal populations are responding to modifications in temperature, humidity, and weather patterns. The distributions of many species are already shifting (or attempting to shift) to higher latitudes and altitudes, and these transitions are expected to intensify. More frequently occurring extreme weather events associated with climate change – droughts, floods, heat

---

2 The research for this chapter was conducted as part of a research project funded by the Netherlands Organization for Scientific Research (NWO), entitled 'Towards Climate Change Proof International Nature Conservation Law: International Law and the Adaptation of Species and Ecosystems to the Effects of Climate Change'.

3 A clear overview and discussion of impacts and adaptation measures is contained in Alison Campbell et al., *Review of the Literature on the Links between Biodiversity and Climate Change: Impacts, Adaptation and Mitigation* (Secretariat of the Convention on Biological Diversity 2009). Attempts by the present author to summarize the scientific literature on the effects of climate change on biodiversity and on recommended adaptation measures were made in Arie Trouwborst, 'International Nature Conservation Law and the Adaptation of Biodiversity to Climate Change: A Mismatch?' (2009) 21 Journal of Environmental Law 419, 419–421 and 426–429.

waves, cold spells – also impact on populations. Nature has adapted to climate changes in similar ways in the past, but the current rate of change is unusually rapid and, moreover, many species are already under substantial stress through habitat fragmentation, over-exploitation and other anthropogenic threats. Overall, climate change is thus anticipated to have increasingly significant adverse consequences for biodiversity.

The scientific literature documenting these effects is huge and growing rapidly. A few recent examples are given here by way of illustration. A review in *Nature Climate Change* finds that bird and butterfly populations in Europe are changing fast, but not fast enough to keep pace with the changing climate.[4] In the last two decades, butterfly and bird communities moved an average 114 and 37 kilometres respectively to the north, but are still lagging well behind the even swifter northward shift of average temperatures. The 'climate debt' built up in 20 years by the studied butterfly and bird communities was found to correspond to a 135 kilometres lag behind climate for butterflies and a 212 kilometres lag for birds. Apparently, even winged creatures have trouble keeping up with changes in climatic conditions, *inter alia* because the base of their food chains does not keep up.[5]

Another publication reports a recent discovery concerning the Ethiopian bush-crow (*Zavattariornis stresemanni*). The curiously restricted range of this globally endangered bird, which has puzzled scientists for decades, turns out to coincide precisely with a 'bubble' of cool, dry climate.[6] Within this area, where the average temperature is below 20°C bush-crows are virtually everywhere. Outside the area, where the average temperature is higher, they do not occur at all, probably because bush-crow chicks cannot survive there. Hence the description of the Ethiopian bush-crow by lead author Paul Donald as 'a cool bird, that appears to like staying that way'.[7] With its suitable climate space shrinking, the bird is thus one of the species in the world most vulnerable to climate change. In other words, the bush-crow is in the same

---

[4] Vincent Devictor et al., 'Differences in the Climatic Debts of Birds and Butterflies at a Continental Scale' (2012) 2 Nature Climate Change 121.

[5] Ibid.

[6] Paul F Donald et al., 'The Restricted Range of the Ethiopian Bush-crow *Zavattariornis stresemanni* is a Consequence of High Reliance on Modified Habitats within Narrow Climatic Limits' (2012) 153 Journal of Ornithology 1031.

[7] Birdlife International, 'Scientists Discover an "Invisible Barrier" that Holds the Answer to One of Nature's Little Mysteries' (16 March 2012) *Birdlife News Posts*, http://www.birdlife.org, accessed on 31 October 2012.

predicament as other species with nowhere else to go as their habitats warm up, including many island and mountain dwellers.

Various proactive measures aimed at helping nature adapt to climate change have been recommended in the scientific literature. Frequently recommended actions are protecting and restoring large, robust natural areas; ensuring adequate connectivity between such areas – thus creating protected area networks; taking management measures to boost the resilience of species and ecosystems to changing conditions and extreme climatic events, *inter alia* by reducing other, 'conventional' threats; and, for certain species, undertaking the active translocation of populations to more suitable areas (also known as 'assisted migration', 'assisted dispersal' or 'assisted colonization').[8] Clearly, with climate change the need for effective international cooperation in nature conservation is on the increase. It should also be noted here that from a biodiversity conservation perspective, the *mitigation* of climate change itself is at least as essential as the aforementioned measures. This is beyond the scope of the present chapter, however, which solely encompasses the *adaptation* of biodiversity to the 'committed' climate change which has already been set in motion.

3 ADAPTING BIODIVERSITY LAW TO CLIMATE CHANGE

3.1 Biodiversity Law and Climate Change

'Biodiversity law' comprises an elaborate and complex amalgamation of legal frameworks at national, regional and global levels. At the lower level, it covers a wide array of (sub)national legislation concerning the protection of natural areas and species. At regional levels, this legislation is complemented by instruments such as the 1968 African Convention on Nature Conservation;[9] the 1940 Western Hemisphere

---

[8] See Campbell et al., above note 3; and Trouwborst, above note 3 at 426–429.

[9] African Convention on the Conservation of Nature and Natural Resources (adopted 15 September; in force 16 June 1969; 1001 UNTS 4); the revised version of this Convention, which was adopted in 2003, has yet to enter into force. See International Union for the Conservation of Nature and Natural Resources (IUCN), *An Introduction to the African Convention on the Conservation of Nature and Natural Resources* (IUCN 2004); and Michael Bowman, Peter

Convention,[10] the 1979 Bern Convention on European Wildlife and Natural Habitats (Bern Convention),[11] the European Union (EU) 1979 Wild Birds[12] and 1992 Habitats[13] Directives, and the 1991 Environment Protocol to the Antarctic Treaty.[14] Yet another level up, nature conservation instruments with a global scope include the 1971 Ramsar Wetlands Convention,[15] the 1972 UNESCO World Heritage Convention,[16] the 1979 Bonn Migratory

---

Davies and Catherine Redgwell, *Lyster's International Wildlife Law* (2nd edn, CUP 2010), 262–296.

[10] Convention on Nature Protection and Wild Life Preservation in the Western Hemisphere (adopted 12 October 1940; in force 1 May 1942; 161 UNTS 193). See Bowman et al., ibid. at 241–261.

[11] Convention on the Conservation of European Wildlife and Natural Habitats (adopted 19 September 1979; in force 1 June 1982; ETS No. 104). See Bowman et al., ibid. at 297–345; and Carolina Lasén Díaz, 'The Bern Convention: 30 Years of Nature Conservation in Europe' (2010) 19 Review of European Community and International Environmental Law 185.

[12] Directive 2009/147/EC of the European Parliament and of the Council on the Conservation of Wild Birds [2010] OJ L20/7; this is the codified version of Council Directive 79/409/EEC as subsequently modified. See Wouter P J Wils, 'The Birds Directive: 15 Years Later' (1994) 6 Journal of Environmental Law 219; Jonathan Verschuuren, 'Effectiveness of Nature Protection Legislation in the EU and the US: The Birds and Habitats Directives and the Endangered Species Act' (2003) 3 Yearbook of European Environmental Law 305; and Bowman et al., above note 9 at 206–212.

[13] Council Directive 92/43/EC on the Conservation of Natural Habitats and of Wild Fauna and Flora [1992] OJ L206/7. See Jonathan Verschuuren, 'Implementation of the Convention on Biodiversity in Europe: 10 Years of Experience with the Habitats Directive' (2002) 5 Journal of International Wildlife Law and Policy 251; Verschuuren, above note 12.

[14] Protocol to the Antarctic Treaty on Environmental Protection (adopted 4 October 1991; in force 14 January 1998; 30 ILM 1461). See Catherine Redgwell, 'Environmental Protection in Antarctica: The 1991 Protocol' (1994) 43 International and Comparative Law Quarterly 599; Bowman et al., above note 9 at 368–374.

[15] Convention on Wetlands of International Importance Especially as Waterfowl Habitat (adopted 2 February 1971; in force 21 December 1975; 996 UNTS 245). See Michael Bowman, 'The Ramsar Convention on Wetlands: Has it Made a Difference?' (2002) 10 Yearbook on International Cooperation on Environment and Development 61; and Bowman et al., above note 9 at 403–450; C Max Finlayson et al., 'The Ramsar Convention and Ecosystem-based Approaches to the Wise Use and Sustainable Development of Wetlands' (2011) 14 Journal of International Wildlife Law and Policy 176.

[16] UNESCO Convention Concerning the Protection of the World Cultural and Natural Heritage (adopted 16 November 1972; in force 17 December 1975;

Species Convention (CMS)[17] and the 1992 Biodiversity Convention (CBD).[18]

Most of the biodiversity law currently in force, including all the aforementioned international instruments, was created before the issue of the adaptation of nature to climate change made its way onto (inter) governmental agendas. The provisions of most instruments, therefore, do not expressly address the issue. Consequently, a growing scholarly literature is devoted to assessing the current capacity of (inter)national nature conservation regimes to facilitate the adaptation of species and ecosystems to climate change, and to examining ways of enhancing that capacity.[19] In the remainder of the present section, several key findings

---

11 ILM 1358). See Francesco Francioni (ed.), *The 1972 World Heritage Convention: A Commentary* (OUP 2008); Bowman et al., above note 9 at 451–482.

[17] Convention on the Conservation of Migratory Species of Wild Animals (adopted 23 June 1979; in force 1 November 1983; 10 ILM 15). See Simon Lyster, 'The Convention on the Conservation of Migratory Species of Wild Animals (The "Bonn Convention")' (1989) 29 Natural Resources Journal 979; Richard Caddell, 'International Law and the Protection of Migratory Wildlife: An Appraisal of Twenty-Five Years of the Bonn Convention' (2005) 16 Colorado Journal of International Environmental Law and Policy 113; Bowman et al., above note 9 at 535–583.

[18] Above note 1; see Daniel Bodansky, 'International Law and the Protection of Biological Diversity' (1995) 28 Vanderbilt Journal of Transnational Law 623; Michael J Bowman and Catherine Redgwell (eds), *International Law and the Conservation of Biological Diversity* (Kluwer Law International 1996); Bowman et al., above note 9 at 587–629.

[19] The relevant literature concerning *international* instruments includes: Michael Bowman, 'Global Warming and the International Legal Protection of Wildlife' in Robin R Churchill and David Freestone (eds), *International Law and Global Climate Change* (Graham & Trotman 1991); Gerard C Boere and Douglas Taylor, 'Global and Regional Governmental Policy and Treaties as Tools Towards the Mitigation of the Effect of Climate Change on Waterbirds' (2004) 146 Ibis 111; Kim Wheeler, 'Bird Protection & Climate Changes: A Challenge for Natura 2000?' (2006) 13 Tilburg Foreign Law Review 283; David Hodas, 'Biodiversity and Climate Change Laws: A Failure to Communicate?' in Michael I Jeffery et al. (eds), *Biodiversity, Conservation, Law and Livelihoods: Bridging the North–South Divide* (CUP 2008); An Cliquet, Chris Backes, Jim Harris and Peter Howsam, 'Adaptation to Climate Change: Legal Challenges for Protected Areas' (2009) 5 Utrecht Law Review 158; Arie Trouwborst, 'International Nature Conservation Law and the Adaptation of Biodiversity to Climate Change: A Mismatch?' (2009) 21 Journal of Environmental Law 419; Saja Erens, Jonathan Verschuuren and Kees Bastmeijer, 'Adaptation to Climate Change to Save Biodiversity: Lessons Learned from African and European Experiences' in

from this literature are highlighted and discussed. Given that attempting a comprehensive instrument-by-instrument approach would not be feasible in the present context, and would in any case have limited added value, the subsequent section will explore a cross-section of national, regional and global instruments from the perspective of a single, albeit crucial, climate adaptation measure: connectivity.

A number of things have become apparent from the legal literature on the adaptation of biodiversity to climate change published to date. It is useful to begin with one crucial reality, which runs the risk of being overlooked in discussions on the adaptation of biodiversity law to climate change, namely that the effective implementation of most legal instruments *as they are* would significantly boost the capacity of species and ecosystems to adapt to climate change. After all, to quote one biological study: '[p]erhaps the most important policy goal should be to encourage large, genetically diverse, populations that will enhance the ability

---

Benjamin J Richardson et al. (eds), *Climate Law and Developing Countries: Legal and Policy Challenges for the World Economy* (Edward Elgar 2009); Daniel Schramm and Akiva Fishman, 'Legal Frameworks for Adaptive Natural Resource Management in a Changing Climate', (2010) 22 Georgetown International Environmental Law Review 491; Jonathan Verschuuren, 'Rethinking Restoration in the European Union's Birds and Habitats Directives' (2010) 28 Ecological Restoration 431; Aline Kühl and Elizabeth Maruma Mrema, 'Impacts of Climate Change on Biodiversity, with a Focus on Migratory Species' in Tuula Honkonen and Ed Couzens (eds), *International Environmental Law-making and Diplomacy Review 2010* (University of Eastern Finland 2010); Arie Trouwborst, 'Conserving European Biodiversity in a Changing Climate: The Bern Convention, the European Union Birds and Habitats Directives and the Adaptation of Nature to Climate Change' (2011a) 20 Review of European Community and International Environmental Law 62; Arie Trouwborst, 'La Adaptación de la Flora y la Fauna al Cambio Climático en un Paisaje Fragmentado, y el Derecho Europeo sobre la Conservación de la Naturaleza' (2011b) 2 Revista Catalana de Derecho Ambiental 1; Arie Trouwborst, 'Transboundary Wildlife Conservation in a Changing Climate: The Adaptation of the Bonn Convention on Migratory Species and its Daughter Instruments to Climate Change' (2012) 4 Diversity 258; Arie Trouwborst, 'Migratory Species Conservation in Warming Polar Oceans, With Particular Reference to Seabirds' in Erik Jaap Molenaar, Alex G Oude Elferink and Don R Rothwell (eds), *The Law of the Sea and Polar Regions: Interactions between Global and Regional Regimes* (Martinus Nijhoff Publishers forthcoming 2013). The literature concerning *national* regimes is rather more extensive, and no effort is made to be exhaustive here. Two examples are Jo E Hossell et al., 'Climate Change and Nature Conservation: Implications for Policy and Practice in Britain and Ireland' (2003) 11 Journal for Nature Conservation 67; and J B Ruhl, 'Climate Change and the Endangered Species Act: Building Bridges to the No-Analog Future' (2008) 88 Boston University Law Review 1.

of … species to adapt to, or exploit, the changes caused by increasing human impact on the global climate'.[20] Such increased adaptability, or resilience, of populations and the habitats on which they depend can be attained precisely 'by reducing conventional threats such as pollution, habitat fragmentation and overexploitation'.[21] And that, in turn, is exactly what existing nature conservation acts and treaties were designed to do. Hence, from a climate change adaptation perspective, improving the effectiveness of existing instruments would seem a much better course of action than discarding them as 'outdated' and throwing them overboard. Likewise, the concrete means required to facilitate the adaptation of nature to climate change chiefly consist of familiar tools, including the legal protection of certain species and certain areas from adverse human impacts, although their application will need to involve a greater degree of adaptability. As compared to pre-climate change nature conservation, to cite another biological study, in many cases 'the management tools will remain largely the same, but the way they are applied will change'.[22]

3.2 Interpretation

Another outcome of prior research highlighted here concerns the considerable potential of *interpretation* as a means of adapting biodiversity law to the current era of climate change.[23] It is useful to briefly recall the principal international law rules governing treaty interpretation in this regard, and then to consider a few concrete instances with those rules in mind.

---

20 Robert A Robinson et al., 'Travelling through a Warming World: Climate Change and Migratory Species' (2009) 7 Endangered Species Research 87, 96.

21 Zoological Society of London, *Climate Change Vulnerability of Migratory Species: Species Assessments – Preliminary Review*, UNEP/CMS/ScC17/Inf.9 (Zoological Society of London 2011).

22 Abraham J Miller-Rushing, Richard B Primack and Cagan H Sekercioglu, 'Conservation Consequences of Climate Change for Birds' in Anders P Møller, Wolfgang Fiedler and Peter Berthold (eds), *Effects of Climate Change on Birds* (OUP 2010) 306.

23 See, in particular, Trouwborst (2011a), above note 19; and Trouwborst (2012), above note 19.

3.2.1 The rules of treaty interpretation

According to the generally accepted main rule, as reflected in the Vienna Convention on the Law of Treaties (VCLT),[24] a treaty 'shall be interpreted in good faith in accordance with the ordinary meaning to be given to the terms of the treaty in their context and in the light of its object and purpose'.[25] As the International Law Commission has clarified: 'When a treaty is open to two interpretations one of which does and the other does not enable the treaty to have appropriate effects, good faith and the objects and purposes of the treaty demand that the former interpretation should be adopted.'[26] In addition to treaty objectives, account must be taken, *inter alia*, of 'any subsequent agreement between the parties regarding the interpretation of the treaty or the application of its provisions'; 'any subsequent practice in the application of the treaty which establishes the agreement of the parties regarding its interpretation'; and 'any relevant rules of international law applicable in the relations between the parties'.[27] Especially relevant for present purposes are the potential influences on the interpretation of treaty provisions of stated treaty objectives ('object and purpose') and of subsequent decisions adopted by treaty bodies.[28]

Both can be illustrated with reference to a notable appeal ruling by the Netherlands Crown of 2007, involving the interpretation of the Ramsar Wetlands Convention.[29] In this case, the Crown ruled that the authorization of the construction of a resort in the proximity of a wetland

---

24 Convention on the Law of Treaties (adopted 23 May 1969; in force 27 January 1980; 1155 UNTS 331). On treaty interpretation generally, see R Gardiner, *Treaty Interpretation* (OUP 2008).

25 Ibid. Article 31(1).

26 *Yearbook of the International Law Commission* (1966, Vol. II) 219.

27 VCLT Article 31(3).

28 On the potential of such decisions serving as 'subsequent agreement' or 'subsequent practice' in the context of treaty interpretation see, *inter alia*, R Churchill and G Ulfstein, 'Autonomous Institutional Arrangements in Multilateral Environmental Agreements: A Little-Noticed Phenomenon in International Law' (2000) 94 American Journal of International Law 623, 641; A Wiersema, 'The New International Law-Makers? Conferences of the Parties to Multilateral Environmental Agreements' (2009) 31 Michigan Journal of International Law 231; Bowman et al., above note 9 at 46.

29 Netherlands Crown Decision (in Dutch) in the case lodged by the Competent Authority for the Island of Bonaire on the annulment of two of its decisions on the Lac wetland by the Governor of the Netherlands Antilles of 11 September 2007, (2007) 347 Staatsblad. For an English summary and commentary, see Jonathan Verschuuren, 'Ramsar Soft Law is Not Soft at All: Discussion of the 2007 Decision by the Netherlands Crown on the Lac Ramsar Site on the

occurring on the Convention's List of Wetlands of International Importance had been rightfully annulled on the basis of Article 3 of the Convention, because an environmental impact assessment (EIA) had not been performed. However, the rather open-ended language of Article 3 does not, as such, appear to require an EIA at all.[30] Various decisions adopted by the Conference of the Parties (COP) to the Ramsar Convention in connection with Article 3 *do* call for EIAs, but these are not by themselves legally binding. After recalling these decisions and citing VCLT, Article 31, the Crown reasoned thus:

> Although Article 3 of the Ramsar Convention does leave the state parties considerable discretionary powers as to the exact procedure, the authorities cannot agree on activities in or nearby a Ramsar site without an EIA … The Ramsar Convention has to be faithfully interpreted and implemented by the state parties so that its aims are achieved … Article 3 of the Ramsar Convention … has to be carried out *in the light of the aim* of conservation and preservation of the special ecological character of wetlands. In addition, … when interpreting the provisions of the convention, *later resolutions, recommendations and guidelines that were adopted by the parties* to the convention, have to be taken into account … Such resolutions, recommendations and guidelines are especially important because Article 3 itself does not offer much to hold on to. In addition, … it is important that the resolutions and recommendations have been adopted unanimously by the Conference of the Parties, in which all state parties, including the Kingdom of the Netherlands, are represented.[31]

Although not every court in every case will necessarily take as generous an approach, this ruling nevertheless provides a clear demonstration of the potential influence of treaty aims and COP decisions on the interpretation of treaty obligations.

Island of Bonaire' (2008), http://www.ramsar.org/pdf/wurc/wurc_verschuuren_bonaire.pdf, accessed 31 October 2012.

[30] Article 3 of the Convention reads: '(1) The Contracting Parties shall formulate and implement their planning so as to promote the conservation of the wetlands included in the List, and as far as possible the wise use of wetlands in their territory. (2) Each Contracting Party shall arrange to be informed at the earliest possible time if the ecological character of any wetland in its territory and included in the List has changed, is changing or is likely to change as the result of technological developments, pollution or other human interference.'

[31] Translation by Verschuuren, note 29 above (emphasis added).

3.2.2 Treaty interpretation and adaptation to climate change

Whereas the Dutch case just discussed did not involve the adaptation of nature to climate change, a parallel with the interpretation of 'outdated' treaty provisions in light of the need for adaptation action is easily drawn. In fact, Article 3 of the Ramsar Wetlands Convention is itself a good example. The obligation of states parties to 'formulate and implement their planning so as to promote the conservation of the wetlands included in the List, and as far as possible the wise use of wetlands in their territory'[32] must now be understood in light of Ramsar Convention COP Resolutions calling on parties to take climate adaptation action, including to 'manage wetlands so as to increase their resilience to climate change and extreme climatic events'.[33] In international nature conservation instruments, broadly phrased conservation obligations like the one in Article 3 of the Ramsar Wetlands Convention abound. Examples from other global instruments include Article 5 of the World Heritage Convention[34] and Article 8 of the CBD.[35] The duties in the latter provision, for instance, are now informed by Conference of the Parties to the CBD Decisions urging parties to 'enhance the integration of climate-change

---

[32] Article 3(1), see above note 30.

[33] Ramsar COP Resolution VIII.3 on Climate Change and Wetlands: Impacts, Adaptation, and Mitigation (adopted 26 November 2002), para. 14; see also Ramsar COP Resolution X.24 on Climate Change and Wetlands (adopted 4 November 2008); and Ramsar COP Resolution XI.14 on Climate Change and Wetlands: Implications for the Ramsar Convention on Wetlands (adopted 13 July 2012).

[34] Article 5(d) commits parties to taking 'the appropriate legal, scientific, technical, administrative and financial measures necessary for the identification, protection, conservation, presentation and rehabilitation' of 'natural heritage'. See also Trouwborst (2009), above note 19 at 432–433.

[35] Article 8(a)-(f): 'Each Contracting Party shall, as far as possible and as appropriate: (a) Establish a system of protected areas or areas where special measures need to be taken to conserve biological diversity; (b) Develop, where necessary, guidelines for the selection, establishment and management of protected areas or areas where special measures need to be taken to conserve biological diversity; (c) Regulate or manage biological resources important for the conservation of biological diversity whether within or outside protected areas, with a view to ensuring their conservation and sustainable use; (d) Promote the protection of ecosystems, natural habitats and the maintenance of viable populations of species in natural surroundings; (e) Promote environmentally sound and sustainable development in areas adjacent to protected areas with a view to furthering protection of these areas; (f) Rehabilitate and restore degraded ecosystems and promote the recovery of threatened species, inter alia, through the development and implementation of plans or other management strategies.'

considerations related to biodiversity in their implementation of the Convention' by taking 'appropriate actions to address' the impacts of climate change on biodiversity,[36] and in particular to:

> Reduce the negative impacts from climate change as far as ecologically feasible, through conservation and sustainable management strategies that maintain and restore biodiversity;
>
> Implement activities to increase the adaptive capacity of species and the resilience of ecosystems in the face of climate change, including, *inter alia*:
>
> (i) Reducing non-climate stresses, such as pollution, over-exploitation, habitat loss and fragmentation and invasive alien species;
> (ii) Reducing climate related stresses, where possible, such as through enhanced adaptive and integrated water resource and marine and coastal management;
> (iii) Strengthening protected area networks including through the use of connectivity measures such as the development of ecological networks and ecological corridors and the restoration of degraded habitats and landscapes…;
> (iv) Integrating biodiversity into wider seascape and landscape management;
> (v) Restoring degraded ecosystems and ecosystem functions; and
> (vi) Facilitating adaptive management by strengthening monitoring and evaluation systems;
>
> Bearing in mind that under climate change, natural adaptation will be difficult and recognizing that *in situ* conservation actions are more effective, also consider *ex situ* measures, such as relocation, assisted migration and captive breeding, among others, that could contribute to maintaining the adaptive capacity and securing the survival of species at risk, taking into account the precautionary approach in order to avoid unintended ecological consequences.[37]

As for a regional example, Article II of the African-Eurasian Waterbirds Agreement (AEWA)[38] – a CMS ancillary instrument – prescribes in general terms that parties 'shall take co-ordinated measures to maintain

---

[36] CBD COP Decision IX/16 on Biodiversity and Climate Change (adopted 30 May 2008), para. A(4).

[37] CBD COP Decision X/33 on Biodiversity and Climate Change (adopted 29 October 2010), para. 8(c)–(e). See also CBD COP Decision VII/15 on Biodiversity and Climate Change (adopted 20 February 2004); CBD COP Decision VII/28 on Protected Areas (adopted 20 February 2004); CBD COP Decision VIII/30 on Biodiversity and Climate Change (adopted 31 March 2006); CBD COP Decision IX/16, note 36 above; and CBD COP Decision X/31 on Protected Areas (adopted 29 October 2010).

[38] Agreement on the Conservation of African–Eurasian Migratory Waterbirds (adopted 16 June 1995; in force 1 November 1999; 6 *Yearbook of International*

migratory waterbird species in a favourable conservation status or to restore them to such a status'.[39] This result-oriented provision ostensibly requires parties simply to do what it takes to secure a favourable conservation status for the species involved, which are listed in Annex 2 to the Agreement. If what it takes includes climate adaptation action, then that is what parties must undertake if they are to implement their treaty obligations in good faith. In other words, Article II, albeit without stating this in so many words, apparently implies a legal duty for AEWA parties to take action to facilitate the adaptation of waterbirds to climate change. Again, this position is reinforced by a series of resolutions dedicated to the topic by the AEWA Meeting of the Parties (MOP).[40] Incidentally, Conservation Guidelines on climate adaptation adopted under the Agreement confirm the conclusion drawn above.[41]

It is not difficult to identify further examples like the ones just discussed. For instance, an analysis of Articles 2 and 4 of the Bern Convention, interpreted in light of the wealth of guidance provided in the Bern Convention Standing Committee Recommendations on measures needed to aid the adaptation of European biodiversity to climate change, rendered comparable conclusions[42] (see also section 4.2 below). Further examples where similar reasoning appears to apply include broad obligations to take the necessary measures to achieve a certain conservation outcome, laid down, for instance, in other CMS ancillary instruments such

---

Environmental Law (1995) 907). See B Lenten, 'A Flying Start for the Agreement on the Conservation of African–Eurasian Migratory Waterbirds (AEWA)' (2001) 4 Journal of International Wildlife Law and Policy 159; and R Adam, 'Waterbirds, the 2010 Biodiversity Target, and Beyond: AEWA's Contribution to Global Biodiversity Governance' (2008) 38 Environmental Law Review 87.

[39] AEWA Article II(1).

[40] AEWA MOP Resolution 3.17 on Climate Change and Migratory Waterbirds (adopted 27 October 2005); AEWA MOP Resolution 4.14 on the Effects of Climate Change on Migratory Waterbirds (adopted 19 September 2008); AEWA MOP Resolution 5.13 on Climate Change Adaptation Measures for Waterbirds (adopted 18 May 2012). For detailed discussion see Trouwborst (2012), above note 19 at 271–274.

[41] Ilya Maclean and Mark Rehfisch, *Guidelines on the Measures Needed to Help Waterbirds Adapt to Climate Change*, AEWA Conservation Guidelines No. 12 (AEWA 2008). With respect to AEWA Article II(1), and also Article III(2)(e), the document submits that '[a]lthough neither clause makes explicit reference to climate change, both suggest that remedial measures are needed to combat adverse effects, including climate change, on waterbirds.'

[42] See Trouwborst (2011a), above note 19.

as the Agreement on Albatrosses and Petrels (ACAP),[43] and in regional instruments addressing the marine environment such as the OSPAR Agreement for the North-East Atlantic.[44] The latter, for example, obliges states parties to take the 'necessary measures' to protect the ecosystems and biodiversity of the marine region in question.[45] Naturally, the effect of interpretation will vary from case to case, depending, *inter alia*, on the wording employed in relevant decisions by parties and in the treaty text itself.

In sum, as a result of interpreting the 'ordinary meaning' of the provisions involved in light of 'object and purpose' and/or subsequent agreements and practice, the aforementioned obligations and others like them can now validly be construed as requiring action to help biodiversity adapt to climate change, even though climate change was not on the drafters' minds at the time the provisions were formulated. In other words, the above analysis clearly demonstrates the considerable potential of interpretation in closing the large gap which would, at first sight, appear to exist between what is currently required by the existing legal instruments on biodiversity conservation and what is needed to enable the adaptation of biodiversity to climate change with minimal losses.

3.3 Law Reform

It would, however, go too far to describe interpretation as a silver bullet in the present context. There are limits to what it can achieve. Different states – and different courts – may take (slightly or substantially) different approaches. For instance, the extent to which, and conditions under which, non-binding decisions by treaty parties can pose as 'subsequent agreement' or 'subsequent practice' remain open to debate. This, together with the general formulation of many of the treaty provisions in question, results in a degree of legal uncertainty as to the precise scope and content of the obligations involved in the field of climate adaptation action. Moreover, treaty provisions can only be stretched so far, and

---

43 Agreement on the Conservation of Albatrosses and Petrels (adopted 19 June 2001; in force 1 February 2004; 2588 UNTS 257), Article II: '(1) The objective of this Agreement is to achieve and maintain a favourable conservation status for albatrosses and petrels. (2) The Parties shall take measures, both individually and together, to achieve this objective.'

44 Convention for the Protection of the Marine Environment of the North-East Atlantic (adopted 22 September 1992; in force 25 March 1998; 32 ILM 1072).

45 Ibid., Annex V, Article 2.

contra legem interpretation – resulting in actual contradiction of the language employed in a treaty – must be avoided.

For these and other reasons, there appear to be plenty of situations which call for actual law reform, that is, the amendment of existing law or the creation of new law. This statement probably applies as much at national levels as it does at international levels. This is easily illustrated with reference to the rather fundamental shift of paradigm which is required of biodiversity law. Whereas, traditionally, nature conservation instruments have precisely focused on conserving species and habitats in their places of origin, they must now become 'moving companies',[46] aiding species and ecosystems in their relocations to higher latitudes and more suitable areas. Conservation instruments, in particular regarding protected areas, will now be expected to facilitate the departure of current species and ecosystems and the arrival of new ones, rather than keeping all things as they are.[47]

Obviously, however, the law reform route has its own drawbacks. Although in theory the advantages of unambiguously phrased provisions concerning concrete climate adaptation measures would seem apparent, the actual amendment of existing legal instruments is typically difficult to achieve and subject to drawn-out procedures. Similar considerations apply to the creation of new ones. Needless to say, both options hinge on the availability of sufficient political will.

It is important to bear in mind, therefore, that the desirability and urgency of climate change-related law reform varies from instrument to instrument. On one end of the scale, there are situations where a lack of express climate adaptation provisions can, in principle, be accommodated by recourse to an evolving interpretation of provisions in light of climate change. The Bern Convention appears to be a good example of this type of situation.[48] On the opposite end of the scale, there are situations where reading commitments concerning climate adaptation action into current obligations is unsatisfactory or plainly will not work. One instance is the legal requirement that, to fully participate as a party (or, in the case of Memoranda of Understanding, as a signatory) in CMS ancillary instruments, a state must be a 'range state' of (one of) the species involved – that

---

46 Term used (in Dutch) by Rolf Roos and Bart van Tooren, 'Flora en Fauna in Rep en Roer' in Rolf Roos and Saskia Woudenberg (eds), *Opgewarmd Nederland* (Stichting Natuurmedia/Uitgeverij Jan van Arkel/Stichting Natuur en Milieu 2004) at 99.

47 See Trouwborst (2009), above note 19 at 441.

48 Trouwborst (2011a), above note 19.

is, a state where the species is *currently* established.[49] This requirement has the potential to interfere with proactive conservation efforts aimed at facilitating expected range shifts, and it will take amendment of the various instruments involved to remove this barrier, as it cannot be 'interpreted away'.[50] Sometimes it is difficult to determine where exactly on the scale particular instruments belong. An example where there is still debate as to whether the existing regime will accommodate the growing need for flexibility and adaptability, concerns the rules regarding the designation and management of protected areas under the EU Birds and Habitats Directives.[51] It is anticipated, however, that the broadly formulated and result-oriented obligations in the EU Directives, in combination with the teleological interpretations often employed by the EU Court of Justice, will go a long way to alleviate these concerns.[52] Finally, mention should also be made of the apparent tensions between 'assisted migration' propositions and existing rules addressing 'invasive alien species' that aim at the prevention and mitigation of harmful consequences following the introduction of non-native species in native ecosystems.[53]

As suggested before, given the necessary political will, there could be considerable merit in the adoption of a Protocol to the CBD on climate adaptation.[54] By setting out concrete criteria regarding measures to facilitate the adaptation of biodiversity to climate change, such a Protocol could act as an overarching framework to guide the adaptation of biodiversity law at regional and national scales.

4 BIODIVERSITY LAW AND THE NEED FOR CONNECTIVITY

Ensuring adequate connectivity, with the latter referring to the 'ease with which organisms move between particular landscape elements',[55] was

---

49 See, e.g., ACAP Article XV.

50 For full discussion see Trouwborst (2012), above note 19 at 281–286.

51 See, *inter alia*, Cliquet et al., above note 19; Verschuuren (2010), above note 19; and Trouwborst (2011a), above note 19.

52 Trouwborst (2011a), above note 19.

53 See Alejandro Camacho, 'Assisted Migration: Redefining Nature and Natural Resource Law under Climate Change' (2010) 27 Yale Journal on Regulation 171.

54 Trouwborst (2009), above note 19.

55 Graeme L Worboys, Wendy L Francis and Michael Lockwood (eds), *Connectivity Conservation Management: A Global Guide* (Earthscan 2010) at xxxi.

always important, given the threats posed to populations of wild flora and fauna by habitat fragmentation. Connectivity can be enhanced through an array of tools, including corridors, stepping stones, and generally through biodiversity-friendly management of landscapes outside/between protected areas, thus creating actual protected area *networks*.[56] Climate change has made such connectivity conservation all the more urgent, given the overall shifts in species' distributions which have been set into motion. Consequently, maintaining and/or restoring connectivity is universally considered a crucial biodiversity adaptation measure. From this point of view, the current section takes a closer look at selected global, regional and national biodiversity conservation instruments.[57] The regional and national parts, respectively, zoom in on the continent with the most fragmented landscape – Europe – and the most fragmented country within this region – the Netherlands.

4.1 Global Biodiversity Law and Connectivity

Article 3 of the Ramsar Wetlands Convention, requiring the conservation of wetlands included in the List of Wetlands of International Importance, and the 'wise use' of wetlands generally, has already been discussed above.[58] It is complemented by a duty to 'promote the conservation of wetlands and waterfowl by establishing nature reserves on wetlands, whether they are included in the List or not'.[59] Wetlands, such as rivers, (can) evidently provide for connectivity. Hence, the duties in the Convention can and do contribute to connectivity conservation. The Convention's parties are well aware of this, and have acknowledged that the 'conservation and wise use of wetlands enables organisms to adapt to climate change by providing connectivity, corridors and flyways along which they can move'.[60] The Ramsar Wetlands Convention is thus a clear example of an instrument the effective implementation of which would patently promote the adaptation of biodiversity to climate change.

Comparable considerations apply to the World Heritage Convention. A significant number of ecologically important sites around the world

---

[56] Generally, see ibid.

[57] For a more extensive review see Barbara Lausche, David Farrier, Jonathan Verschuuren, Antonio GM La Viña, Arie Trouwborst and Charles-Hubert Born, *The Legal Aspects of Connectivity Conservation* (IUCN forthcoming 2013).

[58] See above note 30.

[59] Article 4(1).

[60] Resolution X.24, above note 33, para. 12.

qualify as 'natural heritage' according to the Convention's definition,[61] and some of these are included in the World Heritage List compiled under the treaty. Parties to the Convention are committed to doing everything within their power to ensure the 'identification, protection, conservation, presentation and transmission to future generations' of the natural heritage situated on their territories.[62] Besides warranting the taking of 'effective and active measures' for the protection of the sites concerned, the aforementioned Article 5 of the World Heritage Convention states that each party: 'shall endeavor, in so far as possible, and as appropriate for each country', to 'integrate the protection of that heritage into comprehensive planning programmes' and to 'take the appropriate legal, scientific, technical, administrative and financial measures necessary for the identification, protection, conservation, presentation and rehabilitation of this heritage'. As hinted at above, it is thus possible to construe this last obligation as extending to connectivity measures, where needed in view of climate change. In any event, the very occurrence on the World Heritage List of mountain ranges like the Canadian Rockies and large wetlands like the Wadden Sea is of significance from a connectivity point of view.

A third global conservation treaty considered here, the Convention on Migratory Species, stipulates with respect to the endangered migratory species listed in its Appendix I that parties 'shall endeavour':

> (a) to conserve and, where feasible and appropriate, restore those habitats of the species which are of importance in removing the species from danger of extinction;
> (b) to prevent, remove, compensate for or minimize, as appropriate, the adverse effects of activities or obstacles that seriously impede or prevent the migration of the species; and
> (c) to the extent feasible and appropriate, to prevent, reduce or control factors that are endangering or are likely to further endanger the species ...[63]

It is not hard to envisage circumstances where connectivity conservation action by CMS parties would be essential to give effect to this provision. It suffices to consider the implications of the requirements under (b) for power lines, roads, fences, wind farms and other human infrastructure, which can impair connectivity in respect of migratory wildlife. Furthermore, the Conference of the Parties to the CMS has acknowledged on

---

61 See Article 2 of the Convention.
62 Article 4.
63 CMS, Article III(4).

various occasions that the objectives of the Convention cannot be attained without ensuring adequate connectivity, particularly with a view to climate change. Parties have, for example, been urged to 'strengthen the physical and ecological connectivity between sites, permitting dispersal and colonization when species distributions shift',[64] and thus achieve a 'higher resilience to climate change'.[65] Of particular importance are the following statements, in which the Conference of the Parties:

> *Requests* Parties to promote the identification of the most relevant sites and corridors for migratory species, with an emphasis on those that are trans-boundary and would benefit from international cooperation;
>
> *Invites* Parties to enhance the coverage, quality and connectivity of protected areas as a contribution to the development of representative systems of protected areas and coherent ecological networks that include all taxonomic groups of migratory species;
>
> *Urges* Parties to undertake habitat restoration and management at protected areas and critical sites in order to ensure habitat availability during the different stages of the life cycle of migratory species;
>
> *Urges* Parties to explore actively the potentially suitable areas for cooperation over transboundary protected areas, ensuring that barriers to migration are to the greatest possible extent eliminated or mitigated and that migratory species are managed under commonly agreed criteria;
>
> *Invites* Parties to undertake concerted efforts to integrate protected areas into wider landscapes and sectors, including through the use of connectivity measures such as the development of biological corridors, where appropriate, and the restoration of degraded habitats and landscapes in order to address the impacts of and increase resilience to climate change;
>
> *Encourages* Parties to explore the applicability of ecological networks and corridors to marine migratory species that are under pressure from human activities such as oil and gas exploration, overexploitation, fishing and coastal development.[66]

Needless to say, these and similar statements by the Conference of the Parties to the CMS are of evident interest with a view to the prior discussion on treaty interpretation.[67]

---

[64] CMS COP Resolution 10.19 on Migratory Species Conservation in the Light of Climate Change (adopted 25 November 2011), para. 8(b).

[65] CMS COP Resolution 10.3 on the Role of Ecological Networks in the Conservation of Migratory Species (adopted 25 November 2011), Preamble.

[66] Ibid., paras 1–4, 6 and 9.

[67] For a more elaborate discussion, see Trouwborst (2012), above note 19.

The last major global treaty considered here is the Convention on Biological Diversity. It is noteworthy from a connectivity perspective that Article 8 of the CBD, cited earlier, refers to a 'system' of protected areas and *other* 'areas where special measures need to be taken to conserve biological diversity'.[68] Another example of a relevant provision is the obligation to – as far as possible and as appropriate – integrate biodiversity conservation into other 'relevant sectoral or cross-sectoral plans, programmes and policies'.[69] This must be deemed to apply, *inter alia*, to agricultural and infrastructural policies, which obviously have profound implications for connectivity, that is, for ecological infrastructure. A growing set of (not themselves legally binding) commitments and guidelines adopted by the Conference of the Parties to the CBD demonstrates that parties attach considerable value to connectivity conservation in the context of the Convention's implementation. For instance, in a particularly unambiguous statement, the Conference of the Parties in 2010 called for 'concerted efforts to integrate protected areas into wider landscapes and seascapes and sectors, including through the use of connectivity measures such as the development of ecological networks and ecological corridors, and the restoration of degraded habitats and landscapes in order to address climate-change impacts and increase resilience to climate change'.[70] In another example, the Conference of the Parties provides a convenient bridge to the subsequent discussion of regional biodiversity law by urging states to 'cooperate *regionally* in activities aimed at enhancing habitat connectivity across ecological gradients, with the aim of enhancing ecosystem resilience and to facilitate the migration and dispersal of species with limited tolerance to altered climatic conditions'.[71]

4.2 Regional Biodiversity Law and Connectivity: Focus on Europe

Indeed, it is well understood that much of the required adaptation action, including ensuring adequate connectivity, will need to be undertaken at a regional rather than a global scale. The challenge of regional cooperation to enhance biodiversity's ability to adapt to climate change is especially momentous in Europe, where nature is already struggling to cope with

---

68 CBD Article 8(a), see above note 35.

69 CBD Article 6.

70 CBD COP Decision X/31, above note 37, para. 14(a). See also the other COP Decisions referred to above in note 37.

71 CBD COP Decision VIII/30, above note 37, para. 4 (emphasis added).

severe human pressures in heavily fragmented[72] landscapes extending across a large number of comparatively small states. According to a study conducted for the European Commission: '[i]n most parts of Europe, protected areas are too small to accommodate changes, and the matrix around them is too modified and intensively used'.[73] The major European biodiversity instruments, the Bern Convention and the EU Birds and Habitats Directives, clearly have their work cut out. The focus here is on the pan-European Bern Convention, a treaty that includes virtually all European states as well as the EU among its 50 parties. The complex regime of the EU Directives, despite its great(er) practical significance in the 27 EU member states, is not dealt with here for reasons of space and to avoid undue duplication.[74]

The Bern Convention aims to 'conserve wild fauna and flora and their natural habitats, especially those species and habitats whose conservation requires the cooperation of several States, and to promote such cooperation'.[75] Article 2 states a general but unconditionally phrased obligation to take 'requisite measures' to 'maintain the population of wild flora and fauna at, or adapt it to, a level which corresponds in particular to ecological [and other] requirements'. This is accompanied by a more qualified obligation in Article 3 to 'take steps to promote national policies for the conservation of wild flora, wild fauna and natural habitats, with particular attention to endangered and vulnerable species, especially endemic ones, and endangered habitats'.[76] The same provision stipulates that each party 'undertakes, in its planning and development policies …, to have regard to the conservation of wild flora and fauna'.[77] Of particular importance for connectivity is Article 4, which addresses habitat protection:

---

[72] See European Environment Agency and Swiss Federal Office for the Environment, *Landscape Fragmentation in Europe* (Publications Office of the European Union 2011).

[73] G Tucker and Y de Soye, *Impacts of Climate Change and Selected Renewable Energy Infrastructures on EU Biodiversity and the Natura 2000 Network – Tasks 2b & 3b: Impacts of Climate Change on EU Biodiversity Policy, and Recommendations for Policies and Measures to Maintain and Restore Biodiversity in the EU in the Face of Climate Change* (IEEP/IUCN 2009) 81.

[74] For analysis of the EU Directives in light of climate change see, *inter alia*, Wheeler, above note 19; Cliquet et al., above note 19; Erens et al., above note 19; Verschuuren (2010), above note 19; Trouwborst (2011a), above note 19; and Trouwborst (2011b), above note 19.

[75] Article 1.

[76] Article 3(1).

[77] Article 3(2).

(1) Each Contracting Party shall take appropriate and necessary legislative and administrative measures to ensure the conservation of the habitats of the wild flora and fauna species, especially those specified in Appendices I and II, and the conservation of endangered natural habitats.
(2) The Contracting Parties in their planning and development policies shall have regard to the conservation requirements of the areas protected under the preceding paragraph, so as to avoid or minimise as far as possible any deterioration of such areas.
(3) The Contracting Parties undertake to give special attention to the protection of areas that are of importance for the migratory species specified in Appendices II and III and which are appropriately situated in relation to migration routes, as wintering, staging, feeding, breeding or moulting areas.
(4) The Contracting Parties undertake to co-ordinate as appropriate their efforts for the protection of the natural habitats referred to in this article when these are situated in frontier areas.

Given current knowledge about the consequences of climate change for biodiversity conservation, the strong, result-oriented obligations of Article 4(1) and Article 2, read in light of the Convention's objectives, appear to entail an obligation for Bern Convention parties to ensure the maintenance and/or creation of adequate ecological connectivity for different species groups.[78]

Recommendations adopted by the Bern Convention's Standing Committee (the principal treaty body in which all parties are represented), *inter alia* in the context of the 'Emerald Network' of Areas of Special Conservation Interest set up under the Convention, reinforce this conclusion.[79] Already in 1991, well before the added urgency created by climate change was formally acknowledged, the Standing Committee recommended parties to 'encourage the conservation and, where necessary, the restoration of ecological corridors', *inter alia* through the following measures:

> Authorising agreements between nature conservation authorities and government or public bodies owning or responsible for such areas with a view to maintaining natural plant cover and preserving the sites of rare or endangered plant species, prohibiting or limiting the use of phytosanitary products and of fire in those areas, as well as restricting the use of machinery to the strict minimum necessary for safety reasons.
>
> Taking measures to restore or to compensate for the loss of ecological corridors caused by the building of new roads and other constructions that

---

78 See also Trouwborst (2011a), above note 19.
79 Ibid.

> prevent animals from migrating or interchanging. In these cases, the responsible authority has to safeguard such crossing routes, for example, by building special tunnels for otters and badgers, by building so-called cerviducts for deer, by closing roads during the spring migrational period for amphibians, or by any other appropriate means.
>
> Maintaining certain watercourses or parts thereof in their natural state, and where necessary restoring them, by prohibiting the building of dams, any straightening or canalisation work and the extraction of materials from their beds, and by maintaining or restoring vegetation along their banks. Ensuring that dredging operations, when they prove essential, do not harm the integrity of the aquatic ecosystem or of the banks.
>
> On other watercourses, limiting canalisation and straightening work to whatever is absolutely essential, providing fish passes across dams, maintaining a minimum flow in low-water periods as far as possible, limiting extraction of materials from the bed and maintaining vegetation along the banks.[80]

More recent Standing Committee Recommendations spell out concrete guidance concerning the adaptation of biodiversity to climate change, with a central role reserved for connectivity conservation. For instance, parties are recommended to establish 'networks of interconnected protected areas (terrestrial, freshwater and marine) and intervening habitat mosaics to increase permeability and aid gene flow'.[81] Similarly, in a specific section on reptiles and amphibians, parties are urged to '[f]acilitate in-situ adaptation and natural range shifts by redoubling efforts to maintain or restore large intact habitats and large-scale connectivity'.[82] As a final example, detailed and comprehensive guidance adopted in 2009 calls on parties to:

> Ensure the development of a sufficiently representative and connected network of protected areas so as to allow for species dispersal and settlement in new suitable sites as a consequence of climate change. In a context of great uncertainty, such a network would constitute an insurance policy to provide protection for most endangered species and habitats …
>
> Connect protected areas into functional ecological networks to allow the movement of species between them. Techniques include, as appropriate,

---

[80] Standing Committee Recommendation No. 25 (1991) on the Conservation of Natural Areas Outside Protected Areas Proper (adopted 6 December 1991), Appendix, Part III.

[81] Standing Committee Recommendation No. 135 (2008) on Addressing the Impacts of Climate Change on Biodiversity (adopted 27 November 2008), Appendix, para. II(3)(c).

[82] Ibid. para. I(13).

buffer zones, stepping stones, corridors, and measures to reduce habitat fragmentation.

Carry out integrated management of the wider countryside to alleviate the overall pressure on biodiversity and facilitate movement of species between conservation areas, as species dispersal is likely to be the most important mechanism of species adaptation to climate change.[83]

4.3 National Biodiversity Law and Connectivity: Focus on the Netherlands

All the global and regional instruments discussed in the preceding parts of the current section are in force for the Netherlands. In this crowded country, much of the landscape is heavily modified and fragmented. In respect of connectivity, the Netherlands' pertinent international and European obligations are implemented primarily through the 'Netherlands Ecological Network' scheme, or EHS (*Ecologische Hoofdstructuur* in Dutch, which literally means 'ecological main structure').[84]

The EHS scheme is aimed at the creation and conservation of a coherent national network of natural areas, which has been taking shape since the mid-1990s. This Network consists of, and connects, large and small existing natural areas, (agri)cultural landscapes with notable ecological values, and areas still to be converted into nature, denominated 'nature development areas'. Specifically, the Network comprises 'core areas', 'nature development areas' and 'connectivity zones'. 'Core areas' are existing areas with ecological values of (inter)national significance, with a minimum size of 250 hectares. They comprise protected natural areas, including the 20 National Parks and all Natura 2000 sites in the Netherlands; estates, forests, agricultural landscapes with notable ecological values; and large water bodies including the Ijsselmeer, Wadden Sea, and the entire portion of the North Sea within the Netherlands

---

[83] Standing Committee Recommendation No. 143 (2009) on Further Guidance for Parties on Biodiversity and Climate Change (adopted 26 November 2009), para. 3(4)–(6).

[84] The current subsection draws from a more extensive case study of the EHS performed by the present author for the IUCN as part of the project 'Legal Aspects of Connectivity Conservation'; see above note 57. See also, *inter alia*, Graham Bennett and Kalemani Jo Mulongoy, *Review of Experience with Ecological Networks, Corridors and Buffer Zones* (Secretariat of the Convention on Biological Diversity 2006); and Graham Bennett, *Green Infrastructure Implementation and Efficiency: In-Depth Case Analysis Theme 1: Ecological Networks. Report for DG Environment on Contract ENV.B.2/SER/2010/0059* (IEEP 2012).

Exclusive Economic Zone. 'Nature development areas' are areas with good possibilities for restoring ecological values of national and/or international significance. 'Connectivity zones' are areas interconnecting the areas in the former two categories. The EHS also aims to connect Dutch areas with natural areas across the borders in Belgium and Germany.

A first rough map for the EHS at the national scale was provided in 1995 in a national spatial policy instrument for rural areas.[85] It also set out guidelines, criteria and targets for the Network. The plan envisaged the inclusion in the terrestrial part of the Network of around 440,000 hectares of existing uncultivated land, 200,000 hectares of agricultural land and 50,000 hectares of 'nature development areas' – altogether accounting for one-sixth of the territory of the Netherlands. Based on this national plan, the twelve Dutch provincial governments delineated the EHS components within their respective jurisdictions and incorporated more specific guidelines and criteria in regional planning instruments. In turn, legally binding land use restrictions were laid down by municipal governments in local zoning plans. The designation, management and protection of the areas composing the EHS have been pursued by the national government through a mix of instruments. These include spatial planning, protected areas legislation, the allocation of budgets for land purchase and management subsidies, and measures aimed at the improvement of the environmental quality within the Network, in particular by addressing the problems posed by low water tables, surface water contamination, acidification and nitrogen deposition.

The national government laid down the annual area to be purchased and/or put under subsidized nature management in a timetable, so as to finalize the entire Network by the end date – originally 2018, now 2021 after a recent re-adjustment of the EHS scheme. The acquisition of new areas to complete the pre-designed jigsaw puzzle of the EHS has relied on substantial funds provided by the national government. Once purchased, ownership and management of the areas concerned have typically been transferred to private and semi-public nature conservation organizations.

By the end of the 1990s, it became apparent that the EHS as envisaged would fall short of meeting its ecological objectives: newly restored natural areas were too fragmented, connectivity zones were undersized

---

[85] Ministerie van Landbouw, Natuurbeheer en Visserij and Ministerie van Volkshuisvesting, Ruimtelijke Ordening en Milieubeheer, *Structuurschema Groene Ruimte I* (Ministerie van Landbouw, Natuurbeheer en Visserij and Ministerie van Volkshuisvesting, Ruimtelijke Ordening en Milieubeheer 1995).

and non-ecological infrastructure such as roads and railways posed too many obstacles to connectivity. To counter these shortcomings, a major new component was added to the EHS scheme in 2000, in the form of plans to form seven large-scale, strategic connectivity zones at a regional rather than a local level, denominated 'robust connectivity zones'.[86] In addition, a specific programme was agreed in 2004 to address physical barriers to connectivity posed by human infrastructure such as highways, railways and man-made waterways.[87] Over 215 major obstacles to connectivity were identified, and plans were made and funds allocated to overcome these between 2005 and 2018. By 2011, 59 obstacles had been comprehensively resolved and a further 44 partly resolved. Typical resolution measures include the construction of underpasses for animals such as badgers, martens and foxes, and the construction of large overpasses ('ecoducts') suitable for large mammal species (red deer, wild boar, roe deer), reptiles, butterflies and a host of other organisms.

Climate change did not figure as part of the original motivation for setting up the EHS, but has since reinforced the need for completing and adequately managing the Network. A 2006 study aimed at identifying the best options for Dutch conservation policy to respond to climate change, calls for a good spatial coherence between natural areas in general, and, in particular, for the accelerated implementation of the envisaged 'robust connectivity zones'.[88] The crucial role of these zones for the adaptation of flora and fauna to climate change was confirmed in a specific study commissioned by the national government in 2009, which concluded that the planned robust connectivity zones would make an important contribution to removing major obstacles to the adaptation of species to climate change resulting from habitat fragmentation.[89]

---

[86] Ministerie van Landbouw, Natuur en Voedselkwaliteit, *Natuur voor Mensen, Mensen voor Natuur* (Ministerie van Landbouw, Natuur en Voedselkwaliteit 2000).

[87] Ministerie van Verkeer en Waterstaat, *Meerjarenprogramma Ontsnippering* (Ministerie van Verkeer en Waterstaat 2004).

[88] Claire Vos, Marja van der Veen and Paul FM Opdam, *Natuur en Klimaatverandering: Wat Kan het Natuurbeleid Doen?* (Alterra 2006).

[89] Willemien Geertsema et al., *Robuuste Verbindingen en Klimaatverandering* (Alterra 2009).

5 CONCLUSIONS – CLIMATE PROOFING BIODIVERSITY LAW

It is appropriate in this final section to draw attention to several general conclusions and recommendations flowing from the above analysis of biodiversity law in light of climate change.

First, there appears to be no need to throw existing legal frameworks on biodiversity conservation overboard and start from scratch. Rather, the *effective implementation* of existing instruments would considerably boost the resilience of species and ecosystems to climate change. Second, to make the most of existing biodiversity law in the present context, there is significant potential in the *interpretation* of relevant provisions in legal instruments in light of current knowledge regarding the needs of species and ecosystems under the influence of climate change. These two findings are clearly supported by the analysis of biodiversity law from a connectivity perspective in the preceding section. Third, in certain situations actual law reform is called for, whether in the form of *modification* of existing law, or the *creation of new law*. To illustrate the latter, at the global level there would seem to be significant merit in the development of a Protocol to the Biodiversity Convention on climate change adaptation. In brief, adapting biodiversity law to the new necessities created by climate change entails interpretation of existing law to the extent possible, and law reform where feasible and necessary.

To summarize, in the current context it appears warranted to seek: (a) the effective implementation of existing law; (b) the interpretation of existing law in light of climate change; (c) the modification of existing law where necessary and feasible; and (d) the creation of new law where necessary and feasible. Finding the optimal combination of these four courses of action would seem to be key to maximizing the contribution of biodiversity law to the successful adaptation of species and ecosystems to climate change.

12. Climate change adaptation and land use planning law

*Keith H Hirokawa and Jonathan Rosenbloom**

1 INTRODUCTION

Governance of local and regional communities has always been an experiment in adaptation. New and shifting social, economic, and environmental challenges have pushed communities to adapt to a wide variety of changes involving anything from technology to growing populations. For instance, automobile travel reduced the size of the world by increasing the distance one could travel; elevators opened thousands of feet of airspace to development and human occupancy; drive-through restaurants fundamentally altered the nature of the relationship between restaurants and their surrounding neighborhoods. In each of these examples, new land uses challenged the capacity of local communities to continue protecting present values and ways of life. In response, local communities employed and, in some cases, reconceptualized traditional land use tools – such as infrastructure planning, nuisance regulation, and zoning – to meet emerging challenges with familiar tools and in familiar ways. In large measure, these land use tools represent a sophisticated type of governance that is aimed at minimizing risks and maximizing the opportunities of economic, environmental, and social change. Further, the tools do so in a way that provides local communities with the flexibility to address changes as they arise, and help local communities adapt to projected climate change related conditions.[1]

While land use tools have previously been implemented to help local communities adapt to new challenges, climate change presents a distinct set of issues different from prior issues facing local communities. Traditionally, urban planning has drawn on lessons and observations of

* The authors would like to express their gratitude to Mary Armistead for her excellent research assistance in preparing this chapter.

[1] See Roger Pielke Jr, Gwyn Prins, Steve Rayner and Daniel Sarewitz, 'Lifting the Taboo on Adaption' (2007) 445 Nature 597, 597–598 (arguing for more focus on resilient and sustainable societies).

the past to direct present decision-making and address new challenges.[2] In contrast, climate adaptation planning requires contemporary decision-making to account for changes that are projected to take place in the future. Further, climate change differs both in scale and in type from past challenges faced by communities, and, as such, will compel local governments to explore new concepts and new planning tools and approaches. For instance, because of the velocity of physical changes to coastal areas, coastal local governments may be unable to employ old paradigms used to identify the most appropriate locations for transportation, housing, commerce, industry, natural resource extraction, recreation, and essential public facilities. Likewise, because of the potential for changes in water availability, local governments will need to consider the risks of incremental water demands as seriously as external threats to water supplies.

A growing interest among municipalities in adaptive land use planning illustrates a willingness to face new climate change related conditions. Many local governments are directing agencies to adopt regulatory rules that require consideration of potential or projected climate impacts in public sector decisions with the goal of improving the local government's adaptive capacity to changes stemming from climate conditions.[3] 'Adaptive capacity' has been defined as 'the ability or potential of a system to respond successfully to climate variability and change, and includes adjustments in both behavior and in resources and technologies'.[4] Governance, leadership, foresight, and flexibility are all determinants of adaptive capacity.

Of course, it is important to note that environmental challenges (as well as the cultural perspectives of those challenges) are not consistent across regions and countries, and as such, adaptive capacity (including

---

[2] We recognize that while zoning is not inherently retrospective and that some scholars have argued it should be prospective, the predominant use of zoning is retrospective in nature and reflects changes that have already impacted the local community. See, e.g., M Christine Boyer, *Dreaming The Rational City: The Myth of American City Planning* (MIT 1983); Donald A Krueckeberg, 'The Culture of Planning' in Donald A Kruekeberg (ed.), *Introduction to Planning History in the United States* (Center for Urban Policy Research 1983); Alejandro E Camacho, 'Mustering the Missing Voices: A Collaborative Model for Fostering Equality, Community Involvement and Adaptive Planning in Land Use Decisions: Installment One' (2005) 24 Stanford Environmental Law Journal 3.

[3] ICLEI-Local Governments for Sustainability, *Progress and Challenges in Urban Climate Adaptation Planning: Results of a Global Survey* (MIT 2012) 3.

[4] National Research Council, *Adapting to the Impacts of Climate Change* (National Academies Press 2010) 160.

the ability to govern through environmental changes) has developed unevenly. Local efforts to engage adaptive planning also illustrate that adaptation does not come in a one-size-fits-all package. Some cities, such as New York and London, have conducted detailed climate vulnerability and risk assessments. Other communities have drawn on available regional and local data to project local climate changing scenarios. Further, some cities have conducted comprehensive assessments while others have focused on one or a subset of natural resources and man-made sectors. The fact that climate changes will affect different communities in different ways does not make these challenges facing local governments less real or daunting, and in some ways makes the challenges more complex, as the possibility of repetition or collaboration among local communities is made more difficult.[5]

Throughout these efforts, it is becoming clear that one of the most effective adaptation tools used by local governments is the power to plan communities. Through land use planning, local governments can increase resilience to major climate shifts and ensure that our communities are equipped with built-in mechanisms to face and mitigate such changes. This chapter identifies some of the most innovative planning tools available to local governments that illustrate the potential to plan for communities' resiliency. The chapter begins by setting forth some of the most severe impacts local governments will experience from climate change. Section 2 is intentionally designed to be a parade of 'horribles'. It is a checklist of significant impacts that may be and, in some circumstances, are already being experienced by local governments. This section recognizes that not all local governments will experience climate change impacts the same, and that climate change adaptation is contextual. Section 3 provides an overview and inventory of the traditional tools that local governments used. The nature and flexibility of the tools illustrate that local communities are particularly well-suited to address the contextual nature of climate change impacts. Section 4 looks more closely at specific attempts by local governments to rely on land use planning to adapt to climate change. This section focuses on what we refer to as structural tools, as they begin to form the basic foundation for a wide variety of adaptation approaches and goals. These tools include identifying metrics, establishing baselines, and assessing performance.

---

[5] This complexity and fractionation of the climate change challenges is furthered by the governance structures that often prohibit local communities from proposing innovative solutions. See Jonathan Rosenbloom, 'New Day at the Pool: State Preemption, Common Pool Resources, and Non-Place Based Municipal Collaborations' (2012) 36 Harvard Environmental Law Review 446.

Section 5, expands on the structural tools and explores specific mechanisms that can help local governments achieve adaptation goals and avoid catastrophic unpreparedness in the climate change arena.

2 A ROLE FOR COMMUNITIES IN CLIMATE PREPAREDNESS

Although governments at all levels will be called upon to face the broad spectrum of challenges stemming from climate change, climate circumstances affect different levels of government in unique ways. Resource competition in the Arctic or high seas, for example, calls for the capacity to engage in international diplomacy, while mitigation of storm surges along vast stretches of coastline may be more relevant to national, state, or regional governments. Where domestic water use is relevant to governance capacity, however, governmental policies become more local and must address changes in hydrology and precipitation, as well as water infrastructure and access. As many of the issues relevant to immediate and daily public needs involve local governance, local governments will experience many of these challenges acutely and directly.[6] Local governments have historically acted as first responders to changes in local conditions. They are directly responsible for minimizing and mitigating the threats of local circumstances to local residents. As such, local governments have a high level of motivation to address climate change impacts that threaten their communities.

Mere knowledge that changes are imminent, however, does not constitute climate change preparedness. Climate change does not issue an idle threat, but instead will deeply impact communities by causing significant and often unanticipated changes to essential public facilities and local ecologies. A recent report drafted by ICLEI-Local Governments for Sustainability summarized the 'wide-ranging effects on urban areas' associated with climate change, including '[a]nticipated changes in temperature, precipitation, and natural hazards [that] have the potential to place stress on infrastructure, residential, office, and historic buildings, environmental quality, and both municipal and ecosystem services'. In the meantime, although developing nations will likely see threats to their assets, 'developing countries are expected to encounter significant increases in illness and death among vulnerable populations as a result of

---

[6] See New York City Panel on Climate Change, *Climate Risk Information* (2009) at 5, http://www.nyc.gov/html/om/pdf/2009/NPCC_CRI.pdf, accessed 24 October 2012.

climate impacts'.[7] Specific examples of projected (and existing) changing conditions facing local governments include:

- Increased water and storm hazards (for example, increases in sea levels and precipitation with corresponding decrease in storm control reliability);
- Decreased water availability (for example, drought conditions) and dependability of water infrastructure;
- Increased ecosystem changes (for example, shifts in species' natural range and migration paths) and decrease in maintenance of ecosystem processes that provide valuable ecosystem services (for example, crop pollination, storm water control, and habitat provision);
- Increased risk of species extinction and loss of biodiversity;
- Increased frequency of extreme weather patterns (for example, heat waves, tornados, hurricanes, typhoons);
- Increased local health, safety, and welfare burden from malnutrition (for example, loss of food sources and soil degradation);
- Increased exposure and vulnerability to infectious diseases (for example, changes in the geographic distribution of diseases).

These conditions in turn will impact many of the services that have traditionally been considered the responsibility of local governments. In many cases, physical climatic changes will interrupt the delivery of services or preclude the placement of services and facilities. In some cases, climatic changes may heighten the need for existing services and create the need for new services. Local governments may be forced to rethink the way certain public services are valued and the manner in which public services are provided. Local governments will be particularly challenged by the responsibility to provide the following:

- Local mass transportation. Transportation routes and methods will be affected by the changing geographical landscape, such as

---

[7] ICLEI, above note 3 at 3; National Research Council, above note 4 at 36–37 ('Examples of other sources of climate [change vulnerability] include population shifts and development choices, such as dense urban development in drought-prone areas; and places and communities that are especially dependent on climate-sensitive industries such as agriculture, forestry, and tourism. In the built environment, each city's residents and infrastructures will be affected in unique ways').

alterations in the flood plain and waterline. Transportation infrastructure will be strained from extreme weather conditions and will need repair and maintenance.

- Local utility infrastructure and facilities. An increase in extreme weather can disrupt service and damage utility lines, such as gas, waste, and water lines, costing local governments millions of dollars.
- Water. Providing potable water may become more difficult as water becomes scarce in some regions and as water becomes contaminated (such as through flooding or saltwater intrusion into drinking sources).
- Removal of waste water. Sewer systems may be overloaded through flooding or infrastructure deterioration, leading to increased use of combined sewer outfalls and enhanced pollution and unsafe, unhealthy conditions.
- Removal of waste. Flooding may produce an increase in debris that may be distributed randomly throughout the jurisdiction, creating unsafe and unsanitary conditions.
- Protection of local wetlands from flooding. Loss of wetlands may destroy local species' habitat, enhance the local area's susceptibility to future flooding, and reduce a powerful carbon sink.
- Electricity and telecommunications. Extreme weather patterns may place a high demand on energy consumption and telecommunications, which could lead to disruptions in service, putting vulnerable populations at risk of health problems.
- Health services. Safety of inhabitants from weather patterns and disease may be made more difficult with a lack of potable water, elevated heat levels, and malnutrition.[8]

This broad depiction of potential and projected changes only touches the surface of the possible struggles that will impact communities. However,

---

[8] Sources for this list are IPCC, 2007: Summary for Policymakers in M L Parry et al. (eds), *Climate Change 2007: Impacts, Adaptation and Vulnerability: Contribution of Working Group II to the Fourth Assessment Report of the Intergovernmental Panel on Climate Change* (Cambridge University Press 2007) 18; National Research Council, above note 4 at 36–37; ICLEI, above note 3 at 3. For a contextualized example, such as those discussed in sections 3–5, see City of Chula Vista, *City of Chula Vista Climate Adaptation Plan*, available at http://www.chulavistaca.gov/clean/conservation/Climate/documents/AttA_ClimateActionPlanUpdate_Apr12ProgressReport_FINAL.pdf, accessed 24 October 2012.

the list represents a diverse group of local government services. Adaptation priorities will depend on the context and severity of the threat, combined with a detailed analysis of the specific local government conditions, including geography, culture, and governance structure. For example, coastal communities will be more vulnerable to rapidly changing sea levels and salt water intrusion on drinking water supplies, while many landlocked areas will be more vulnerable to drastic drought conditions similar to those experienced throughout the Midwest in the United States and parts of Australia since 2003. Although means will differ, ideally localities situated in geographic regions vulnerable to water shortage will have the foresight to secure their water infrastructure capacity against challenges to collection, storage, and use. Likewise, local agricultural communities around the world have been suffering from changing temperatures and shorter growing seasons and may, at some point, be unable to adapt their growing practices. It is to be hoped that these communities are devising the means to survive temperature shifts and climatic variation through changes in farming practices and geographical dependencies. In times of water shortage and increased storm surges, wildlife and trees may become a liability and security threat. On the other hand, some local governments will rely on natural resource resiliency more than ever to provide ecosystem services. To this extent, local governments should be engaged in the process of creating and enhancing green infrastructure assets, including local and urban forests, habitats, and watersheds.

3 TRADITIONAL TOOLS TO PROTECT THE COMMUNITY

Although climate change adds a new dimension and scale to the types of challenges that local governments face, it was noted above that local governments are particularly experienced at accommodating changes in their economic, environmental, and social circumstances. Local governments have previously adapted to a wide variety of changing conditions, such as accounting for new transportation patterns, new travel technologies, trends in nightlife and commodities, rapid economic development, and emerging resource needs. From these struggles, local governments have proven to be agile, flexible, and innovative in addressing these changes. This is particularly important as many local governments provide critical infrastructure necessary for education, security, public health and safety, and recreation. As the Department for Environment, Food and Rural Affairs (DEFRA) in the United Kingdom stated:

'Under localism, councils have the opportunity to play a critical role in helping communities by adapting their services (for example in the provision of social care, public health and emergency planning), the management of their estates and through their role as community leaders by increasing resilience of local spaces (for example through planning, community and businesses resilience)'.[9]

This section explores a few of the more common tools traditionally employed by local governments to adapt to changes in economic, social, and environmental circumstances. It also addresses the ways in which the traditional tools may be inadequate, by themselves, to equip communities to become more resilient to climate change. The section then compares the traditional tools to city-building approaches that draw on the principles of sustainability and preparedness. It may be relevant to this discussion that local land use authorities are often identified as culprits in uncontrolled pollution and pollution prevention failures. As Dan Tarlock states:

> The regulation of private land use to achieve environmental protection objectives remains the weakest link in modern environmental law. Many of the major environmental challenges such as the control of non-point source water pollution, the conservation of biodiversity, and the limitation of automobile emissions, including carbon dioxide, are at the core of land use regulation problems. Yet, in the main, we continue to develop and abuse land, regardless of environmental stresses that development causes. As environmental protection once again rises on the political agenda, the need to address the gap between land use regulation and environmental protection is becoming more critical; the regulatory gap impedes or cancels much of the progress that we have made to improve the conditions of our air sheds and watersheds – let alone confront the linked challenges of biodiversity conservation and adaptation to global climate change.[10]

Local involvement in environmental quality has been based on (and limited by) the idea that environmental quality is a welfare issue. Hence, the first doctrine of relevance is nuisance, as '[t]here is no question that modern environmental law finds its roots in common law nuisance'.[11] Most land uses cause unmitigated impacts across property boundaries.

---

[9] Department for Environment, Food and Rural Affairs, *Climate Ready, Co-creation Progress Update and an Invitation to Respond* (July 2012), http://www.defra.gov.uk/publications/files/pb13801-climate-change-progress-check.pdf, accessed 24 October 2012.

[10] A Dan Tarlock, 'Land Use Regulation: The Weak Link in Environmental Protection' (2007) 82 Washington Law Review 651, 654–657.

[11] *State v Schweda*, 736 N.W.2d 49, 56 (Wisc. 2007).

Nuisance prevents the use of property to injure others under the canon: *sic utere tuo ut alienum non laedus* (one should use one's own property in such a manner as not to injure that of another).[12] Nuisance law has provided courts with some reprieve in the circumstance of complaining property owners, in a way that affords courts the opportunity to minimize the adverse impacts on a community scale. In particular, the doctrine of nuisance elevates location and context as it authorizes courts to choose among competing land uses; ultimately, some uses are more appropriate in particular locations, and a nuisance may merely be a valuable thing that does not fit at a particular location, 'like a pig in the parlor instead of the barnyard'.[13] Moreover, nuisance law requires courts to accommodate private needs in a property system – a system that allows us to decide how to use concepts like property to reach the ends and goals of social welfare.

Common law doctrines like nuisance have the potential to accommodate the needs of public control over land uses even following major shifts in ecology.[14] Due to the flexibility in the application of the doctrine, combined with its insistence that property rights be legitimized with reference to context and intensity, nuisance is a suitable foundation to lead property into the uncertainty of climate change law. Yet nuisance has shortcomings. In particular, although nuisance law may be an effective (albeit inefficient) method of allocating liabilities for land uses, it: a) delegates the arrangement of communities to the courts; and b) does not accommodate an interest in preventing nuisances by planning. Hence, although the employment of nuisance remains an effective tool used by local governments, changes in the law have concentrated the local power to regulate the use of land through zoning.

In the United States, New York adopted the first comprehensive zoning ordinance in 1916. Six years later, a federal advisory commission published the Standard Zoning Enabling Act (SZEA), a model act for state governments that authorized the progressive control of land use and its impacts by local governments. The SZEA suggested the creation of homogenous land use districts 'to lessen congestion in the streets', and 'to secure safety from fire', 'to provide adequate light' and 'to avoid

---

12 *Black's Law Dictionary* (6th edn, West Publishing 1990) 1380.

13 *Village of Euclid v Ambler Realty Co*, 272 U.S. 365, 388 (1926).

14 J B Ruhl, 'Making Nuisance Ecological', (2008) 58 Case Western Reserve Law Review 753, 756; J B Ruhl, 'The "Background Principles" of Natural Capital and Ecosystem Services – Did Lucas Open Pandora's Box?' (2006) 22 Journal of Land Use and Environmental Law 525, 541–543.

undue concentration of population'.[15] States adopting the SZEA allowed their local governments to regulate to protect property values by 'conserving the value of buildings', and by 'encouraging the most appropriate use of land throughout [the] municipality'.[16]

Zoning survived its first significant legal challenge in 1926 with the Supreme Court's decision in *Village of Euclid v Ambler Realty Co*.[17] The *Euclid* court held that private property expectations can be legitimate against the needs of the public welfare only where they are consistent with public needs in the particular context. The court stated: 'the question whether the power exists to forbid the erection of a building of a particular kind or for a particular use, like the question whether a particular thing is a nuisance, is to be determined, not by an abstract consideration of the building or the thing considered apart, but by considering it in connection with the circumstances and the locality'.[18]

By borrowing from nuisance, the court allowed local governments to envision the future of a community and anticipate social, economic, and environmental developments. The *Euclid* court was equipping local governments with the authority to evolve with a 'changing world'.

Until recent years, urban life was comparatively simple but, with the great increase and concentration of population, problems have developed, and are constantly developing, which require, and will continue to require, additional restrictions in respect of the use and occupation of private land in urban communities. Regulations, which a century ago, or even half a century ago, probably would have been rejected as arbitrary and oppressive, are now seen as wise, necessary and valid when applied to existing conditions. Such regulations are sustained under the complex conditions of our day, for reasons analogous to those that justify traffic regulations, which, before the advent of automobiles and rapid transit street railways, would have been condemned as fatally arbitrary and unreasonable. And in this there is no inconsistency. For, while the meaning of constitutional guaranties never varies, the scope of their application must expand or

---

15 Advisory Committee On Zoning, Department of Commerce, *A Standard State Zoning Enabling Act: Under Which Municipalities May Adopt Zoning Regulations* §3 (1926). See Ruth Knack, Stuart Meck and Israel Stollman, 'The Real Story Behind the Standard Planning and Zoning Act of the 1920s' (1996) (Feb) Land Use Law 3.

16 Ibid.

17 272 U.S. 365 (1926).

18 Ibid. at 388.

contract to meet the new and different conditions which are constantly coming within the field of their operation.[19]

To some extent, traditional tools, such as zoning and nuisance abatement, have proven effective in controlling land use impacts and protecting local resources. Zoning controls have been called upon in the direct assertion of control over environmental matters by local governments. Communities have experimented, for example, with: large lot zoning districts to control fire and disease and protect aquifers; urban growth boundaries to preserve open space, rural lifestyles, and agricultural land uses; environmental overlay districts and other performance-based zoning tools to ensure environmental standards in the land use process; building setbacks to control congestion and encourage storm water control and groundwater recharge; and environmental quality districts that control erosion and the transportation of sediments through urban areas.[20]

However, such tools have also been problematic in stabilizing communities to the point of stagnancy.[21] For instance, because nuisance requires courts to determine the reasonableness of particular impacts by reference to context and location, early landowners in a neighborhood had an opportunity to establish baseline levels of reasonableness of land use impacts. Similarly, well-established communities may be resistant to community efforts to adapt to shifts in economic, environmental, or social needs. In other words, the traditional land use tools solidify local land use practices, thereby making communities vulnerable to the changing impacts of climate change.[22]

---

[19] Ibid. at 387.

[20] See, e.g., Ohio: *Meisz v Village of Mayfield Heights*, 92 Ohio App. 471 (Ohio Ct. App. 1952); *Ketchel v Bainbridge*, 52 Ohio St. 3d 239 (Ohio 1990); *Cash v Cincinnati Zoning Board of Appeals*, 117 Ohio App.3d 319 (Ohio Ct. App. 1996); City of Tucson, Land Use Code § 2.8.6 (adopting an overlay zoning district to protect natural vegetation and sensitive habitat from storm water events); § 33:6, *infra* (discussing the use of minimum lot sizes to preserve agricultural land); § 9:58, *infra* (discussing setback regulations).

[21] This chapter primarily concerns the methods of maintaining an adaptive regulatory structure, rather than adaptive forms of property. As such, the remainder of this chapter omits consideration of nuisance and other opportunities for adaptive approaches to property rights. However, it is worth noting that the flexibility of nuisance allows the doctrine to champion stability amidst changing baselines and values, and as such, should be considered an available tool for adaptive communities.

[22] For a discussion of stationarity and its problems, see Robin Kundis Craig, '"Stationarity is Dead" – Long Live Transformation: Five Principles for Climate Change Adaptation Law' (2010) 34 Harvard Environmental Law Review 9, 27.

Given these shortcomings, many local governments have taken a more proactive approach to city design and its relationship to how the community copes (and interacts) with the environment. The recent planning successes suggest that any feasible local adaptation planning effort must include a multi-disciplinary and (in some sense) generational perspective of land use planning. John Nolon has described land use planning as a convergence of art and science, politics, and place. In this description, land use planning is marked by maintaining governance and culture through the flexible adaptation of the city to the changing needs of communities:

> City planning is a science and an art concerned primarily with the city's ever-changing pattern. As a pure science, it examines causes (history and etiology) and reciprocal influences of man and environment (urban geography and ecology). As applied science, it synthesizes these findings with those of the economic, sociological, and political sciences as well as the technological branches of statistics, civil and sanitary engineering, architecture, landscape architecture, and other pertinent branches of human knowledge, in an attempt to thoroughly understand conditions and their contexts and trends. As an art, it utilizes these materials, instructs or organizes citizens, molds events, and thwarts or guides trends to bring about the changes in city design which it contemplates.[23]

With the aid of planning tools, local governments can prospectively integrate resiliency and environmental protection into the community-building process.[24] Planning is designed to bring 'extensive and often uncontrollable development'[25] to the forefront by compelling local governments to envision affordable and efficient governance. Recent planning trends, such as 'Smart Growth' and New Urbanism, encourage local governments to design sustainably and govern responsibly without the need to rely on nuisance abatement.[26] These planning approaches seek to

---

[23] John R Nolon, 'Comparative Land Use Law: Patterns of Sustainability', (2005) 37 The Urban Lawyer 807, 818–819.

[24] See generally James A Kushner, 'Smart Growth: Urban Growth Management and Land-Use Regulation Law in America' (2000) 32 The Urban Lawyer 211.

[25] Patrick Gallagher, 'The Environmental, Social, and Cultural Impacts of Sprawl' (2000) 15 Natural Resources & Environment 219.

[26] See generally James E Holloway and Donald C Guy, 'Smart Growth and Limits on Government Powers: Effecting Nature, Markets and the Quality of Life Under the Takings and Other Provisions' (2001) 9 Dickinson Journal of Environmental Law & Policy 421 (providing an analysis of the implications of changing our way of thinking toward Smart Growth programs); Patricia E

engage existing challenges and projected changes by protecting natural resources in the planning process.

What makes land use planning compelling and valuable is its resilience and flexibility to tailor it to specific local conditions. The process of formulating land use plans requires the public, open and transparent engagement of a community in envisioning the identity and values of the community. It is a process of achieving consensus on the most pressing challenges facing the community. Citizens collaborate with land use professionals to anticipate public and private needs; formulate governmental responses to those needs; and interrupt trends that would cause those needs, all with an eye on the public health, safety, morals, and general welfare of the community. It is an opportunity to assess the new challenges that may arise 10, 20, or 30 years into the future.[27]

4 METRICS, BASELINES, AND ASSESSMENTS: UNDERSTANDING SPECIFIC CLIMATE CHANGE VULNERABILITIES

This section explores three land use planning initiatives – metrics, baselines, and assessments – that help local governments understand their climate change vulnerabilities. Identifying metrics allows for an effective measurement of climate change impacts relevant to the local community.[28] Establishing baselines for each metric helps communities track ecological, economic, and social changes: 'Collecting the evidence', notes DEFRA, is the first step to effectively creating an adaptation plan

---

Salkin, 'The Smart Growth Agenda: A Snapshot of State Activity at the Turn of the Century' (2002) 21 St. Louis University Public Law Review 271 (comparing the effectiveness of various state Smart Growth programs).

[27] See, e.g., Or. Rev. Stat. § 197.015(5) (requiring local governments to adopt comprehensive land use maps and policies addressing 'sewer and water systems, transportation systems, educational facilities, recreational facilities, and natural resources and air and water quality management programs').

[28] In supporting local planning efforts, DEFRA has recognized the importance of identifying metrics, establishing baselines, and monitoring progress as an initial step for local entities. 'The first step has to be accepting that some form of climate change is happening, assessing the risks and then taking action within your own community to reduce your vulnerability to these changes'. DEFRA, *Adapting to climate change: A guide for local councils* (January 2010) at 11, available at http://archive.defra.gov.uk/environment/climate/action/documents/adapt-localcouncilguide.pdf, accessed 24 January 2013.

and creating 'a foundation for delivering adaptation actions'.[29] Finally, assessing changes in the metrics will require communities to put the data to work by comparing opportunities, threats, and assets.[30] Through regular and methodical assessment, local conditions can be contextualized, while changes can be confronted through the local government's most effective tools. In short, crafting a strategic adaptation plan builds upon a good understanding of local climate change impacts and the local culture and fabric. Metrics, baselines, and assessments are relevant as *threshold initiatives* in adopting a substantive planning strategy because they can determine effective directions and priorities for adaption goal-setting that target the most serious local climate change vulnerabilities.

Identification of key metrics should be the starting point for the development of an effective local adaptation plan. The metrics should include indicators of community values that will illustrate and measure the changes from climate change that are relevant to a particular community asset. For example, the City of Chula Vista, California, recognized rising temperatures as a relevant climate change related condition and proposed 11 adaptation strategies to address this condition. The City identified 'potential metrics that could be tracked and reported on an ongoing basis to quantify the [City's] performance' in preparing for changing conditions. Some of those metrics included:

- Actual temperature reductions;
- Total square footage of 'cool' paved surfaces;
- Number of wildfire education materials distributed;
- Percentage increase in community awareness from outreach and education;

---

29 Ibid.

30 In the process of identifying metrics and establishing baselines, communities will be compelled to distinguish between measures that further adaptation goals, and those which mitigate the impacts of climate change. In the context of adaptation the relevant baselines are directed at the impacts stemming from climate change, whereas in mitigation the relevant baselines are directed at measuring the source of climate change. It is the difference between measuring the change in sea level versus the change in carbon dioxide emitted into the atmosphere. While the two are related, the latter (carbon dioxide emissions) is a greenhouse gas contributing to climate change, while the former (sea levels) is an effect of climate change. For purposes of adaptation, planning bodies would seek to measure the changes in vital natural resources or community characteristics, as opposed to the causes of those changes, notwithstanding overlaps and redundancies.

- Number of surveys performed to monitor shifts in local species' range and diversity;
- Number of habitat restoration areas where climate change impacts were incorporated into designs.[31]

Of course, the decision to use an individual metric corresponds to the location-specific impacts discussed in section 2 and should be intended to establish a solid foundation for a culture of sustainability.[32] Hence, the City of Dubuque, Iowa, in the United States illustrates ways by which a community should consider an array of metrics and select those that reflect the community's values. In arriving at its metrics, Dubuque, in conjunction with the University of Iowa School of Urban and Regional Planning, reviewed over 1,200 different metrics and ultimately settled on approximately 60 that now drive the community's response to climate change and, more generally, the community's vision for a sustainable community. The Dubuque metrics include measurements of carbon dioxide emissions, urban forest levels (tree canopy cover), areas covered with prairies and wetlands, drinking water contamination levels, and asthma and obesity rates.[33] Such impacts are identified by using indicators as assessment mechanisms, organized into the following categories: Regional Economy; Smart Energy Use; Smart Resource Use; Community Design; Green Buildings; Healthy Local Food; Community Knowledge; Reasonable Mobility; Healthy Air; Clean Water; and Native Plants and Animals.

---

31 See City of Chula Vista, above note 8.

32 The adaptation plan drafted by the City of Lewes, Delaware, provides another good example of how communities can benefit from the use of metrics, baselines, and assessments. The City identified key metrics, established baselines, and then honed in on two key vulnerabilities through an assessment process. Those vulnerabilities, saltwater intrusion into the aquifer and property damage from increased flooding, were then given priority in seeking methods to adapt to these changing conditions. This process provided a basis upon which Lewes could measure indicators of local importance (thereby keeping adaptation planning relevant to the community) and fashion response plans in the most effective way. The City of Lewes, *Hazard Mitigation and Climate Adaptation Action Plan* (June 2011), http://www.deseagrant.org/sites/deseagrant.org/files/attachments/Lewes%20Hazard%20Mitigation%20and%20Climate%20Adaptation%20Action%20Plan.pdf, accessed 24 October 2012.

33 University of Iowa School of Urban and Regional Planning, Iowa Initiative for Sustainable Communities, *Sustainability Progress Report 2012* (2012) 76–77, http://www.cityofdubuque.org/DocumentCenter/Home/View/15189, accessed 24 October 2012.

Dubuque's program is far-reaching because each metric is measured at a given time to establish a baseline to monitor future changes. The baseline provides a controlled indicator from which the community can evaluate climate change related conditions and the community's ability to adapt to those conditions. This requires the community to employ a monitoring system to evaluate whether the metric is generating useful information. The result of the Dubuque planning process is a cutting edge implementation tool that demonstrates the best of climate preparedness and allows the city to grasp the total cost of continuing climate change monitoring – a model for other local governments. The excerpt from the Resource Use section of the Dubuque plan (Appendix B) (see box) suggests the effectiveness of the tool both in terms of usefulness in cost–benefit analysis and budgeting for the City, but also in terms of its use as a planning and assessment tool for identifying climate change vulnerabilities. Dubuque's assessment methodologies were reproduced across the 11 categories and almost 60 separate indicators to create a comprehensive and deep look at the ways the City can respond to climatic changes.

SMART RESOURCE USE

Indicator: Groundwater Conservation

Time: 1 hour

The net water withdrawal of the Jordan Aquifer near Dubuque was determined by acquiring standing water level data from … the Water Plant Manager at the City of Dubuque Water Department and the Iowa DNR Geological Department's GEOSAM database of the four primary deep water wells utilized by the City of Dubuque. Deep water sources are less resilient to changing climate and human impact and provided a better measure of overall water supply sustainability, and therefore the alluvial well measures were not incorporated. Data from the Iowa DNR's GEOSAM database provides historical information on water levels of local Dubuque water supplies, of which the static water levels of the four deep water wells were obtained for the earliest years available. The earliest static water level available was for Well 6 in 1935; therefore Well 6 was used as the baseline measure as it provided the most comprehensive measure to determine long-term groundwater withdrawal trending.

1. More recent, monthly measures of local static water levels of the Jordan Aquifer, at Well 6, below Dubuque were acquired from Dubuque's Water Department.
2. These monthly static water levels were averaged for each year from 2005–2011.
3. The change in the static water level from 1935 was determined for each year as a net change.
4. The data for this indicator was displayed as the net water extracted, in feet, since the 1935 static water level of Well 6.

Indicator: Sustainable Materials Management

Time: 1 hour

Landfill diversion was determined based on the percent of solid materials that were recycled, reused, or composted through municipal collection services only. This data was acquired from … the Resource Management Coordinator of the City of Dubuque's Public Works Department. The data consisted primarily of household solid materials collected per month, which includes refuse, recyclables, and yard waste.

1. The monthly amount, in tons, of recyclables and yard waste diverted to the local recycling center and composting facility were classified as 'solids diverted' as they did not go to the landfill, which was averaged for each year available.
2. The amount of solids diverted is then taken as a percent of the total tons of all residential refuse, recyclables, and yard waste collected for each year from 2006 to 2011.
3. The same process was completed for the City of Oshkosh, where the city data was received from … the Winnebago County Landfill.

Indicator: Trash/Refuse Generation

Time: 30 minutes

Trash/refuse generation was measured based on the amount of solid discards produced per household, which is based on municipal collection services only. The measure of solid discards generated includes routinely produced discards, or refuse, that enters the landfill.

1. The solid discards data was collected from the City of Dubuque Public Works Department as the tonnage collected within Dubuque by city services.
2. The total tons of material were then converted to pounds of solid discards by multiplying the amount in tons by 2,000 (1 ton = 2,000 pounds).
3. Data regarding the annual number of households within Dubuque was acquired through the City of Dubuque as the number of households who are customers of the city curbside collection service. The number of customers estimated on an annual basis is calculated from the total base monthly fee revenues divided by the monthly base fee and adjusted slightly for customers subscribing to tipper carts and not paying a monthly base fee.
4. The annual number of household customers from 2006 to 2010 was used to normalize the poundage of solid discards generated by household in the City of Dubuque.
5. Data was processed the same for the City of Oshkosh, which was received from … the Winnebago County Landfill.[1]

Note:

1 University of Iowa School of Urban and Regional Planning, Iowa Initiative for Sustainable Communities, *Sustainability Progress Report 2012* (2012) 76–77, http://www.cityofdubuque.org/DocumentCenter/Home/View/15189, accessed 24 October 2012.

The manner in which local governments identify their own metrics and inventories of local climate change vulnerabilities is a reflection of local values, and as a result, the inventory process will likely include considerations of cost and public participation. Part of the assessment process also includes identifying the natural and built assets that are critical to the community, including health, social, communication, fire, and transportation services. An evaluation of the specific vulnerabilities of these services is helpful, as they may be impacted by climate change conditions. Poor or weak housing stock, vulnerable populations, and hazardous waste sites and their location to natural resources are just a few community characteristics that may be particularly susceptible to climate change.

In the United States, STAR Communities recently publicized an innovative program that is centered on seven sustainability goals and 46 objectives.[34] The 20 cities which volunteered to serve as the initial STAR Communities will have their sustainability performance measured against progressive assessments and an objective rating system. The process is designed to identify inefficiencies and facilitate reductions in some climate change vulnerabilities on a contextual basis. Not surprisingly, the first 'Local Government Action' under the program's 'Climate Adaptation' objective is to adopt a climate adaptation plan which establishes a monitoring system for climate change vulnerabilities. One of the initial STAR Communities cities, Des Moines, Iowa, has been engaged in a complex process to identify key metrics that will help illuminate whether the city is effectively adapting to climate change related conditions. Des Moines has engaged a wide range of interested parties that bring various levels of climate sophistication and community values to the process. Based on the information obtained through identifying metrics, establishing baselines, and assessing changes in the metrics, the STAR Communities will have a foundation to understand the changes occurring in their communities and draft corresponding goals to adapt to those changes.

Another interesting development took place in the summer of 2012, when the Local Government Association in England launched the 'Climate Local' initiative to enable collaboration among local councils on tackling climate change adaptation strategies. The initiative supports local efforts to conduct risk assessments to understand how global challenges will affect local entities, and suggests the implementation of those assessments though adaptation policies. Several local councils in England have also begun the adaptation process, starting with an assessment tool created by the UK Climate Impacts Programme (UKCIP)[35] called the Local Climate Impacts Profile ('Profile'). The Profile is designed to help the local entity identify key metrics, establish baselines, and assess how climate change will (or is) impact(ing) the community. The Profile 'demonstrate[s] not only the potential impacts of climate change but also the extent to which the authority, community or organization is prepared

---

34 STAR Communities refers to Sustainability Tools for Assessing and Rating Communities, a project that fosters a 'voluntary, self-reporting framework for evaluating, quantifying, and improving the livability and sustainability of U.S. communities'. See http://www.starcommunities.org, accessed 24 October 2012.

35 According to its web page, UKCIP 'was established in 1997 by the UK government, when it was then known as the UK Climate Impacts Programme. Since it began, UKCIP has been based at the Environmental Change Institute at the University of Oxford', http://www.ukcip.org.uk/lclip/why-do-lclip, accessed 24 October 2012.

and able to respond now (i.e., to deal with current weather events)'.[36] In performing its summary of its community's Profile, Nottingham, England, for example, used a variety of metrics to obtain baselines upon which it then developed a separate action plan. Some of its key findings involved its vulnerability to severe weather events, such as flooding, heavy snow, and heat waves and associated damage to infrastructure.

A final yet essential consideration is that the assessment process is not a one-time activity in the climate change context. Rather, communities may find that once they have identified key metrics and baselines and have begun a system of ongoing monitoring, the chosen metrics need to be revised to capture key (and better) data. The assessment and adaptation process should also incorporate this flexibility, as adaptation is an attempt to prepare for the future without the benefit of certainty as to future conditions. Efficiency and accuracy can be improved through an internal evaluative process to ensure the relevance of the process to actual climate change related conditions.

5 OBSTACLES, INCENTIVES, AND MANDATES: ACHIEVING GOALS THROUGH COMMUNITY BASED PREPAREDNESS

Once communities have gone through the process of identifying metrics, establishing baselines, and assessing changes, they have established a solid foundation upon which to begin the arduous task of implementing the substantive adaptation responses made compelling in the assessment process. At the very least, communities have three legislative tools that are relevant here as adaptive land use planning opportunities: (1) remove obstacles located in existing land use planning laws that hinder the adaptation process; (2) embed incentives into land use laws to encourage private and public sector local adaptation planning; and (3) mandate land use standards to prepare for climate change related conditions and alleviate further deterioration stemming from climate change. In a powerful and succinct article, Chris Duerksen describes the following tools as 'three paths to sustainability':

---

36 Ibid.

- Removing obstacles: Most modern [zoning] codes create barriers to sustainability, often unintentionally. For example, small wind turbines and solar panels are often prohibited by residential zoning regulations, height controls, or design standards.
- Creating incentives: Some sustainable technologies are relatively new and experimental – like green roofs. Zoning codes can foster increased density and other incentives to encourage use of such technologies.
- Enacting standards: While removing obstacles and creating incentives will be important, no zoning code can succeed without mandatory regulations that require certain actions or prevent harm. As Teddy Roosevelt once said, a smile and a six-shooter sometimes work better than a smile alone.[37]

Communities and their local governments have adopted innovative substantive adaptation policies that implement each of Duerksen's three paths, either individually or in any variety of combination. Some local sustainability policies approached adaptation in a comprehensive manner, covering a wide variety of land use planning topics and contemplating reliance on all three tools. The National Research Council, for example, has suggested following a broad framework. Vulnerabilities can be reduced by limiting the magnitude of climate change through actions to limit greenhouse gas (GHG) emissions, reducing sensitivity (the underlying social, cultural, economic, geographic, ecological, and other factors that interact with exposures to determine the magnitude and extent of impacts), or improving coping capacity (the ability to avoid, prepare for, and respond to an impact so that it is not seriously disruptive). Actions to reduce sensitivity and increase coping capacity are key to effective adaptation to climate change.[38]

Premised on a similar structure of reducing vulnerability and maximizing resiliency a number of local governments have adopted extensive and

---

[37] Chris Duerksen, *Saving the World through Zoning* (American Planning Association 2008). While Duerksen was primarily focused on the relationship between zoning and climate mitigation, the three tools are equally applicable to the use of zoning and land use controls to facilitate climate change adaptation. Because the tools do not have an inherent substantive commitment – meaning they do not imply any particular form or principle of implementation and can be flexible – they can be used to explore opportunities relevant to mitigation and/or adaptation.

[38] National Research Council, above note 4 at 29.

comprehensive adaptation strategies. In the United States, the communities of Marion County, California; Worcester, Massachusetts; Durham, North Carolina; Portland, Oregon; and Pittsburgh, Pennsylvania, represent local governments that have undertaken comprehensive strategies to adapt to projected climate change related conditions. New York City's *plaNYC* integrates an extensive adaptation strategy through its far-reaching comprehensive plan, which is driven by removing obstacles, incentivizing changes, and mandating standards on a variety of essential areas of concern, including:

- Precipitation, flooding, and storm water run-off to ensure potable water and to protect the wastewater infrastructure;
- Increased energy demand during heat waves and designating public buildings as emergency cooling centers;
- Emergency power generators to supply energy when peak demands arrive during heat waves or in case of electric outages;
- Underground energy and telecommunications infrastructure to withstand flooding and sea level rises;
- Energy, building, and sewer codes to adapt buildings to high wind conditions, flooding, and high temperatures;
- Transportation infrastructure to accommodate flooding, saltwater damage, increased power demands and power outages, overheating of subway platforms, and increased stress on infrastructure.[39]

Adopting these overarching strategies, while bold and far-reaching, can be somewhat cumbersome for local communities; perhaps due to the political capital needed to succeed in this arena,[40] but also because of the demand these strategies place on governmental resources. As discussed above, however, local governments are familiar with a variety of legal tools that may be used to prepare for and adapt to economic, environmental, and social changes, and local governments may be able to employ these seasoned strategies effectively for adaptation purposes.

Below, we set forth individual, targeted policies that have been employed to help further adaptation policies. The policies cover a wide

[39] The City of New York, *PlaNYC: A Greener Greater New York* (2011), http://nytelecom.vo.llnwd.net/o15/agencies/planyc2030/pdf/planyc_2011_planyc_full_report.pdf, accessed 24 October 2012.

[40] See Jared Lang, 'The Politics of Greening Governmental Operations' in Keith H Hirokawa and Patricia E Salkin (eds), *Greening Local Government: Legal Strategies for Promoting Sustainability, Efficiency, and Fiscal Savings*, (ABA 2012) 3.

array of land use planning approaches but need not be adopted as part of a larger planning strategy – individual policies may be easier to implement and may be more directed to a community's assessment of its adaptation needs. In addition, it may be easier to convince local planning bodies to address individual policies which lessen the collective action problems often associated with comprehensive plans or attempts to reduce greenhouse gas emissions. The policies respond directly to local conditions, even though they may also have a positive impact on the global community.[41]

5.1 Planning and Energy

Specific and local land use planning efforts have the ability to facilitate the adaptation of local communities to unforeseen fluctuations in energy use and production. The localization of energy production, for example, can help reduce community reliance on outside and uncontrolled institutions and can help the community become more resilient to climate change related conditions. Land use planning can foster the localization of energy production through a variety of methods, including by establishing community based energy production facilities, such as community owned large solar arrays or wind farms. Planning for local energy production may include the removal of obstacles (for example, height limitations prohibiting wind turbines or limitations on roof installations prohibiting solar panels), the provision of incentives for producing local energy, and the establishment of mandates concerning on-site or local energy production. Further, land use planning may help prepare communities for changes in energy use and production by siting single, small energy generation resources near important energy demand centers, such as hospitals, which can help reduce peak demand and avoid service disruptions. Additional planning techniques include requiring energy audits and/or upgrades at the time of real property interest transfers; designation of special energy zones, which establish minimum standards of performance pertaining to energy use, efficiency, and production; and zoning for alternative roof types to reduce energy usage.[42]

---

[41] For a description of the inefficiencies involved with mitigating climate change through a reduction in greenhouse gas emissions at the local level see Rosenbloom, above note 5 at 462–468.

[42] See, e.g., The Hepburn Wind Project, http://hepburnwind.com.au, accessed 24 October 2012 (the first community-owned wind farm in Australia, the past summer Hepburn won the World Wind Energy Award, as it produced its 10 millionth kilowatt hour of clean, local energy for 2,300 households); The

5.2 Planning and Emergency Centers

Local land use planning can help governments accommodate the needs of their citizens in times of flooding, heat waves, and other severe weather events by establishing emergency centers. This may include the zoning or designation of high, dry areas for sleeping and shelter, as well as the identification of cooling centers during heat waves. Designating these locations before they are necessary should reduce confusion and loss of life and property at the time they become necessary.[43]

5.3 Planning and Transportation

Multiple transportation modes may be at risk from flooding, damage to infrastructure, strain on materials (bridges, roads), increased electrical demand, and power outages. Available transportation at relevant scales for getting people to safe areas and bringing in supplies during heat waves and floods will be critical to serve cities in times of emergency, which are accompanied by increased and immediate needs for the movement of people and services. Land use planning, which can be accomplished by planning more compact communities, and transit-oriented development (TOD), can help by decreasing the community's reliance on vehicular travel over long distances.[44] TODs are implemented through the planning process to promote mixed-use development, social equity, and achieve the high density critical mass essential for the

---

Hull Wind Turbine, Hull, Massachusetts, http://www.town.hull.ma.us/Public_Documents/HullMA_Light/Hull%20Wind%20Turbine, accessed 24 October 2012 (describing Hull Wind One, a 660 kilowatt wind turbine owned by the local government). For successful examples of cooperatives in Europe see http://www.chelseagreen.com/Wind/articles/NEurope96.htm, accessed 24 October 2012, and Canada, http://www.windshare.ca and http://www.ontario-sea.org, accessed 24 October 2012.

[43] See, e.g., SPURREPORT, *Climate change hits home: Adaptation strategies for the San Francisco Bay Area* (SPUR 2011), http://www.spur.org/files/SPUR_ClimateChangeHitsHome.pdf, accessed 24 October 2012; City of Chula Vista, above note 8.

[44] For information on TODs see http://www.transitorienteddevelopment.org/transportation.html, accessed 24 October 2012; see also Smart Growth Network, *About Smart Growth*, http://www.smartgrowth.org/about/default.asp, accessed 24 October 2012. For the American Planning Association's model codes to promote smarts see http://www.planning.org/research/smartgrowth, accessed 24 October 2012.

long-range success of local and regional public mass transit. In considering TODs, local communities have mandated and incentivized density increases, parking space reductions, and design guidelines to promote walking, biking, and mass transit. TODs are also believed to help stabilize the local economy by providing employment opportunities and affordable and diverse housing.

5.4 Planning and Urban Farming

Removing existing land use planning obstacles to urban farming (such as, minimum grass frontage requirements), incentivizing urban residents to plant, grow, and harvest crops, and, in some cases, mandating certain vegetation growth may help address a number of climate change related conditions. As discussed above, scarcity of food – particularly in urban areas – is likely to be one climate related change, resulting in poor diets and malnutrition. Land use planning to encourage urban farming could help address this by assisting the community in becoming more resilient to projected commodity market changes. In addition, depending on the type of vegetation, an urban garden may reduce run-off and serve as a greenhouse gas sink, alleviating additional climate change related conditions.[45]

5.5 Planning and Water

The State of Arizona has recognized that temperature changes will affect surface water levels and water supplies, decreasing hydroelectric output and agricultural production and increasing the potential for wildfires and air pollution.[46] Yet water conservation, potability, and access to water, are vital components of sustainability and economic, environmental, and social adaptation to climate change in any region.[47] For instance, all local governments are compelled to consider how and where to locate water and sewerage facilities to minimize climate vulnerabilities and service

---

[45] For in depth discussion of legal challenges and regional perspectives on urban farming, see Sarah B Schindler, 'Of Backyard Chickens and Front Yard Gardens: The Conflict between Local Governments and Locavores' (2012) 87 Tulane Law Review (forthcoming).

[46] Arizona Department of Environmental Quality, http://www.azclimatechange.gov, accessed 24 October 2012.

[47] See, Keith H Hirokawa, 'The Relevance of Land Use Law to Climate Change Preparedness: The Case of Sustainable Water Practices' (2009) 40(5) Trends (ABA Section of Environment, Energy, and Resources).

interruptions caused by climatic tragedies. Local governments may be concerned with aquifer and surface water protection, and so may establish minimum requirements for pervious surfaces or require setbacks and buffer zones along rivers and within important watershed areas to control flooding. Local governments have embraced the goals of water conservation and are actively seeking change in both their regulations and operations. Some municipalities are adopting mandatory, restrictive regulations for domestic and industrial water usage, while others are engaging in water planning or incorporating water efficiency standards into the land use process. Local governments are also becoming quite efficient at incorporating innovative land use tools in the building design process, such as implementing green roof requirements to control run-off; water efficiency plans to decrease resource reliance and use; effective grading strategies so that storm water flows to intended areas; and locational requirements that minimize population vulnerabilities and direct critical infrastructure (such as, nurseries and hospitals) to areas less prone to floods.

5.6 Planning and Permitting Processes

Many land use permitting processes may provide legitimate and effective mechanisms for addressing adaptation needs. Requests for variances from the zoning code, for example, often require the applicant to show a hardship under the existing zoning framework.[48] A community could revise the variance process to recognize the relevancy of climate change to land use hardships. Variance approvals could be conditioned upon implementing approved adaptation strategies that would integrate adaptation into the community by continuous monitoring and reporting or flexible performance standards. Such principled flexibility could be incorporated into any land use permitting process (rezoning, special and conditional use permits, building and occupancy permits) to require proof that the resulting land use demonstrates resiliency principles, such as decreased reliance on gray infrastructure (for example, multi-modal transportation or on-site electricity generation), increased adaptability to new technologies (for example, rainwater capture and gray water reuse),

---

48 *Erickson v City of Portland*, 496 P.2d 726, 728–29 (Or 1972) ('Variances traditionally have been considered escape valves to allow property owners relief from zoning restrictions which, when applied to particular land, have the result of making that land completely unusable, or usable only with extraordinary effort').

decreased natural resource impacts (for example, water conservation and materials reuse), or decreased climate vulnerability.

The challenge in this context is identifying processes where adaptation is appropriate and where the perceived burdens of adaptive measures are not too onerous. In theory, conditioning permits on adaptive tools would benefit landowners by directing land uses toward more adaptive and resilient manifestations. However, even where the law contemplates adaptive measures, of course landowners may invoke property doctrines that protect the prerogatives of owners to preserve the status quo.[49] One area in which adaptation will be challenging for planning entities is in promoting certain types of improvements in coastal and other flood-prone areas.[50]

5.7 Planning and Brownfields

In preparing for unforeseen changes, local planning efforts can identify hazardous waste sites in accessible areas, including flood plains and low-elevation areas that will be vulnerable to rising sea and freshwater levels. Most local planning efforts do not track the circumstances or require remediation of contaminated sites that may release hazardous materials when water levels rise. Adaptation policies can focus efforts to control contamination risks by requiring removal of hazardous materials from areas particularly susceptible to flooding. Communities can also properly zone hazardous uses for the future to separate them from vital natural resources.

5.8 Planning and Wetlands

Wetlands provide ecosystem services that will be essential for coping with climate change.[51] Wetlands can efficiently capture, store, and transport water from significant storm events, making wetlands an important adaptation tool for absorbing storm and flood energy. Wetlands

---

[49] For instance, existing land uses may be protected as legally nonconforming uses in which rights are extended to uses that were lawful when initiated, but later became unlawful due to changes in zoning regulations. *See Clackamas County v Holmes*, 508 P.2d 190, 192 (Or 1973). Often nonconforming rights may be continued and even altered, but may also be subjected to a reasonable amortization. *Cope v City of Cannon Beach*, 855 P.2d 1083, 1084 (Or 1993).

[50] For analysis of the property challenges in this context, see Timothy M Mulvaney, 'Exactions for the Future' (2012) 64 Baylor Law Review 101.

[51] The term 'wetland' has been defined to mean those 'areas that are inundated or saturated by surface or ground water at a frequency and duration

provide structure and shelter for plants and wildlife. They constitute ecological transition areas between uplands and watercourses, and they capture sediments and pollutants before they can reach streams.[52] Wetlands also serve as a protective barrier from flooding waters and provide a natural habitat for wildlife. Proper land use protections for these services will continue to be critical to maintaining a functioning ecosystem fostering human well-being amidst changing climatic conditions. Local governments can most effectively implement adaptation strategies in wetlands by focusing on protecting both wetland structure and functionality.[53] Hence, Clark County, Washington, protects wetlands through the collaboration of its Wetlands Protection Ordinance and Habitat Conservation Ordinance. The county requires development applicants to 'demonstrate that the activity: [s]ubstantially maintains the level of habitat functions and values as characterized and documented using best available science' and '[m]inimizes habitat disruption or alteration beyond the extent required to undertake the proposal'.[54] Clark County allows mitigation where '[d]isrupted functions and values shall be mitigated on-site as a first priority, and off-site thereafter'.[55] Such regulations, that impute value to wetlands beyond mere acreage, may result in better distribution and interconnectedness of wetland services throughout a community.

5.9 Planning and Urban Forests

Trees provide a wide array of services that are essential to community resiliency and adaptation to climate related changes. Communities that invest in urban forests through land use planning are better positioned to

---

sufficient to support and that under normal circumstances do support, a prevalence of vegetation typically adapted for life in saturated soil conditions'; 40 CFR § 230.3(t).

[52] Robert A Gleason et al. (eds), *Ecosystem Services Derived from Wetland Conservation Practices in the United States Prairie Pothole Region with an Emphasis on the U.S. Department of Agriculture Conservation Reserve and Wetlands Reserve Programs* (USGS 2008), http://www.fsa.usda.gov/Internet/FSA_File/pp1745.pdf, accessed 24 October 2012.

[53] See, generally, Jeanne Christie and Jon Kusler, *Recommendations for a National Wetlands and Climate Change Initiative* (2009), http://aswm.org/pdf_lib/recommendations_2008_112008.pdf, accessed 24 October 2012.

[54] Clark Cnty, Wash, Code § 40.440.020(a)(2)(a)–(b).

[55] Clark Cnty, Wash, Code § 40.440.020(A)(3)(b).

benefit from the services provided by trees, including biomass production, provision of wildlife habitat, soil stability, storm water control, and shade.

Communities can capitalize on urban forest services by preserving open spaces and park lands, planting and maintaining street trees, providing access to community gardens, and regulating human interaction with trees. Hence, the City of Vancouver, Washington, notes that: 'Green infrastructure provides important ecological and social functions that translate into direct cost savings to local government and indirect stimulation of the local economy. Unlike traditional gray infrastructure capital improvements, such as transportation and water systems …, green infrastructure accrues value [and] provides greater services as time passes'.[56] These efforts benefit both common and private areas with tree services.

When local governments commit to investments in the urban forest, they are likely to realize the continuing benefits of ecosystem services. These benefits rely upon an understanding of the relationship between tree management strategies and the services we want to receive: 'A fundamental premise behind the management of urban forest resources is that humans can manipulate the structure of vegetation to enhance various forest benefits.'[57] Tree protection can be implemented through the use of traditional tools, such as nuisance abatement. However, local governments seeking to maximize the benefits of tree services typically employ planning principles to protect inventories and needs of urban forest canopy throughout communities.[58]

---

[56] City of Vancouver, Washington, *Urban Forest Street Management Plan* (2007) 7, http://www.cityofvancouver.us/parks-recreation/parks_trails/urban_forestry/ufmanplan.asp, accessed 24 October 2012.

[57] John Dwyer et al., *Connecting People with Ecosystems in the 21st Century: An Assessment of Our Nation's Urban Forests* (USDA Forest Service 2000), http://www.fs.fed.us/pnw/pubs/gtr490/pnw_gtr490.pdf, accessed 24 October 2012 ('this mechanism occurs in the context of societal values (what the community wants) and ecological principles (what is ecologically possible)').

[58] See generally, Keith H Hirokawa, 'Urban Forests as Green Infrastructure' in Keith H Hirokawa and Patricia Salkin (eds), *Greening Local Government: Legal Strategies for Promoting Sustainability, Efficiency, and Fiscal Savings* (ABA 2012).

6 CONCLUSION

The fact that we are discussing adaptation indicates a commitment to the idea that change is coming, and local communities must make a concerted effort to brace for new challenges by increasing adaptive capacity and improving climate change resiliency. Land use control is relevant because society's vulnerability to major shifts is indicated by the manner in which we design cities, including the locations at which we facilitate the construction of homes, the placement of shopping centers, and the direction that water and sewerage are transported through and away from town. As such, the land use planning actions described in this chapter acknowledge local governments' important role in leading the struggle toward more sustainable, climate-smart decisions. The actions also indicate that while geographical and jurisdictional restrictions may limit the reach of local adaptation efforts, local governments nonetheless wield important powers in the climate change context, particularly as the capacity to cope with climate change rests in the challenges society may face to receive essential services or trade in the marketplace. These challenges are largely determined by our priorities in land use planning.

Climate change will provoke fundamental changes that are specific to particular communities. Environmental insecurity, economic instability, and social uncertainty due to shifts in the global climate present significant environmental problems to local governance. Vital community services are vulnerable to climate contingencies, whether the issue involves transportation, air and water quality, communication, energy transmission or generation, or waste disposal. Although local governments alone cannot affect all of the changes necessary for successful local adaption, the employment of the strengths and authority of local governments, as well as the drive of communities, are essential in formulating a coherent response to forthcoming climate challenges.

13. Climate change adaptation and green building

Keith H Hirokawa and Aurelia Marina Pohrib

1 INTRODUCTION

Climate change poses one of the most significant challenges ever faced in terms of the stability of civil society, the security of access to natural resources, and the ecological integrity of the human environment. Committed warming and the projected rapid rate of impacts due to climate change will result in an unpredictable and unfamiliar environment, requiring both civil society and natural systems to adapt.[1] Increases in air and water temperatures, reduced frost days, increased frequency and intensity of heavy downpours, a rise in sea levels, reduced snow cover, glaciers, permafrost, and sea ice, a longer ice-free period on lakes and rivers, lengthening of the growing season, and increased water vapor in the atmosphere have also been observed. These climate related changes are expected to continue while new ones develop. Likely future changes for the United States (US) and surrounding coastal waters include more intense hurricanes, with related increases in wind, rain, and storm surges (but not necessarily an increase in the number of these storms that make landfall), as well as drier conditions in southwest regions of the US and the Caribbean. These changes will affect human health, water supply, agriculture, coastal areas, and many other aspects of society and the natural environment.[2]

Climate change is also expected to cause substantial impacts on the built environment and physical infrastructure. Sea level rise is expected to impair the security of coastline generation and transmission facilities,

[1] US Global Change Research Program, *Global Climate Change Impacts in the United States* (Cambridge University Press 2009) 9–12, http://downloads.globalchange.gov/usimpacts/pdfs/climate-impacts-report.pdf, accessed 26 October 2012.

[2] Ibid. at 9.

accelerate erosion, and adversely affect fresh water supplies.[3] Heat waves and high summer temperatures will result in high demands for electricity for temperature control. An increased frequency and severity of storm events will challenge the physical integrity of pipelines and transmission lines. It is critical to address these changes, as the built environment has historically served as protection for humans against the unexpected consequences of ecological changes; humans and civil society are more vulnerable with increases in population and growing uncertainty regarding the security of housing and infrastructure. In addition, any significant and rapid changes to the built environment will undeniably affect humans in sociological, psychological, and economical ways: 'The built environment literally "houses" economic activities, individuals and families, as well as society's cultural heritage. Its purpose is to protect life, health and psychological and social welfare of its inhabitants, host economic activities and sustain aesthetic and cultural values.'[4] The Census Bureau projects that the United States' population in 2050 will be 439 million, a 46 percent increase from 2007,[5] suggesting that climate change vulnerabilities include challenges of scale.[6]

---

[3] See generally, National Research Council, Board on Earth Sciences and Resources (BESR), Ocean Studies Board (OSB), *Sea-Level Rise for the Coasts of California, Oregon, and Washington: Past, Present, and Future* (2012).

[4] Frank Henning Holm, Towards a sustainable built environment prepared for climate change?, Presentation to Global Policy Summit on the Role of Performance-Based Building Regulations in Addressing Societal Expectations, International Policy, and Local Needs, National Academy of Sciences, Washington, DC (2003) 1,http://www.irccbuildingregulations.org/pdf/3–05.pdf, accessed 26 October 2012.

[5] US Census Bureau, *National Population Projections* (2008), http://www.census.gov/population/www/projections/downloadablefiles.html, accessed 26 October 2012.

[6] Other issues arising from population increase deal with individuals and families, particularly the make-up of the family. If population increases and climate change issues are not significantly addressed, and as such society has not found a way to adapt to changes, what will happen to the structure of the household? The Census Bureau, in their calculating projection accounted the birth rate at 2.1 per woman; will climate change increase or decrease the birth rate? If land available for development significantly decreases will families congregate together such that a household size will increase from about a family of 2–4 persons to include extended families? Thus, social issues are also important aspects of green building and, like climate change and carbon dioxide emissions, should be taken into account when moving towards a sustainable built environment.

Preparing for the effects of climate change will require an adaptation strategy that demands shifts from current practices and demands, adjustments to the carbon-dependent economy, and a transformation of our values.[7] Adaptation 'requires a constructive legal and social response to continuous interacting, often unpredictable, and perhaps irreversible changes'.[8] As human behavior has become a primary focus of strategies to address climate change, the public dialogue has turned to building practices in response to evidence of the overwhelming pressure that today's buildings place on energy and natural resource supplies, especially compared to other sectors of the economy.[9] Because 'the building sector contributes the largest single source of emissions to our environment',[10] building construction activities significantly contribute to the human drain on adaptive capacity. As such, the design and construction practices that created the approximately 81 million resource, energy, and water inefficient buildings in the US[11] 'are a logical place to address many of today's environmental issues, particularly those concerning energy use and climate change'.[12]

Green building[13] practices address building design, construction, and operation in a comprehensive manner to reduce the overall environmental

---

[7] Roberto Danino, 'Foreword' in David Freestone and Charlotte Streck (eds), *Legal Aspects of Implementing the Kyoto Protocol Mechanisms* (Oxford University Press 2005) v.

[8] Robin Kundis Craig, '"Stationarity is Dead" – Long Live Transformation: Five Principles for Climate Change Adaptation Law' (2010) 34 Harvard Environmental Law Review 9, 27.

[9] US Dept. of Energy Buildings Technology Program, 'Obama Administration Launches New Energy Efficiency Efforts' (2009), http://www1.eere.energy.gov/buildings/news_detail.html?news_id=12607, accessed 26 October 2012.

[10] The American Institute of Architects, *Local Leaders in Sustainability: Green Building Policy in a Changing Economic Environment* 6 (2009), http://www.aia.org/aiaucmp/groups/aia/documents/document/aiab081614.pdf, accessed 26 October 2012.

[11] Michael B Gerrard, 'Preface' in J Cullen Howe and Michael B Gerrard (eds), *The Law of Green Buildings: Regulatory and Legal Issues in Design, Construction, Operations, and Financing* (ABA 2010) xv.

[12] Ibid.

[13] US Environmental Protection Agency, 'Green Building, Basic Information', http://www.epa.gov/greenbuilding/pubs/about.htm, accessed 26 October 2012 (The Environmental Protection Agency defines green building as 'the practice of creating structures and using processes that are environmentally responsible and resource-efficient throughout a building's life-cycle from siting to design, construction, operation, maintenance, renovation and deconstruction.

impact of the built environment, increase the efficiency of energy, water, and other resource use, and reduce waste, pollution, and environmental degradation.[14] In large part, green building practices diverge from conventional building codes by identifying standards of acceptable building performance, rather than setting design measurements, and accounting for the life cycle of materials that comprise the structure.[15] Green building champions the idea that 'a more informed approach to the built environment can result in buildings that complement the natural environment, instead of conflict with it'.[16] Informed building practices can help to mitigate the effects of the built environment on climate change.[17] However, although mitigation is an important first step in the fight against climate change, greater attention to adaptation is needed because mitigation efforts alone are unlikely to solve these problems.[18] The question asked in this chapter, therefore, relates to the potential of green building practices to serve as a measure for adaptation to climate change impacts through technology, land use planning, building construction, and design.

This chapter identifies adaptation principles that are relevant to the built environment and the impacts to and from the built environment that are most suitable for adaptation planning. This review of green building principles and strategies shows that green building practices support adaptation by reducing reliance on produced energy and physical infrastructure (flexibility for uncertain future circumstances) and providing

---

This practice expands and complements the classical building design concerns of economy, utility, durability, and comfort. Green building is also known as a sustainable or "high performance" building').

[14] Ibid.

[15] Life cycle 'provides an assessment of a product or building component over its lifecycle. Instead of focusing on a single aspect of a product's sustainability, such as recycled content, it measures a product's entire life including extraction/harvesting of materials, processing/manufacturing, transportation, installation, operations & maintenance, and reuse/recycling/disposal.': Scott DeGaro, 'Life Cycle Cost Analysis vs. Lifecycle Analysis: The Green Building Implications', http://www.green-buildings.com/content/781534-life-cycle-cost-analysis-vs-lifecycle-analysis-green-building-implications, accessed 26 October 2012.

[16] Keith H Hirokawa, 'At Home with Nature: Early Reflections on Green Building Laws and the Transformation of the Built Environment' (2009) 39 Environmental Law 507, 510.

[17] Global Stewards, 'Sustainable Building Solutions for Green Buildings', http://www.globalstewards.org/buildings.htm, accessed 26 October 2012.

[18] Daniel H Cole, 'Climate Change, Adaptation, and Development' (2008) 26 UCLA Journal of Environmental Law and Policy 1, 2.

standards that are accountable to locational challenges and other physical circumstances (flexibility instead of uniformity).

2 SUSTAINABLE BUILDING PRACTICES

Building construction has resulted in far-reaching impacts to the natural and built environment.[19] According to the US Energy Information Administration, '[i]n 2009, the residential and commercial building sector was responsible for over 50% of total annual U.S. energy consumption'.[20] The US Department of Energy likewise reported for 2009, that the residential and commercial building sector was accountable for '74 percent of total U.S. electricity consumption'[21] and that the buildings comprising this sector were 'collectively responsible for 39% of total U.S. greenhouse gas emissions and 7.7% of worldwide emissions'.[22] Moreover, in addition to their climate change contributions, buildings emit other air pollutants, and the materials used in their construction and operation contribute to indoor air pollution.[23]

Green building has become the industry and regulatory response to the adverse impacts of building[24] and presents an opportunity to make enormous strides toward a sustainable future. Specifically, there is some

---

19 Hirokawa, above note 16 at 511.

20 Gerrard, above note 11 at xv (author included in his calculations 42% of building operations, 8% building construction, and the embodied energy of building materials). If only strictly looking at residential and commercial building consumption the percentage would be just under 40% for 2009, just above 40% for 2010, and just above 40% for 2011. See US Energy Info Admin, 'Annual Energy Outlook 2012 Early Release Overview, Table 2: Energy Consumption by Sector and Source' (2012), http://www.eia.gov/forecasts/aeo/tables_ref.cfm, accessed 1 November 2012.

21 Gerrard, above note 11 at xv.

Ibid.

22 Ibid.

23 Hirokawa, above note 16 at 511 ('Also, many indoor building materials release hazardous toxins, impairing indoor air quality and reducing occupant health and productivity').

24 Sarah Fox, 'A Climate of Change: Shifting Environmental Concerns and Property Law Norms Through the Lens of LEED Building Standards' (2010) 28 Virginia Environmental Law Journal 299, 300 ('With an increasing recognition of the amount of energy that goes into creating and using buildings of every type, one of the strategies currently being implemented to stem the tide of climate change is emissions reduction in the construction industry').

certainty that we will continue to build (and rebuild[25]) in the foreseeable future. As of 2010, the total US building stock is approximately 275 billion square feet. During normal economic times, approximately 1.75 billion square feet of buildings are torn down each year. Every year, approximately 5 billion square feet of new buildings are renovated and approximately 5 billion square feet of new buildings are built. By the year 2035, approximately three-quarters (75%) of the built environment will be either new or renovated.[26]

Construction at this scale can be sustainable by reducing building materials and waste and making building operations more efficient.[27] By applying construction standards to the millions of new homes constructed annually in the US, there is an opportunity 'to minimize the impacts of construction on the depletion of natural resources, greenhouse gas emissions, human health, and generation of waste'.[28]

The green building industry has made significant strides toward sustainable building practices, often in cooperation with federal, state, and local government encouragement to reduce greenhouse gas (GHG) emissions.[29] Several green building standards have been developed by third-party private entities,[30] including the most widely employed strategies in the US: Energy Star,[31] Green Globes,[32] and, in particular, the US

---

[25] Note that 'almost three-quarters of our nation's 81 million buildings were built before 1979 [many of which] were designed and constructed for limited service, and … will eventually require either significant retrofits or replacement', have led to an increase in building regulations and policies. The White House, 'Obama Administration Launches New Energy Efficiency Efforts' (29 June 2009), http://www.whitehouse.gov/the-press-office/obama-administration-launches-new-energy-efficiency-efforts, accessed 26 October 2012.

[26] Architecture 2030, 'A Historic Opportunity', http://architecture2030.org/the_solution/buildings_solution_how#, accessed 26 October 2012.

[27] Hirokawa, above note 16 at 510.

[28] Ibid. at 513.

[29] Even before the Obama administration promoted green growth as a strategy for stimulating the economy, several states, counties and municipalities were committing to green building ordinances. See The White House, above note 25.

[30] See generally Sarah B Schindler, 'Following Industry's LEED®: Municipal Adoption of Private Green Building Standards' (2010) 62 Florida Law Review 285.

[31] See Energy Star, 'About Energy Star', http://www.energystar.gov/index.cfm?c=about.ab_index, accessed 26 October 2012 (A program between EPA and the US Department of Energy).

[32] See Green Building Initiative, 'About GBI', http://www.thegbi.org/about-gbi/, accessed 26 October 2012.

Green Building Council's (USGBC) Leadership in Energy and Environmental Design (LEED).[33] The LEED rating system has 'become the default standard for determining what makes a building "green"'.[34] The LEED program is intended to define green building criteria by providing a rational standard for measurement and allowing for 'green' building claims to maintain integrity by siphoning off false and exaggerated claims.[35] The LEED program has developed a point system for project certification which recognizes sustainable use of materials, land utilization, and design considerations.[36] Points earned under various criteria lead to four shades of green: certified, silver, gold, and platinum.[37] As the certification level increases, the qualifying building becomes more environmentally friendly. LEED awards these points to construction 'projects that implement strategies for better environmental and health performance',[38] since 'it is generally accepted that the built environment (homes,

---

[33] Geoffrey M White, Joshus Nichols and Jeff York, 'Green Building Rating Systems and Green Leases' in J Cullen Howe and Michael B Gerrard (eds), *The Law of Green Buildings: Regulatory and Legal Issues in Design, Construction, Operations, and Financing* (ABA 2010) 15.

[34] Jesse W Abair, 'Green Buildings: What It Means to Be "Green" and the Evolution of Green Building Laws' (2008) 40 The Urban Lawyer 623, 623–624.

[35] Hirokawa, above note 16 at 514–515.

[36] *See* US Green Building Council, *LEED for New Construction and Major Renovations* (2005) 6–7, http://www.usgbc.org/ShowFile.aspx?DocumentID=1095, accessed 26 October 2012 [hereinafter LEED-NC 2005].

[37] Compare US Green Building Council, *LEED for New Construction and Major Renovations* (2009) 8, http://www.usgbc.org/ShowFile.aspx?DocumentID=8869, accessed 26 October 2012 [hereinafter LEED-NC 2009] with LEED-NC 2005, above note 36 at 6–7 (LEED-Certified (26–32 credits); LEED-Silver (33–38 credits); LEED-Gold (39–51 credits); and LEED-Platinum (52–69 credits)). See also US Green Building Council, *LEED for Homes* (2008) iv, http://www.usgbc.org/ShowFile.aspx?DocumentID=3638, accessed 26 October 2012 [hereinafter LEED for Homes] ('The level of performance is indicated by [the same] four performance tiers – Certified [45–59 points], Silver [60–74 points], Gold [75–94 points], and Platinum [90–136 points] …').

[38] US Green Building Council, *LEED Green Building Rating Systems*, http://www.usgbc.org/DisplayPage.aspx?CMSPageID=222, accessed 26 October 2012 (The rating systems give points based on the following categories: Sustainable Sites, Water Efficiency, Energy and Atmosphere, Materials and Resources, Indoor Environmental Quality, and Innovation and Design Process, and LEED for Homes also considers Location and Linkages and Awareness and Education).

[residential and] office buildings, schools, roads, dams, etc.) imposes adverse but avoidable impacts on the natural environment'.[39]

Building green means making environmentally conscious and sustainable decisions throughout the building process, mainly by incorporating resource efficiency into every step in order to minimize environmental impacts. The implementation of a variety of green building practices produces aggregate effects toward natural resources conservation and air and water quality protection.[40] As such, green building systems generally implement a 'whole systems' approach by directing changes in building design, construction, and operation, focusing on energy and water efficiency, resource efficient building design, eco-conscious materials, and indoor environmental quality.[41]

For example, green building systems seek significant improvements in the range and availability of green building materials, such as 'low carbon footprint materials, more efficient end-use electrical equipment and lighting, heat/power recovery, renewable energy supply systems and manufacturing plant location'.[42] Likewise, the siting of a building, whether residential, commercial, or industrial, is just as important as the materials used to build it. Sustainable sites take into consideration location and the land uses pertaining to that particular site, focusing 'builders on minimizing damage to the environmental characteristics of a given building site'.[43]

Green building systems integrate reductions from conventional building standards of water use and waste generation. Efficient use of water in building operations by installing high efficiency fixtures and appliances such as low-flow showerheads, faucets, and toilets, Energy Star dishwashers and washing machines, leads to an overall conservation of water.[44] Other ways in which water conservation can be achieved is

---

39 Hirokawa, above note 16 at 508.

40 See Build it Green at http://www.builditgreen.org/building-design-construction/, accessed 26 October 2012.

41 White, Nichols and York, above note 33 at 15.

42 Architecture 2030, 'Energy', http://architecture2030.org/the_solution/solution_energy, accessed 26 October 2012.

43 Hirokawa, above note 16 at 515.

44 See Energy Star, 'Qualified Products', http://www.energystar.gov/index.cfm?fuseaction=find_a_product, accessed 26 October 2012. See also Larissa Larsen et al., *Green Building and Climate Resilience: Understanding Impacts and Preparing for Changing Conditions* (University of Michigan 2011) C-121 (The need to use of high efficiency fixtures is high throughout all the regions of the US. 'Water efficient fixtures, including faucets, showerheads and toilets, reduce the flow rate of water though the fixture').

through programmed, low-volume irrigation systems, rainwater collection systems, wastewater treatment systems, and hot water recirculation systems.[45] Water efficiency 'seeks to change water use practices, but not the purposes to which water is put; [it] encourages the capture of rainwater, on-site treatment and gray-water reuse, and installation of water-conserving fixtures throughout buildings'.[46]

Green building features include efficient use of natural resource materials. Green building standards encourage the use of renewable materials, including rapidly renewable wood species such as bamboo and recycled-content materials in carpets, tiles, and concrete. LEED awards points for the 'diver[sion of] reusable materials from collection in landfills, [which] reduces the burden on natural resources by focusing on materials reuse',[47] and 'incorporate[s] the salvage or reuse of building materials, products, and furnishings'.[48] The new, technologically improved, energy-efficient features of a building are the building blocks of curbing GHG emissions. For example, if all buildings in the US were covered by the most recent version of the international energy conservation code – upon which many state energy codes are based – 61 trillion BTUs[49] of energy would be saved and four million metric tons of GHG emissions would be avoided in the first year.[50]

Energy-efficient features include appliances, windows, and water heating systems that have Energy Star ratings.[51] Efficient lighting fixtures and bulbs and on-site renewable energy sources, such as photovoltaic electricity, can also further decrease the building's overall energy consumption.[52]

Finally, indoor air quality features are also employed in order to 'ensure that green buildings have optimal lighting, thermal comfort, and healthy indoor air quality for their occupants'.[53] Ventilation effectiveness

---

45 See Adaptive Building Solutions, http://www.adaptivebuilding.com/green-building-key-features.php, accessed 26 October 2012.

46 Hirokawa, above note 16 at 516.

47 Ibid.

48 Ibid. See also White, Nichols and York, above note 33 at 21 ('Major renovation projects are eligible for points for reusing the existing walls, floors, and roof, and for reusing interior non-structural elements').

49 1 BTU = 1055 joules.

50 Gerrard, above note 11 at xv.

51 See Energy Star, above note 44 ('Save energy and fight climate change with ENERGY STAR qualified products. They use less energy, save money, and help protect the environment').

52 Ibid.

53 White, Nichols and York, above note 33 at 21.

and air quality are important aspects of buildings. The heating, air conditioning, and ventilation system (HVAC) must be appropriately sized for an efficient and properly ventilated building.[54] Low-emitting materials, such as Low-VOC paints,[55] finishes, and wallpaper create a healthier living and working environment.[56]

3 GREEN BUILDING AS AN ADAPTIVE MEASURE

Adaptive building strategies are built upon an understanding of how climate change will impact the human and built environment. Climate change will directly impact cultural, social, economic, and political life in the US through increasing pressure on water resources, crop and livestock production, coastal areas, and human health and safety. Climate change will interact with many social and environmental stresses[57] in global and complex ways, reaching even the basic functioning of the economy,[58] and will expose increasing uncertainties in both the state of the environment and the goals that might be accomplished through regulation. Even strategies to curb GHG emissions and reduce the now enormous ecological footprint of developed countries will lack predictability due to future contingencies relating to social needs, practices, and demands.[59] Alyson Flournoy summarizes the challenge of uncertainty as follows: 'any ecosystem or natural resource for which we choose an objective or target is changing; we are unlikely to have complete information on the present status of the resource in relation to the target;

---

54 Ibid.

55 Eartheasy Solutions for Sustainable Living, 'Non-Toxic Paints', http://eartheasy.com/live_nontoxic_paints.htm, accessed 26 October 2012 ('Indoor air is three times more polluted than outdoor air, and according to the EPA, is considered one of the top five hazards to human health. Paints and finishes are among the leading causes. Paints and finishes release low level toxic emissions into the air for years after application. The source of these toxins is a variety of volatile organic compounds (VOCs) which, until recently, were essential to the performance of the paint').

56 See White, Nichols and York, above note 33 at 21.

57 US Global Change Research Program, above note 1.

58 Craig, above note 8 at 27.

59 J B Ruhl, 'General Design Principles for Resilience and Adaptive Capacity in Legal Systems – with Applications to Climate Change Adaption' (2011) 89 North Carolina Law Review 1373, 1392 ('Climate change will cause numerous natural and social systems to depart from settled patterns and ranges of variability, and thus to behave in ways beyond the control or prediction of any legal instruments and institutions').

and if we design a management strategy, change and surprise preclude accurate prediction of whether the strategy will achieve our objectives'.[60] Of course, science and technology are constantly evolving to expand our knowledge of climate change impacts and, as such, our understanding and approaches will be constantly changing.[61]

3.1 Regulatory Uncertainty

Notwithstanding the remaining uncertainties regarding the scale and scope of climate change, there is agreement with respect to both the likelihood of climate change impacts to the built environment and the nature of those impacts.[62] Shifts in average temperatures and other climate events across the globe will cause shifts in the demand and the supply of energy. Decreasing demand for energy supply, such as by altering the character of energy needs in the built environment, will lower vulnerability to changes in the ability to produce and deliver energy. Shifting trends in precipitation and water runoff place the reliability of water supplies in jeopardy, suggesting the need to increase resiliency to changes in water supplies and delivery, such as increasing capacity for multiple uses and water reuse. Operation and maintenance of transportation systems will be challenged by changing climate patterns and the resulting new challenges to established transportation systems, suggesting a necessity to decrease our reliance on transportation and recognize the importance of the efficiencies of dense populations.

Improving adaptive capacity and resiliency requires that we incorporate these shifts and the remaining uncertainties into the goals of regulatory control.[63] As Robin Craig notes, the most pragmatic response

---

[60] Alyson C Flournoy, 'Protecting a Natural Resource Legacy while Promoting Resilience: Can it be Done?' (2009) 87 Nebraska Law Review 1008, 1026.

[61] Craig, above note 8 at 41 ('Lack of knowledge about the nature, scope, and extent of climate change effects, particularly at the level of specific resources and ecosystems and local communities, limits citizens' and governments' abilities and willingness to make rational choices regarding adaptation strategies, thus undermining adaptive capacity'); Flournoy, above note 60 at 1026 ('Because our knowledge about ecosystems is limited, we do not even have complete information about the present status of the resources we are trying to protect or conserve').

[62] Larsen et al., above note 44.

[63] Adaptive capacity can be defined as 'the idea that a system might sense threats to system equilibrium and respond by changing resilience strategies without changing fundamental attributes of the system': Ruhl, above note 59 at 1387. Adaptive systems, including legal systems, must 'manag[e] the mix of

to such uncertainty is not regulatory nihilism but principled flexibility: 'the more we acknowledge pervasive uncertainties regarding what climate change actually means … the more we should restructure environmental and natural resources law to give as many species and systems as possible the best chance to survive and adapt to whatever changes come'.[64] By taking a flexible approach, regulation can build adaptive capacity by allowing the conceptualization of regulatory strategies in a moving context.

The built environment presents multiple complex and expensive difficulties related to these adaptation challenges. Homes and businesses need to be protected from major climatic events and shifts. Of course, buildings necessarily become outdated or need structural repairs on a schedule largely determined by building design and constructive methods. Studies report that in approximately 24 years about 75 percent of the built environment will be either new or renovated.[65] Thus, the building sector has been recognized, both nationally and internationally, for its potential in achieving GHG emission reductions.[66] Built infrastructure presents the same complexities and challenges: 'it is difficult and expensive to alter or replace infrastructure designed to last for decades (such as buildings, bridges, roads, airports, reservoirs, and ports) in response to continuous and/or abrupt climate change'.[67]

Our understanding of the relationship between climate change and the local environment needs help in identifying how building practices can be flexible and responsive to the future uncertainties of climate change. Other building aspects that are targets for resiliency improvements include building size, orientation, and layout, so that buildings can take advantage of natural daylight and strategies to reduce heat gain in the summer and heat loss in the winter. The landscapes surrounding buildings can feature regional vegetation that may be appropriate for the

---

resilience properties such as reliability, efficiency, scalability, modularity, and evolvability'. Ibid. at 1388.

[64] Craig, above note 8 at 39. It has been pointed out that environmental law and ecology have been captured by a tendency toward thinking in terms of 'stationarity', which is composed of the notion 'that natural systems fluctuate within an unchanging envelope of variability: P C D Milly et al., 'Stationarity Is Dead: Whither Water Management?' (2008) 319 Science 573.

[65] Architecture 2030, above note 26.

[66] Since the housing sector is a major contributor to climate change by singlehandedly emitting large amounts of carbon dioxide, it also presents a major opportunity to change it so that it is responsible for less amounts of emissions.

[67] US Global Change Research Program, above note 1 at 10.

climate and grouped according to water needs.[68] New developments can incorporate pervious surfaces composed of gravel, permeable block pavers, grids, or other permeable systems.[69] Such improvements have impacts beyond the walls of a structure, including 'the psychological well-being of individuals, promoting community self-image and fostering community pride'.[70]

Regulatory systems are designed to exceed the demands of the status quo,[71] to accommodate immediate needs for management actions, to facilitate consensus around clearly articulated goals,[72] and to monitor the development of norms within the governed practices to ensure that outcomes are met or actions re-evaluated. Hence, although we know that reducing reliance on the existing energy infrastructure and curbing GHG emissions are necessary to avoid climate change impacts, in reality 'we have little understanding of "the complex, multivariable, nonlinear, cross-scale, and changing [socio-ecological systems]" that exist even prior to climate change impacts'.[73] Current efforts to reduce the demand for natural resources, such as motion sensor faucets to ensure against water waste and over consumption, the use of sensors for lighting to reduce electricity consumption and waste, and the use of digital control systems to monitor boilers, mitigate energy usage and present cost reductions and adaptive improvements at a relatively small investment.[74]

---

[68] See LEED-NC 2005, above note 36.

[69] Ibid.

[70] Alexandra Dapolito Dunn, 'Siting Green Infrastructure: Legal and Policy Solutions to Alleviate Urban Poverty and Promote Healthy Communities' (2010) 37 Boston College Environmental Affairs Law Review 41, 50 ('Community-based organizations can promote green projects as well. For example, the New York Restoration Project (NYRP) funds green improvements and upkeep in economically and environmentally burdened areas in New York. One NYRP effort, the Target East Harlem Community Garden, successfully fused community gardening with the installation of solar panels, wind turbines, and a drip irrigation system').

[71] Schindler, above note 30 at 317.

[72] See generally Craig, above note 8 at 9.

[73] Ibid. at 40–41.

[74] Many of the techniques seen as mitigating factors can nonetheless fall under the reactive adaptation theory, but in a way that is just name calling. Various types of adaptation can be distinguished, including anticipatory and reactive adaptation, private and public adaptation, and autonomous and planned adaptation.

In like manner, building efficiency standards present opportunities to set and strive for identifiable adaptation goals.[75] Reduction of energy needs through passive energy designs or reducing grid reliance through on-site wind and solar generation contribute to achieving 'net zero energy' buildings,[76] illustrated by the 'zero energy capable' requirements in the City of Austin, Texas.[77] By Resolution, the Austin City Council authorized the Mayor to form a Zero Energy Capable Homes Task Force and develop a strategy mandating that all new single-family homes are zero energy capable by 2015.[78] Austin's plan includes an aggressive approach to retrofits and upgrades that may include point-of-sale regulations and incentives.[79] Likewise, California continues to contribute to sustainability strategies and is now promoting its 'Zero Net Energy Action Plan' for commercial buildings.[80] The Plan encourages commercial building owners to 'increasingly embrace zero net energy performance (including clean and distributed generation), reaching 100 percent penetration of new starts in 2030'[81] and 50 percent of existing building

---

75 In general, green buildings can reduce energy use by 20–50% (compared to conventional counterparts) by attention to design principles and building orientation. However, energy savings beyond 50% typically rely on more innovative strategies in design and technology: Greg Kats, *Greening Our Built World: Costs, Benefits, and Strategies* (Island Press 2010) 26–32.

76 California Public Utilities Commission, Press Release, 'CPUC and Business Leaders Launch Zero Net Energy Action Plan; Goal of Efficient, Clean Energy Powered Buildings by 2030' (1 September 2010), http://docs.cpuc.ca.gov/PUBLISHED/NEWS_RELEASE/122911.htm, accessed 31 October 2012.

77 Although 'zero energy homes' are equipped to draw from the utility grid, they are intended to produce the energy needed for consumption. Zero net energy capable homes encompass residential homes that are 65% more efficient than those built under the standard code and utilize on-site energy generation: Memorandum from Zero Energy Capable Homes Task Force to Mayor and Council of City of Austin Tex. (2007) 1, http://www.austintexas.gov/edims/document.cfm?id=109500, accessed 31 October 2012.

78 Ibid.

79 City of Austin, Tex. Res. No. 20070215–023, § (3)–(3)(a), http://www.austintexas.gov/edims/document.cfm?id=100723, accessed 31 October 2012.

80 California Public Utilities Commission, above note 76. For more information on California's Energy Efficiency Strategic Plan and the Zero Net Energy actions plan, see also California Public Utilities Commission, *Energy Efficiency Strategic Plan* (2011), http://www.cpuc.ca.gov/PUC/energy/Energy+Efficiency/eesp/, accessed 31 October 2012.

81 *Californian Energy Efficiency Strategic Plan, Zero Net Energy Action Plan: Commercial Building Sector 2010–2012* at 7, http://www.cpuc.ca.gov/NR/rdonlyres/6C2310FE-AFE0–48E4-AF03–530A99D28FCE/0/ZNEActionPlan

retrofits.[82] California's zero net energy community is growing and includes public and private educational facilities, recreational facilities, and commercial centers that boast energy and other natural resource achievements in both new construction and retrofits.[83]

Climate change adaptation requires that we regulate for an uncertain future. Green building strategies provide tools to advance adaptation concerns and needs, including the flexibility gained by reducing infrastructure and resource reliance.

3.2 Locational Flexibility as to Specific Climatic Conditions

Due to the uncertainties of climate change, it 'is not possible to provide a list of actions to be taken now to adapt in each region and sector'.[84] The most appropriate adaptation strategies for buildings will vary from region to region. As local governments know very well, 'adaption is fundamentally implemented at local and regional levels and needs to consider the socioeconomic and political factors'.[85] At the least, climate change will impact the way new land developments are planned, sited, and constructed. For example, sea level rise will increase risks of erosion, storm surge damage, and flooding for coastal communities, especially in the southeast regions of the US and parts of Alaska. Reduced snowpack and earlier snow melt will alter the timing and amount of water supplies, posing significant challenges for water resource management in the western regions of the US.[86]

Responding to environmental impacts by reducing vulnerability to climate change needs to be accompanied by 'prudent planning [which] requires that steps be taken to respond to the changes that are on the horizon, and that real estate development activities adapt to the new

---

FINAL83110.pdf, accessed 31 October 2012; California Public Utilities Commission, above note 76.

82 *Californian Energy Efficiency Strategic Plan*, above note 81 at 17.

83 Ibid. See also *Massachusetts Zero Net Energy Buildings Task Force, Getting to Zero: Final Report of the Massachusetts Zero Net Energy Buildings Task Force* (2009), http://www.mass.gov/eea/docs/eea/press/publications/zneb-taskforce-report.pdf, accessed 31 October 2012 (making recommendations for reductions in building energy consumption and increases in on-site renewable energy generation).

84 National Academy of Sciences, *Adapting to the Impacts of Climate Change* (2010) ix.

85 Ibid.

86 US Global Change Research Program, above note 1 at 10.

realities of the world with a changing climate'.[87] Making buildings more adaptable requires strategies that address whole buildings and comprehensive designs.[88]

The incorporation of resiliency strategies into building standards 'will allow a system to absorb disturbances such as increased precipitation or flooding while maintaining its structure and function'.[89] Regional differences are an important aspect to incorporate into green building practices because a failure to match implementation standards with regional climatic needs will produce redundancies and superfluous gains. A body of green building standards that accounts for regional demands and circumstances will be more successful in addressing the challenges facing the built environment.[90] In addition, employing specific strategies based on regional climatic differences can provide adaptive benefits by increasing resiliency in the built environment.

For example, water conservation may be the most significant goal in one region, while energy conservation is the most significant goal in another. Building envelope, siting, and landscaping strategies depend on exposure to sun and availability of water. Building operational needs in relation to heating, cooling, lighting, water, and waste will depend on regional circumstances. Climate-appropriate landscaping[91] to accommodate future climate change is more important in northeastern, northwestern, and southwestern regions of the US, but less so in the midwestern and southeastern regions.[92] In contrast, the need to minimize

---

87 J Kevin Healy, Margery Perlmutter and L Margaret Barry, 'Environmental Review and Climate Change Adaptation' in J Cullen Howe and Michael B Gerrard (eds), *The Law of Green Buildings: Regulatory and Legal Issues in Design, Construction, Operations, and Financing* (ABA 2010) 313, 330–331 ('Making adjustments in urban planning and building design to compensate for already rising sea levels is a new paradigm that is just beginning to make its way into American public policy, building, and zoning codes').

88 J B Ruhl argues that adaptive capacity is more likely obtained through a de-emphasis on front-end decision making, such as comprehensive planning. Ruhl, above note 59 at 1394; J B Ruhl, 'Climate Change Adaptation and the Structural Transformation on Environmental Law' (2010) 40 Environmental Law 363, 416–423. He suggests that governance principles must shift towards more inclusive, collaborative, and flexible models: Ruhl, above note 59 at 1395.

89 Larsen et al., above note 44 at 7.

90 Schindler, above note 30 at 321.

91 Larsen et al., above note 44 ('To create landscapes that will maintain their design and function in response to changes in precipitation, temperature, and pest types').

92 Ibid.

impervious surfaces,[93] employ building orientation strategies,[94] or use retention ponds[95] or solar zoning[96] may be high throughout all regions of the US. Regional differences also account for various climate change impacts, as landlocked states are not necessarily worried about rising sea levels. Attention to equipment, process, and operation will also reap rewards in terms of resource efficiencies with high building performance.[97]

In addition to the regional climatic demands relating to geographical location, adaptive building approaches can account for the locational demands of particular adaptive buildings or particular built environmental constraints. For instance, the volume of GHG emissions from automobile traffic depends on building location and relates directly to a successful reduction in GHG emissions.[98] Adaptive building can integrate transportation information to improve adaptive capacity, specifically by incorporating locational needs to increase density and walkability, intensifying a diversity of land uses and multiple uses to improve access to public transportation and pedestrian-friendliness of the infrastructure, and reducing driving and related transportation costs. In addition, adaptive building approaches can be regionally responsive by reducing the effects of, or potential for, storm water runoff and pollution; preserving and restoring open space and habitats; increasing materials, energy, and water efficiency in buildings and infrastructure; and supporting more diverse, economically sustainable communities.[99] These types of adaptive capacity improvements require flexibility across a variety of standards and site-specific accountability for building impacts.

---

93 Ibid. at Appendix C C-61 ('To reduce erosion and stormwater runoff from sites and to limit increased surface temperature. To reduce the risk of contaminated sewer overflow and flooding in response to increased precipitation or severe storm events').

94 Ibid. at Appendix C C-63 ('To limit an increase in energy use as a building responds to changing climate conditions. To allow for optimal use of passive heating and cooling strategies').

95 Ibid. at Appendix C C-65 ('To collect and hold stormwater, even in the case of severe storm events').

96 Ibid. at Appendix C C-73 ('To increase energy efficiency by maximizing the ability of a building to utilize solar energy for space heating, water heating, and daylighting').

97 Larsen et al., above note 44 at 7.

98 *See* US Dept. of Transportation, 'Transportation and Greenhouse Gas Emissions', http://climate.dot.gov/about/transportations-role/overview.html, accessed 1 November 2012.

99 Kats, above note 75 at 94.

Insights into a sufficiently flexible building approach might be found in the adoption of the 2010 California Green Building Standards Code, also known as 'CALGreen'. CALGreen went into effect in 2011 and has been hailed as the first statewide green building code in the nation. The CALGreen code applies to newly constructed low-rise residential (three stories or less) and most non-residential buildings, but not to building additions, alterations, or repairs.[100] Like its voluntary counterparts, the CALGreen Code includes site selection considerations, construction practices, building operation efficiencies, materials selection, and natural resource conservation measures. Although the intent of CALGreen is to create uniform and enforceable standards for new California buildings, it does not eliminate jurisdictional prerogatives. Local jurisdictions may customize their own codes where circumstances of climate, topography, or geology justify a departure. Third-party programs, such as LEED, may be used where consistent with the demands of the CALGreen program.

A second important development is seen in the intensely collaborative efforts of professionals across the green building industry. In 2008, various industry groups initiated a national, collaborative effort to develop minimum green building standards to accelerate commercial green building practices throughout the country. Standard 189, which covers criteria including sustainability, water and energy efficiency, indoor environmental quality, and materials and resource use, was developed through the partnership of the American Society of Heating, Refrigerating and Air Conditioning Engineers (ASHRAE), the Illuminating Engineering Society (IES), and USGBC.[101] Standard 189 was intended to provide baseline, minimum standards for building design, construction, and operation practices for states and local governments.

Soon thereafter, the International Code Council developed the International Green Construction Code (IGCC) in cooperation with the American Institute of Architects (AIA) and the ASTM International

---

[100] CalRecycle, 'Local Government Construction and Demolition (C&D) Guide: Frequently Asked Questions', http://www.calrecycle.ca.gov/LGCentral/Library/canddmodel/instruction/faq.htm#GreenBuild, accessed 1 November 2012.

[101] ANSI/ASHRAE/USGBC/IES, *Standard 189.1, the Standard for the Design of High-Performance Green Buildings, Except Low-Rise Residential Green Buildings* (February 2010), www.ashrae.org/greenstandard, accessed 1 November 2012.

(ASTM).[102] The IGCC is intended to be consistent and coordinated with the other International Code Council Codes, and to this extent recognizes certain accepted and existing practices and protocols. However, the IGCC surpasses other green building codes on two critical features. First, the IGCC is *designed* to serve as a mandatory, regulatory overlay, and therefore can set goals in ways previously unattainable:

> The IGCC is poised to produce environmental benefits on a massive scale: a scale impossible to attain with purely voluntary green building programs and rating systems. Voluntary programs have pioneered the Green movement, and with the foundation they have laid, the IGCC is positioned to achieve significant market transformation in those segments that are not likely to react to voluntary programs.[103]

Second, woven into the mandatory nature of the IGCC are pathways and options that allow buildings to seek green certification in context. The IGCC contains the flexibility necessary to allow for regional perceptions of green building needs. In addition to the jurisdictional flexibility afforded in the IGCC for climatic and other regional circumstances, the IGCC also includes provisions for recognition of Standard 189.1 as an alternative path to compliance. Chapter 3 of the IGCC, which serves as the 'core' of the program, is 'designed to allow the local jurisdiction to meet regional goals and priorities by determining whether certain provisions are to be mandatory and whether enhanced energy performance or reduced plumbing fixture flow rates will be required for compliance with … [the] code'.[104] All other code provisions are mandatory.[105]

Climate change adaptation is subject to the lesson that 'vulnerability associated with climate change is based on underlying social and ecological stresses, and these stresses tend to vary dramatically from

---

[102] International Code Council, *International Green Construction Code (IGCC) (Public Version 2.0)* (2010) ii, http://www.iccsafe.org/cs/igcc/pages/default.aspx, accessed 1 November 2012.

[103] International Code Council, *International Green Construction Code (IGCC) Synopsis* (2010) 2, http://www.iccsafe.org/cs/IGCC/Documents/PublicVersion/IGCC_PV2_Synopsis.pdf, accessed 1 November 2012.

[104] Ibid.

[105] Moreover, recognizing the mutually beneficial value of the several years of expert content in ASHRAE with Standard 189.1 and IGCC's unparalleled delivery model to all 50 states and more than 22,000 jurisdictions, ICC, AIA and ASTM worked together with ASHRAE, IES and USGBC to launch a joint effort in March of 2010 to release the IGCC with Standard 189.1 as an alternate path to compliance.

place to place'.[106] Local governments have thus become relevant to adaptation, and they have participated by seeking to have local values reflected in green building codes and standards. The productive collaboration resulting among industry and governmental interests has drawn thousands of commercial and individual participants (and their buildings) into a sustainable construction scheme[107] that can be responsible to the regional and locational demands that will accompany climate changes.

3.3 Adaptive Buildings

As noted above, adaptation requires a substantial amount of flexibility in both the physical structures and the regulatory process. Chuck Hoberman, a 'transformable-design guru',[108] has joint ventured the Adaptive Building Initiative (ABI) in 'an effort to equip big buildings everywhere with skins that move'.[109] His motto is that '[i]f a building can change, it can respond to the environment and self-optimize'.[110] ABI 'designs and produces adaptive façades and building envelopes'.[111] It is currently 'designing a new generation of buildings that optimize their configuration in real time by responding to environmental changes'.[112]

---

106 Ibid.

107 See US Environmental Protection Agency, *Celebrating a Decade of Energy Star Buildings: 1999–2009* (2009) 7, http://www.energystar.gov/index.cfm?c=business.bus_ES_bldgs, accessed 1 November 2012 ('Over the past decade, the ENERGY STAR Buildings program has experienced staggering growth by every measure. As the end of 2009 approaches, the energy performance of more than 120,000 buildings representing nearly 14 billion square feet has been measured through ENERGY STAR. More than 5,000 organizations have joined the ENERGY STAR buildings program as partners. Nearly 9,000 buildings have earned the ENERGY STAR across all 50 states').

108 Tim McKeough, 'Chuck Hoberman's Buildings Adapt to the Environment' (*Fast Company*, 1 April 2010), http://www.fastcompany.com/1588653/chuck-hobermans-buildings-adapt-environment, accessed 1 November 2012.

109 Ibid.

110 Ibid.

111 Hoberman – Transformable Design, *Adaptive Building Initiative*, http://www.hoberman.com/abi.html, accessed 1 November 2012 ('By controlling light levels, solar gain, and thermal performance, ABI's adaptive systems reduce energy usage, enhance comfort, and increase the flexibility of the built environment').

112 Ibid.

Adaptive buildings, either newly constructed or retrofitted, 'can adapt their shape and function in real time to environmental changes'.[113] The benefits are far reaching: 'Buildings that continuously attune their configurations in accordance with changing environmental conditions use less energy, offer more occupant comfort, and feature better overall space efficiency than static buildings.'[114] Modular buildings, flexible designs, and building designs that adjust to climatic conditions are appearing both in design and on the ground. The adaptive approach to building façades is innovative and fundamentally rethinks architectural constructs.[115] Adaptive designs also provide some reprieve to the current struggle cities face in reducing carbon footprints. Recent carbon inventories for major US cities indicate that 60 to 70 percent of carbon emissions in densely urban areas result from the construction, operation, and demolition (often, without recycling) of buildings and their associated infrastructure. Cities that aim to reduce carbon emissions are facing this seemingly insurmountable challenge. For instance, New York City's *PlaNYC* calls for 30 percent carbon reductions by 2030 based on today's emissions rates.[116]

Adaptive buildings are underway around the world. ABI is currently engaged in Madrid, Spain, where it 'is installing 27,000 square feet of Adaptive shading in two glass-roofed Foster + Partners buildings on Madrid's City of Justice campus, the new home for Spain's Supreme Court'.[117] The design takes advantage of the location and movement of the sun throughout the day; 'a computer system continually adjusts the shades' and light filters into the building 'but the technology deploys enough shades to prevent offices from overheating'.[118] When the system is closed, '[t]he hexagonal shading units, made of steel and aluminum, fold flat and vanish into the roof's structure'.[119] Hoberman and ABI have also committed to designing a 14-story building façade in Tokyo for

---

113 Chuck Hoberman and Craig Schwitter, 'Adaptive Structures: Building for Performance and Sustainability' (*Design Intelligence*, 11 August 2008), http://www.di.net/articles/archive/2881/, accessed 1 November 2012 ('This field remains far less developed than other areas of practice, but the logic of adaptive performance – which is time-based, responsive, and dynamic – is compelling').

114 Ibid.

115 Ibid.

116 Ibid. See also Chapter 12, section 5.

117 McKeough, above note 108.

118 Ibid.

119 Ibid.

POLA Cosmetics.[120] The façade was designed as 'a custom kinetic mechanism by which the façade panels respond to environmental stimuli with an opening and closing effect'.[121]

4 AWARENESS, ENCOURAGEMENT, AND PROGRESS TOWARD ADAPTIVE BUILDINGS

As noted elsewhere in this book, adaption to climate changes will require more than strategies that seek only minimization of damage and ecological recovery.[122] Adaptation means 'making adjustments in physical, ecological, economic, and social systems to compensate for climate impacts' to make 'human and natural systems more *resilient* to the impacts of climate change'.[123] Resiliency has been an elusive goal in the past, but trends in sustainable construction have resulted in 'the large numbers of green construction projects that they were designed to encourage'.[124] Successes and interest have grown considerably, enough to support a reasonable observation that 'the green building movement is no longer a fringe concept espoused solely by environmentalists'.[125]

In November 2010, green building certification (measured in LEED terms) reached a significant milestone: 'the total footprint of commercial projects certified under the U.S. Green Building Council's (USGBC) LEED Green Building Rating System worldwide surpassed one billion square feet. Another six billion square feet of projects are registered and currently working towards LEED certification around the world.'[126] The US Green Building Council projects continued growth in the green

---

[120] Terri Peters, 'Chuck Hoberman Designs Transformations' (2010) 29 Mark Magazine, 200, 204, http://www.hoberman.com/news/Mark_Magazine_December2010.pdf, accessed 1 November 2012.

[121] Ibid.

[122] Ruhl, above note 88 at 383.

[123] Healy, Perlmutter and Barry, above note 87 at 331 (emphasis in original) ('A resilient system has the capacity to "absorb and rebound from weather extreme, climate variability or change and continue functioning"').

[124] Abair, above note 34, at 628.

[125] Ibid. at 632.

[126] US Green Building Council, Press Release, 'One Billion Square Feet of LEED Certified Green Building Projects Worldwide. Achievement Manifests Building Industry Progress, Market Transformation' (10 November 2010), http://www.usgbc.org/Docs/News/One%20Billion%20Release%202010.pdf, accessed 1 November 2012. This achievement is distributed among over 36,000 commercial projects and 38,000 single-family homes.

building industry to a contribution of USD 554 billion to the US gross domestic product by 2013.[127] In part, awareness of green building concepts has resulted from the performance of green buildings. Green building code improvements relating to building maintenance and operation have attracted new consumers into the green building market[128] because of the ease of capturing substantial value both in building operation costs and in property values over the lifetime of the building. Green building costs are quickly recoverable and growing closer to parity with conventional building costs.[129]

Another reason for green building growth has been governmental support. Although in some cases communities have been slow to champion the need to increase building resiliency,[130] governments have recognized that 'slowing the growth rate of GHG emissions and then reversing it is the key to addressing climate change'.[131] Governments have successfully advocated for principled changes, in which green building appears as a key to addressing climate change and environmental quality, as well as human health and social and economic issues. Today, 'governments at every level are engaged in an effort to enact environmentally-friendly laws in the hope of reducing long-term costs,

---

[127] Ibid. See also Ariel Schwartz, 'U.S. Green Building Market Will Balloon to $173.5 Billion by 2015' (*Fast Company*, 6 July 2010, http://www.usgbc.org/News/USGBCInTheNewsDetails.aspx?ID=4456, accessed 1 November 2012 ('U.S. green building market value will balloon from $71.1 billion now to $173 billion by 2015. Commercial green building is expected to grow by 18.1% annually during the same time period from $35.6 billion to $81.8 billion. In this case, green building is defined as building with resource use and employee productivity in mind.'); Franklyn Cater, 'Green Building, A Real Estate Revolution?' NPR *All Things Considered* (30 September 2010), http://www.usgbc.org/News/USGBCInTheNewsDetails.aspx?ID=4501, accessed 1 November 2012 ('Green building now accounts for nearly one-third of new construction in the U.S. That's up from 2 percent in 2005, according to McGraw-Hill Construction, which tracks the industry.'); Mary Ellen Podmolik, 'Housing's Silver Lining is Green' (*Chicago Tribune*, 10 December 2010), http://www.usgbc.org/News/USGBCInTheNewsDetails.aspx?ID=4569, accessed 1 November 2012 (identifying reports of residential green-construction market generation at $12 billion to $17 billion this year).

[128] Kats, above note 75 at v.

[129] Green Building Council, 'Making the Business Case for High Performance Green Building' (2003), http://www.wgba.org, accessed 1 November 2012.

[130] Edna Sussman, 'Case Study: Climate Change Adaptation Planning Guidance for Local Governments in the United States' (2009) 9 Sustainable Development Law & Policy 31.

[131] Architecture 2030, above note 26.

drawing "green collar" jobs and businesses, and establishing sustainable design and construction procedures'.[132] In 2011, President Barack Obama launched the 'Better Buildings Initiative', a program intended to 'make commercial buildings 20 percent more energy efficient over the next decade by catalyzing private sector investment through a series of incentives to upgrade offices, stores, schools and other municipal buildings, universities, hospitals, and other commercial buildings'.[133] President Obama's plan attempts to create these incentives through tax breaks and additional financing opportunities, and has committed 'over USD 20 billion in funding building energy efficiency initiatives through the American Recovery and Reinvestment Act'.[134]

CALGreen has been characterized as 'the most stringent, environmentally-friendly building code in the United States'[135] and is expected to achieve significant reductions in GHG emissions, energy consumption, as well as an increase in water conservation for the state.[136] CALGreen requires the use of environmentally friendly features, such as efficient plumbing and fixtures to cut indoor water use, efficient heaters and air conditioners, and the diversion of 50 percent of construction waste to recycling.[137] In California, architects, developers, and officials involved in approving new constructions now have to adhere to new

---

[132] Abair, above note 34 at 628.

[133] The White House, 'President Obama's Plan to Win the Future by Making American Businesses More Energy Efficient through the "Better Buildings Initiative"' (3 February 2011), http://www.whitehouse.gov/the-press-office/2011/02/03/president-obama-s-plan-win-future-making-american-businesses-more-energy, accessed 1 November 2012.

[134] Laura Clapper, 'Better Buildings Initiative to Improve Building Efficiency in 10 years' (7 February 2011), http://www.constructiondigital.com/tags/energy-efficiency/better-buildings-initiative-improve-building-efficiency-10-years, accessed 1 November 2012.

[135] California Building Standards Commission, 'CALGreen: The 2010 California Green Building Standards Code, Are You Ready?', http://www.documents.dgs.ca.gov/bsc/CALGreen/The-CALGreen-Story.pdf, accessed 1 November 2012; Carolyn Allen, 'CAL Green California Green Building Code for 2011' *California Green Solutions* (2010), http://www.californiagreensolutions.com/cgi-bin/gt/tpl.h,content=3438, accessed 1 November 2012 (that will apply to new commercial buildings, hospitals, schools, shopping malls and homes).

[136] See California Building Standards Commission, *2010 California Green Building Standards (CALGreen) Code,* http://www.bsc.ca.gov, accessed 1 November 2012.

[137] Ibid.

mandatory guidelines on issues ranging from water efficiency and conservation to indoor air quality.[138]

Local governments are implementing such standards by encouraging, and in some cases mandating, the construction of green buildings.[139] For instance, the City of Baltimore has self-identified as among the local governments that 'stepped up to demonstrate leadership on climate policy amidst federal inaction'.[140] In 2006, the City of Baltimore adopted its Comprehensive Master Plan to codify the four themes that illustrate the essential aspects of the City: live, play, earn, and learn.[141] The City's Sustainability Plan was intended to serve as an element of the Comprehensive Plan and to 'expand upon and complement the recommendations contained in the Comprehensive Plan'.[142] The themes added through the sustainability framework included: cleanliness, pollution prevention, resource conservation, greening, transportation, environmental education, and green economy. Baltimore identifies reduced access to natural resources and infrastructure capacity problems resulting from changes in the global climate system.[143] Baltimore relies upon green building as a strategy for pollution prevention (both in the improvement of the health of indoor environments and the reduction of GHG emissions) and to reduce reliance on energy, and plans to upgrade efficiency in Baltimore schools and increase the popularity of green building standards in the construction industry.[144]

---

138 Ibid.

139 Schindler, above note 30 at 287. Mayor Bloomberg of New York City has championed 'many of the initiatives proposed as part of the Better Buildings Program, which will help us achieve our PlaNYC goal of a 30 percent reduction in greenhouse gas emissions by 2030': News From the Blue Room, 'Statement of Mayor Michael R. Bloomberg on President Obama's Better Buildings Initiative' (3 February 2011), http://www.nyc.gov/html/om/html/2011a/pr045-11.html, accessed 1 November 2012.

140 City of Baltimore, *The Baltimore Sustainability Plan* (2009) 15, http://www.baltimoresustainability.org/uploads/files/Sustainability_Plan.pdf, accessed 1 November 2012.

141 Ibid. at 14.

142 Ibid.

143 Ibid. at 56.

144 Ibid. at 42, 56, 100, and 116. The City did not identify green building as a strategy to address water conservation or waste reduction. Ibid. at 61–67.

Although the building industry faced some troubled beginnings in sustainable building practices,[145] a growing body of evidence supports the promises of green building to deliver efficient and adaptive benefits. In the meantime, the building industry has embraced[146] the contribution of green building through the emergence of a viable market to support green building products and practices. This will prove essential in the employment of green building practices to build adaptive capacity.

5 CONCLUSION

Climate change science predicts significant challenges to the social, economic, political, and environmental conditions in which we live.[147] Mitigation efforts, although successful in the short term, are an inadequate response to the coming threat of a constantly changing environment. In the meantime, social and cultural realities have aligned climate needs with adaptive technologies. Due to persistence, individuals are regularly faced with information about the carbon footprints of consumer goods; governments provide (and sometimes mandate) alternatives to conventional energy and water uses; and markets and the media have been focusing heavily on emission trading systems at the international, national, regional, and local levels as a means of reducing (GHG) emissions. The dialogue on climate change has turned to adaptive capacity, and one obvious path is green building practices.

Buildings emit large amounts of GHGs, are responsible for large amounts of electricity consumption, and result in large amounts of construction, demolition, and water waste. We need to take note of the

---

[145] See Darren A Prum, 'Green Building, High Performance Builds, and Sustainable Construction: Does It Really Matter What We Call Them?' (2010) 21 Villanova Environmental Law Journal 1.

[146] Building Design & Construction, *Green Buildings and the Bottom Line* (2006) 1, http://www.bdcnetwork.com, accessed 1 November 2012 ('What started out as a charismatic crusade had matured into an established sector of the U.S. construction industry').

[147] See John R Nolon, 'The Law of Sustainable Development: Keeping Pace' (2010) 30(4) Pace Law Review 1246, 1248 ('That economic development is linked to the quality of the environment is undeniable. The Commission noted that "there has been a growing realization in national governments and multilateral institutions that it is impossible to separate economic development issues from environmental issues; many forms of development erode the environmental resources upon which they must be based, and environmental degradation can undermine economic development"').

likelihood that the impacts of climate change on the built environment will be devastating. However, the emerging regime of green building laws and regulations addresses several areas of climate change adaptation concern by making buildings more complementary to the local environment. Green building, as a means of adaptation, can direct new aspects of the built environment to create healthier and more resource efficient models of construction, renovation, operation, maintenance, and demolition.[148] Buildings can be sustainable, and what is sustainable about green buildings can also be adaptive.

Efforts to facilitate effective adaptation technologies are proving successful in improving the existing information-gathering capacity, reducing reliance on natural resource production, and increasing flexibility in the built environment. Green building codes can further each of these goals while making a substantial contribution to the overall quality of human health in the built environment. Moreover, the development of green building standards illustrates the power of consensus and the potential offered by public–private collaboration in sustainability matters, reflecting the likelihood that 'climate change adaptation policy will be profoundly influenced by strategies and initiatives from the private sector taking autonomous initiatives'.[149] Adapting to climate change with individualized and industry-wide strategies and initiatives will go a long way toward reducing the threats from climate change. Many state and local lawmakers throughout the US have adopted mandatory green building policies as one element in a comprehensive adaptation scheme.[150]

Green building is becoming the 'standard in communities across America',[151] as the explosion of models and rating systems have made their way to the forefront of the green building market. Development of green building rating systems by the National Association of Homebuilders, the International Green Construction Code, and the State of California add to the more mainstream models and illustrate the depth and breadth of the green building industry's commitment to face climate change impacts and the technological strides towards providing adaptable/flexible mechanisms that protect against unavoidable climate change circumstances.

---

148 US Global Change Research Program, above note 1 at 10.

149 Ruhl, above note 88 at 382.

150 Abair, above note 34 at 628–632 (noting that as environmental concerns become more common, green construction has become more popular and more cost-efficient).

151 The American Institute of Architects, above note 10 at 5.

Of course, green building alone will not secure our adaptation needs, and cities that arrive at such a conclusion will do so erroneously: 'A city, once having adopted a … green building ordinance, will falsely believe that it has sufficiently addressed its environmental concerns. In reality, this "solution" sticks a band-aid on a major wound, calls the process successful, and stops there'.[152] Sustainability illustrates that 'we need more savings, and faster, in order to reduce total emissions at the necessary scale, scope and speed'.[153] A thoughtful, measured approach will avoid future difficulties in the green building program and ensure long term adaptive needs are also taken into consideration. Sustainable building represents a commitment to engage that planning process.

---

152 Schindler, above note 30 at 293.

153 Rob Watson, *Green Building Market and Impact Report 2010* (Greenbiz 2010) 34.

14. Climate change adaptation and environmental and pollution control law

Jonathan Verschuuren

1 INTRODUCTION

Environmental law, including pollution control law, is relevant for adaptation in three ways. First, pollution control law helps to make the natural environment (land, water, air) resilient to change. A natural environment that is already weakened by contamination will suffer greater negative impacts by climate change than a pristine environment. The combination of an influx of nutrients into surface waters, for example, and rising water temperatures can lead to massive harmful algal blooms in both marine and freshwater aquatic environments.[1] Adaptation policies should, therefore, be aimed at further reducing or preventing the emission of pollutants into the environment.

Second, environmental law instruments will have to be applied for adaptation purposes. Installations vulnerable to extreme weather events, flooding or droughts have to be made resilient to these impacts of climate change so as to prevent pollution. Flooding of a chemical industry plant or a waste management site, for instance, can lead to serious pollution in a large area. Therefore, permits and other legal instruments aimed at regulating these activities, have to force the owners of these installations to be prepared for such impacts. Environmental impact assessments can be used to discover whether industrial installations, infrastructure projects or other large-scale human activities are, from a long term perspective, 'climate proof'.

Third, many adaptation measures can or will have negative environmental side-effects. Coastal adaptation measures, such as sea walls and coastal armouring, can impact coastal biodiversity, as will water diversions to combat droughts. Spraying pesticides for disease vector removal may negatively impact the environment. New recreational facilities will

[1] S K Moore et al., 'Impacts of Climate Variability and Future Climate Change on Harmful Algal Blooms and Human Health' (2008) Environmental Health 7 (Suppl 2):S4 doi:10.1186/1476–069X-7-S2-S4.

be created with a potential negative impact on biodiversity, for instance, in coastal areas that, because of temperature rise, are becoming more attractive for tourists, or ski slopes at higher altitudes as lower-lying slopes are increasingly witnessing a decrease in snowfall.

In this chapter, the three roles of environmental law for adaptation will be briefly discussed, focusing on such essential legal instruments as environmental quality standards, environmental permits and environmental impact assessments.

2 APPLYING ENVIRONMENTAL LAW INSTRUMENTS TO ADAPTATION

2.1 Environmental Quality Standards

As mentioned in the introduction, the overall environmental quality determines the level of resilience of the environment to change. Therefore, it is generally thought that adaptation strategies in the field of pollution control law should aim at eliminating particular kinds of discharges and emissions to reduce pollution stressors.[2] In many countries, environmental quality standards have been set to make sure that a certain minimum quality level is achieved. These quality standards may become harder to achieve, leading to increased pressure on norms and requiring quality standards to be set or adapted in order to reduce emissions from existing sources. Higher temperatures, for example, lead to elevated concentrations of ground level ozone, which is harmful to human health as well as to vegetation. Adaptation, then, would at least be aimed at reducing existing emissions, so as to keep achieving the original targets. Preferably, though, stricter targets for ground level ozone and other pollutants should be set so as to increase resilience. The same is true for water quality standards. To prevent or reduce harmful algal blooms, for example, stricter targets for the maximum level of nutrients in aquatic systems need to be set and achieved. Soil quality laws, such as laws that aim to address desertification and erosion,[3] are also affected by climate change, as higher temperatures and changes in precipitation have

---

[2] R K Craig, '"Stationarity is Dead"– Long Live Transformations: Five Principles for Climate Change Adaptation Law' (2010) 48 Harvard Environmental Law Review 9, 45.

[3] Implementing the UN Convention to Combat Desertification in those Countries Experiencing Serious Drought and/or Desertification, Particularly in Africa (UNCCD), Paris, 17 June 1994, UN GA, Distr.Gen. A/AC.241/27.

an impact on the quality of the soil.[4] Hence, quality standards for soils may have to be set (in case they do not yet exist) or adapted to take climate change impact into account.

Environmental quality standards have to be achieved through reducing emissions. This can be done in a variety of ways, ranging from setting emission limit values in environmental permits for individual industries, to generic emission limits for cars or prescribing certain agricultural practices. All of these instruments need to be reviewed with the aim of tightening emission controls. By way of example, we will focus on environmental permits as one of the main instruments to achieve an improvement in the overall quality of the environment.

2.2 Environmental Permits

The Fukushima incident in 2011 showed the catastrophic impact an extreme weather event can have on a nuclear power station. The tsunami that hit the Japanese coast caused large scale power outages and a complete loss of external power on the site of the Fukushima nuclear power station. The station's own diesel generators were flooded and could not be used, which then led to the melting of nuclear fuel and to explosions, causing radioactivity to spread across Japan and the Pacific Ocean.[5] Nuclear power stations – but also any industrial plant where toxic chemicals are stored or used, or storage areas of petroleum – are typically installations that may need additional permit requirements to make them more resilient to climate change. Such requirements could include measures against flooding, or against the spread of toxic chemicals in case of flooding; limitations as to the maximum quantities of chemicals present on the premises; resilient back-up power systems; and water saving and storage facilities to ensure the availability of cooling or process water in times of water scarcity. In extremely vulnerable situations, it may be necessary to repeal the permits altogether and look for alternative locations for the installation, for instance removing it from the floodplain to a higher elevated area.

Hazardous waste sites are other important examples of sites that may be vulnerable to climate change. Sussman and others show that in the US, regulators have the authority to conduct new kinds of surveys and

---

4 See more extensively Chapter 8.

5 A detailed account of the disaster and the findings of subsequent inquiries into the causes of the disaster is reported in a Wikipedia article 'Fukushima Daiichi nuclear disaster', http://en.wikipedia.org/wiki/Fukushima_Daiichi_nuclear_disaster, assessed 1 November 2012.

mapping to identify hazardous sites at the shoreline and low-elevation areas, as well as waste sites likely to be at greatest risk for leaching as a result of sea level changes.[6] When such vulnerabilities are detected, permits can be updated or repealed.

As already mentioned, environmental permits also need to be reviewed with the aim of improving the overall environmental quality to create resilience. For the US, Craig suggests switching from prescribing 'best conventional control technology' to 'best available technologies'.[7] In the EU, the Industrial Emissions Directive is already aimed at setting emission limits that require the application of best available techniques.[8] When reviewing the way the provisions of this Directive define best available techniques and legislate how regulators are to determine what constitutes best available techniques and emission limit values associated with these techniques in a specific case, it is obvious that the Directive does not require taking the potential consequences of climate change into account. The Directive does state that the emission of greenhouse gases should not be regulated in the permit, because those emissions are regulated under the EU Emissions Trading Scheme (ETS).[9] That, however, does not in any way reduce the potential role of the Industrial Emissions Directive in an adaptation policy. In the author's view, the provisions of pollution control laws like the Industrial Emissions Directive should be reviewed in order to check whether they are 'adaptation proof'. Do the provisions require the permitting authorities to take potential impacts of climate change into account? Do they aim at increased resilience? Do they provide the opportunity to regularly assess permit conditions with a view to changing weather conditions or changing climatic circumstances? The EU Industrial Emissions Directive does require the authorities periodically to reconsider all permit conditions, but it does not explicitly state that expected or experienced climatic changes are a reason to do so.[10] More generally, it must be criticized that this important Directive, centrepiece of EU environmental law, does not in any way refer to adaptation.

---

[6] E Sussman et al., 'Climate Change Adaptation: Fostering Progress through Law and Regulation' (2010) 18 N.Y.U. Environmental Law Journal 55, 116.

[7] Craig, above note 2 at 45–46.

[8] Articles 11 and 14 of Directive 2010/75/EU on industrial emissions, (2010) OJ L334/17.

[9] Directive 2010/75/EU, Article 9.

[10] Ibid. Article 21.

2.3 Environmental Impact Assessments

Environmental law relies heavily on impact assessments carried out before taking a decision. Taking the EU as an example, its Member States have to perform:

- environmental impact assessments of certain projects, such as large infrastructure projects, large industrial activities, power stations, waste disposal sites etc.;[11]
- strategic environmental assessments of certain plans and programmes that are prepared for agriculture, forestry, fisheries, energy, industry, transport, waste management, telecommunications, tourism, town and country planning or land use etc.;[12]
- appropriate assessments of the impact of plans and projects on so-called Natura 2000 sites, i.e., protected areas under the Birds and Habitats Directives;[13]
- reviews of the impact of human water use on the status of surface and ground water;[14]
- assessments of significant effects of industrial emissions on the environment.[15]

Assessments like these become increasingly difficult because our predictive capacity cannot deal with the many uncertainties involved in climate change. Or, as Ruhl puts it: 'nonlinearities in change dynamics, environmental feedback properties, and the interactions of social and ecological responses will soon exceed the boundaries of environmental stationarity that have allowed environmental impact assessment and cost–benefit analysis to maintain what reliability and credibility they have'.[16]

This, however, should be seen as a challenge that needs to be overcome. Impact assessments do play a very important role in any

---

[11] Directive 2011/92/EU on the assessment of the effects of certain public and private projects on the environment, (2012) OJ L26/1 ('EIA Directive').

[12] Directive 2001/42/EC on the assessment of the effects of certain plans and programmes on the environment, (2001) OJ L197/30.

[13] Article 6 of Directive 92/43/EEC on the conservation of natural habitats and of wild flora and fauna, (1992) OJ L206/7. See extensively, Chapter 11.

[14] Article 5 of Directive 2000/60/EC establishing a framework for community action in the field of water policy, (2000) OJ L327/1.

[15] Article 12 of Directive 2010/75/EU on industrial emissions, (2010) OJ L 334/17.

[16] J B Ruhl, 'Climate Change Adaptation and the Structural Transformation of Environmental Law' (2010) 40 Environmental Law 363, 414.

adaptation policy because they are explicitly designed not only to look into short-term effects, but also into long-term impacts of an activity.[17] Environmental impact assessment legislation is also well-poised to evaluate complex and cumulative impacts.[18] Impact assessments enable the authorities to take into account changing climatic conditions when assessing the potential or likely impact of an activity on the environment and, thus, enable them to require measures to be taken to make the activity resilient to these changing conditions. Impact assessments also allow the authorities to review alternatives that may be less vulnerable to the impact of climate change. There are many examples of how environmental impact assessments can play a role here. In an environmental impact assessment, various locations of new infrastructure, for instance, may be assessed on their vulnerability to the effects of climate change, such as flooding. Or, when flooding is inevitable in a given area, the vulnerability of infrastructural designs can be assessed, including, for instance, an elevated transport system. Similar considerations are true for any installation that poses a risk to the environment.

Often, environmental impact assessment legislation is formulated in a broad way, so as to include all future potential effects of the activity on the environment. The wording used in impact assessment laws and regulations may already allow for climate change adaptation needs to be included in the process.[19] It seems, however, advisable to explicitly include the obligation to take into account climate change in impact assessments in the relevant laws and regulations, in order to ensure that this is actually done and, when necessary, it can be tested in court.

In the EU, a major revision of the EIA Directive is being used to insert a range of provisions aimed at making the Directive 'climate proof'.[20] These revisions include the duty to:

---

[17] K M Baldwin, 'NEPA and CEQA: Effective Legal Frameworks for Compelling Consideration of Adaptation to Climate Change' (2009) 82 Southern California Law Review 769, 805.

[18] C W Christopher, 'Success by a Thousand Cuts: The Use of Environmental Impact Assessment in Addressing Climate Change' (2008) 9 Vermont Journal of Environmental Law 549, 606.

[19] For instance, this is the case in New York; see Sussman et al., above note 6 at 80.

[20] Proposal of 26 October 2012 for a Directive amending Directive 2011/92/EU, COM(2012) 628 final.

- assess the effect of a project on climate change and on the 'exposure, vulnerability and resilience' to climate change;[21]
- select projects subject to EIA using the following characteristic: 'impacts of the project on climate change (in terms of greenhouse gas emissions including from land use, land-use change and forestry), contribution of the project to an improved resilience, and the impacts of climate change on the project (e.g. if the project is coherent with a changing climate)';[22]
- describe the aspects of the environment likely to be significantly affected by the proposed project, such as 'climatic factors, climate change (greenhouse gas emissions, including from land use, land use change and forestry, mitigation potential, impacts relevant to adaptation, if the project takes into account risks associated with climate change)' (…) 'as well as the exposure, vulnerability and resilience of [these] factors to natural and man-made disaster risks';[23]
- describe the likely significant effects of the proposed project, such as 'the use of natural resources, in particular land, soil, water, biodiversity and the ecosystem services it provides, considering as far as possible the availability of these resources also in the light of changing climatic conditions' and 'the greenhouse gas emissions, including from land use, land use change and forestry …'.[24]

However, an important shortcoming of impact assessments is that they are all carried out prior to deciding on a new project. Existing installations, infrastructure or other activities that are vulnerable to climate change, therefore, are not subject to these assessments. Sometimes, especially in the case of installations that can only operate after having obtained an environmental permit, legislation requires regular reviews and updates of existing regulations (see section 2.2 above). There may also be sectoral legislation that requires an assessment of the resilience of existing infrastructure, power stations, and other installations, as is shown in the chapters in this book on water management, land use, and electricity. If, however, such sectoral legislation is not in place, or does not cover everything, this gap in the legislative framework for adaptation needs attention from the legislature.

---

[21] Amended Article 3(b) and (e) respectively.

[22] Amended Annex III under 1(g).

[23] Amended Annex IV under 1(4).

[24] Amended Annex IV under 1(5)(b) and (f) respectively.

Environmental impact assessments, thus, may help in discovering the weaknesses of human activities from the perspective of climate change and in creating resilience. Some attention should, however, also be paid to the negative impact of adaptation measures. Here, impact assessments have an important role to play as well. Avoiding or reducing negative side-effects of adaptation measures can be achieved through the instruments of environmental impact assessment and strategic environmental assessment. Such assessments offer an instrument to search for the most effective adaptation measure with the least negative side-effects; for instance, sea defences that have the least impact on coastal ecosystems and coastal biodiversity; water diversion projects that do not, or only to a limited extent, harm other legitimate water users; or new recreational facilities located and designed in such a way that they do not harm local biodiversity.

3 CONCLUSION

In adaptation literature, relatively little attention is paid to the role of basic environmental law instruments such as environmental quality standards, environmental permits and environmental impact assessments in adaptation law. In this chapter, I have reviewed the three roles these basic instruments can play in adaptation law: (1) create an overall environmental quality that makes the environment more resilient to climate change; (2) reduce the risk of (increased) pollution as a consequence of extreme weather events or of climate change more generally; (3) reduce the negative side-effects of adaptation measures on the environment.

In general it can be concluded that current environmental laws already possess features that are relevant to adaptation, such as the requirement to regularly assess the permit conditions under pollution control law and the requirement to look at the long-term impacts of a project on the environment under environmental impact assessment law. However, it seems that the legislature should explicitly force regulators to take into account changing climatic conditions and extreme weather events when setting environmental quality standards or when requiring certain conditions to be met in environmental permits. Impact assessment schemes aimed at both specific projects and strategic plans should explicitly require the assessors to take long-term climate change adaptation requirements into account. The potential contribution of these environmental law instruments to climate change adaptation is simply too large not to embrace the opportunities they encompass.

15. Climate change adaptation and electricity infrastructure

Rosemary Lyster and Rebekah Byrne

1 INTRODUCTION

The threats of climate change to electricity infrastructure have been well documented and may be considered in terms of climatic variables, impacts and implications.[1] The climate variables include: increase in intensity of extreme wind; increased frequency and intensity of storms and bushfires; increases in temperature, number of hot days and electrical storms; and decrease in rainfall. The impacts are predicted to be: damage to transmission and distribution above-ground assets resulting in increased blackouts; reduced network capacity along with accelerated deterioration of assets; and an increase in the number and length of blackouts due to electricity demand exceeding supply. The implications considered as economic impacts include: increased capital and maintenance expenditure on electricity transmission and distribution infrastructure; increased demand for skilled staff leading to shortages in this area; short-term lost revenue to transmission and distribution companies; accelerated depreciation and deterioration of assets; and increased power prices to consumers. Additional demand-side implications, particularly in

---

[1] Australian Academy of Technological Sciences and Engineering (ATSE), *Assessment of the Impacts of Climate Change on Australia's Physical Infrastructure* (2008); Energy Networks Association, *Energy Network Infrastructure and the Climate Change Challenge report by Parsons Brinkerhoff to ENA* (2009); Ross Garnaut, *The Garnaut Climate Change Review* (Cambridge University Press 2008) 445–465; Cleo Paskal, *The Vulnerability of Energy Infrastructure to Environmental Change* (Chatham House and Global EESE 2009); Mladen Kezunovic, Ian Dobson and Yimai Dong, 'Impact of Extreme Weather on Power System Blackouts and Forced Outages: New Challenges' (online paper 2009), http://www.ece.wisc.edu/~dobson/PAPERS/kezunovicBPC08.pdf, accessed 1 November 2012; Maunsell Australia, *Impact of Climate Change on Infrastructure in Australia and CGE Model Inputs* (Garnaut Climate Change Review 2008), http://www.garnautreview.org.au/CA25734E0016A131/WebObj/02-A Infrastructure/$File/02-A%20Infrastructure.pdf, accessed 1 November 2012.

developed economies, include: community security risk; increased insurance costs as a result of business and food loss; traffic congestion/accidents/no traffic lights and short term unemployment leading to hindered emergency responses; train service disruption; life support/health risks; deterioration in quality of home life and building services disruption including elevators and lifts.[2]

This chapter sets out the scientific case for legal and technology adaptation responses in the electricity infrastructure sector in the face of climate extremes and disasters. It proposes the adoption of the Smart Grid as one of the technological adaptation measures and relies on the experience of the Queensland flood disasters in Australia in 2010 to provide some generic examples of how governments and utilities might respond to protect electricity infrastructure.

2 MANAGING THE RISK OF EXTREME CLIMATE EVENTS AND DISASTERS TO ELECTRICITY INFRASTRUCTURE AND SUPPLY

In 2012, the Intergovernmental Panel on Climate Change (IPCC) launched its latest scientific report entitled *Managing the Risks of Extreme Events and Disasters to Advance Climate Change Adaptation* (SREX Report).[3] This Report provides the most recent assessment of climate change impacts on electricity infrastructure as well as discussing the economic costs of adaptation measures and global and regional climate disasters; implications for development, recovery and reconstruction; and technologies for extreme events.

2.1 Extreme Events and Extreme Impacts on Electricity Infrastructure

The SREX Report states with high confidence that extreme impacts on humans and social systems will depend on the degree of exposure and vulnerability to that extreme event, in addition to the magnitude of the physical event. As such, even non-extreme events can result in extreme impacts on human systems where vulnerability and exposure are high. For example, an extreme event with a large spatial scale (as in an ice

---

2 Maunsell, above note 1 at 4.

3 Ch B Field et al. (eds), *Managing the Risks of Extreme Events and Disasters to Advance Climate Change Adaptation: Special Report of the Intergovernmental Panel on Climate Change* (Cambridge University Press 2012).

storm or windstorm) can have an exaggerated, disruptive impact on electricity transmission and distribution networks due to the systemic societal dependence on the networks.[4] The Report notes that electricity transmission infrastructure is particularly vulnerable to extreme storm events, particularly wind and lightning, drought, extreme rainfall events and floods, and in some cases heat waves.

2.1.1 Extreme storm events

The SREX Report notes that the passage of storms across France in 1999 caused the greatest devastation to an electricity supply network ever witnessed in a developed country. During the storms, 120 high-voltage transmission pylons were toppled and 36 high-tension transmission lines were lost, representing one-quarter of the total lines in France.[5] Another example is the 2008 Chinese ice storm, which affected a hundred million people for days, weeks or even a month, depending on their location.[6] It inflicted direct economic losses exceeding USD 20 billion, demonstrating that an extreme event of this nature can undo socio-economic and ecological structures that have taken decades to establish. The electricity grid was the first critical infrastructure to fail across the affected region with 80 per cent of the power supply facilities damaged in the province of Guizhou. In the province of Hunan, the electricity grid supplying electricity to the city of Chenzhou was almost completely destroyed and had to be rebuilt, leaving more than four million people in the dark for several weeks. The failure of the grid triggered a cascade of breakdowns of other essential services, such as water utilities, and cooking and heating for millions of families became impossible. Hospitals and factories were forced to close and mines were submerged because of inoperable pumps. China's principal mode of transportation, the railway system, was disrupted by powerless electric trains. Highways and airports were also closed because of treacherous conditions. With all transportation systems compromised, supply chains of energy, food and other vital goods were severely compromised. Coal reserves reached emergency levels and power plants had to be shut down. Sudden and unexpected food shortages occurred resulting in soaring food prices, in some cities by 1,000 per cent.[7] The failure of China's transportation hubs affected the

---

[4] Ibid. at 42, 235.

[5] Ibid. at 249.

[6] See Benzhi Zhou et al., 'The 2008 Chinese Snow Storms: The Socioeconomic-Ecological Impacts and Sustainability Lessons Learned' (2011) American Meteorological Society 47.

[7] Ibid. at 52.

entire country with the consumer price index nationwide increasing by 34 per cent between December 2007 and February 2008.[8]

2.1.2 Drought

According to the SREX Report, severe droughts may also affect electricity supplies if the supply of cooling water to power plants is disrupted or hydroelectricity schemes are compromised.[9] For example, drought has a serious impact on electricity production in New Zealand where around two-thirds of supply is from hydroelectricity and low rainfall results in increased fossil fuel electricity generation, which the Report regards as a maladaptation to climate change.[10] In December 2011, the Balkans experienced an acute region-wide electricity shortage as a result of dramatic falls in water levels to hydroelectric dams during a month-long drought. The hydroelectric power plants only managed to produce 10 kWh, which was the lowest production level since 1926.[11] A severe drought between 2009 and 2010 in southwest China crippled the region's hydroelectric power generation sector. Hydroelectric power was the preferred energy option in the water-abundant and mountainous southwest regions of China, including Yunnan, Guizhou, Guangxi and Sichuan Provinces and Chongqing District. Overall, drought reduced the hydroelectricity capacity of the southwest power plants by 40 per cent. In Yunnan Province, the generated power decreased by 20 per cent while in Guangxi Province 90 per cent of the hydro capacity was reported to be 'paralysed'.[12]

2.1.3 Extreme rainfall events and floods

Risk management and adaptation methods are important in reducing the effects of flooding on electrical infrastructure and the resultant exposure and vulnerability of communities. One of the most recent case studies of the impact of floods on electricity infrastructure is the devastating flood

---

[8] Ibid. at 53.

[9] Field et al., above note 3 at 249.

[10] Ibid. at 261.

[11] Georgi Mitev Shantek, 'Serious Electricity Shortage Hits the Balkans' (*Southeast European Times*, 21 December 2011), http://www.setimes.com/cocoon/setimes/xhtml/en_GB/features/setimes/features/2011/12/22/feature-04, accessed 7 November 2012.

[12] Xina Xie and Michael J Economides, 'China's Hydroelectric Sector Crippled by Severe Drought' (*Energy Tribune*, 12 May 2010), http://oilprice.com/Alternative-Energy/Hydroelectric/Chinas-Hydroelectric-Sector-Crippled-By-Severe-Drought.html, accessed 7 November 2012.

events that struck Queensland, Australia, in 2010 to 2011. There was widespread damage to the electricity network, with power outages experienced even where the local electricity infrastructure was not damaged. This was either the result of damage occurring elsewhere to connection parts of the network or because of precautionary disconnections to prevent electric shock.[13] During the floods, 300,000 customers lost power in two major towns, Ipswich and Brisbane. In the Lockyer Valley, one of the most seriously affected rural areas in Queensland, much of the electricity infrastructure was destroyed, causing 5,000 people to lose power.[14] The Australian government established the Queensland Floods Commission of Inquiry which released its Final Report[15] in March 2012. The Report includes important recommendations on increasing the flood resilience of electricity infrastructure and will be relied upon in this chapter.

2.1.4 Heat waves

There is increased likelihood of disruption of electricity supplies during heat waves.[16] This is because both generation and transmission generate heat and on hot days this can be reinforced by positive feedback loops as demand for electricity rises with greater use of air conditioners. This compromises the ability to dissipate heat from parts of the system, which is crucial for the continued effectiveness of the infrastructure. The heat related risks to coal-fired power stations also include the following: work on coal mines is minimized on high risk days due to coal's combustibility; the potential danger of bushfires to power stations and, by implication, transmission infrastructure; and the warming of water that is supposed to cool the steam turbines.[17] The impact of heat waves on electricity infrastructure was evident during the 2003 extraordinary heat wave in Europe. In France, electricity became scarce, construction

---

[13] Queensland Floods Commission of Inquiry, *Queensland Floods Commission of Inquiry: Final Report* (2012) 239, http://www.floodcommission.qld.gov.au/publications/final-report, accessed 7 November 2012.

[14] Ibid. at 239.

[15] Ibid. at 240.

[16] Field et al., above note 3 at 259.

[17] National Climate Change Adaptation Research Facility, *Impact and Adaptation Response of Infrastructure and Communities to Heatwaves: the Southern Australian Experience of 2009* (NCCARF 2010) at 49.

productivity fell, and the cold storage systems of approximately 20 to 30 per cent of all food-related establishments were found to be inadequate.[18] A similar phenomenon was recently witnessed in the southeastern metropolitan regions of Australia, which experienced a record-breaking heat wave in 2009 where maximum daily temperatures were 12 to 15°C above the seasonal average. Night-time temperatures were also unusually high, with the heat build-up exacerbated by exceptionally low surface moisture resulting from an extended drought. The heat wave occurred as an intense, long-lived Phase I, which caused the most apparent human distress from high temperatures, followed by Phase 2, where extreme heat wave conditions combined with severe bushfires across the region.[19] Such experiences demonstrate that there is little resilience to unexpected perturbations such as extreme heat waves if an electricity system is already operating with little spare capacity. The vulnerability of electricity infrastructure can be increased by a combination of driving forces such as ageing infrastructure; rapidly increasing demand (especially from air conditioning); and the predicted increase in the frequency and severity of climate-related extreme events.[20] One of the adaptation responses might be the undertaking of scenario testing for potentially hotter and more prolonged extreme events on continuity by infrastructure and electricity service providers.[21]

2.1.5 Extreme events, impacts, and development

There is general consensus that, when compared with developed countries, developing countries are more economically vulnerable to climate extremes. This is largely because developing country economies have:

- less resilience as they depend more on natural capital and climate-sensitive activities, such as cropping and fishing;
- they are often poorly prepared to deal with existing climate variability and physical hazards;
- maladaptation due to the absence of financing, information, and techniques in risk management, as well as weak governance systems is likely to exacerbate the damage;
- in regions with a fast-growing population and asset stocks (such as in coastal areas) little attention is given to climate-proof investments;

---

18 Field et al., above note 3 at 257.

19 National Climate Change Adaptation Research Facility, above note 17 at v.

20 Ibid. at 75.

21 Ibid. at ix.

- low levels of economic development mean that there is an adaptation deficit and a lack of ability to transfer costs through insurance and fiscal mechanisms;
- developing countries have large informal sectors.

However, in some cases developed countries also suffer severe disasters because of social vulnerability and inadequate disaster protection as witnessed with Hurricane Katrina in New Orleans, United States.[22]

2.2 Strategies for Electricity Supply During Extreme Events

The SREX Report draws together experiences from a number of recent disasters to indicate strategies that might be adopted during extreme events. Disaster events may compromise electricity supplies so that partial or total blackouts are experienced. The strategies that might be adopted during extreme events include conserving energy, relying on alternative forms of energy, including alternative forms of generation, the rescheduling of activities to a future date, or focusing on the low- or no-energy elements of the business operation. All of these demand-side management responses might be combined with advance electricity storage, for example, when low precipitation reduces hydroelectricity production. As discussed above, this has already occurred in many jurisdictions but may also be increasingly problematic due to climate change projections. Yet another strategy could include rationing electricity supply in accordance with a legally sanctioned process which recognizes a list of priorities between various sectors such as industry, agriculture, electricity production, and domestic water supply.[23] This legislative technique is already adopted in Australia, for example, where water for 'critical human needs' enjoys priority over all other claims on a water source.[24] Also various types of water access licences are prioritized, with water utilities enjoying the highest protection. In the event of drought where water allocations need to be reduced, the access licences enjoying the least legislative protection will have their allocations reduced first.[25]

---

22 Field et al., above note 3 at 265.
23 Ibid. at 307.
24 Water Act 2007 (Cth) Part 2A.
25 Water Management Act 2000 (NSW) s. 58.

3 FRAMING THE COSTS OF EXTREMES AND DISASTERS

The economic costs associated with climate extremes and disasters can be categorized as impact or damage costs and adaptation costs. Impact costs may be either direct or indirect costs. For example, a direct impact will occur when electricity transmission lines are destroyed by wind. However, where this causes businesses to cease operations, putting many people out of work and creating other socio-economic problems, these may be classified as indirect impacts. Adaptation costs are those associated with: adaptation planning; actual adaptation such as risk prevention, preparedness, and risk financing; reactive adaptation such as emergency disaster responses, rehabilitation, and reconstruction; and, finally, the implementation of adaptation measures. The benefits of adaptation include the value of avoided impacts and damages as well as the co-benefits generated by the implementation of adaptation measures.[26]

Although the losses due to the failure of electricity infrastructure during weather and climate extremes are not specifically itemized in the SREX Report, the overall losses build the case for the urgent adoption of climate adaptation strategies and legal responses. Estimates of annual losses from global weather- and climate-related disasters have ranged from a few billion USD in 1980 to above USD 200 billion in 2005 (the year of Hurricane Katrina), not counting indirect and intangible losses. On a global scale, annual material damage from large weather- and climate-related events has increased eight-fold between the 1960s and the 1990s, while the insured damage has increased seventeen-fold. Between 1980 and 2004, the total cost of extreme weather events was USD 1.4 trillion, of which only one-quarter was insured. The human impact of natural disasters is unequally distributed across regions, with Asia experiencing the highest number of weather- and climate-related disasters in the period 2000 to 2008. Economic loss was distributed as follows: Americas 54.6 per cent of total loss, Asia 27.5 per cent, and Europe 15.9 per cent. Africa accounted for only 0.6 per cent of global economic losses, but economic damages from natural disasters are under-reported in this region compared to other regions.

Middle-income countries with rapidly expanding asset bases have borne the largest burden. During the period from 2001 to 2006, losses in middle-income countries amounted to about 1 per cent of gross domestic product (GDP), compared with 0.3 per cent in low-income countries and

[26] Field et al., above note 3 at 265.

less than 0.1 per cent in high-income countries, based on limited evidence. In small, exposed countries, particularly small island developing states, these wealth losses, expressed as a percentage of GDP and averaged over both disaster and non-disaster years, can be considerably higher, exceeding 1 per cent in many cases and 8 per cent in the most extreme cases between 1970 and 2010. This indicates a far higher vulnerability of the economic infrastructure in developing countries.

It is also important to note that the number of weather- and climate-related disasters has increased more rapidly than losses from non-weather disasters, which could indicate a change in climate extremes, although there are other possible explanations. Drought and flood losses may have grown due to a number of non-climatic factors, such as increasing water withdrawals and a decrease in storage capacity in catchments as a result of urbanization, deforestation, the sealing of natural surfaces, and channelization. These factors adversely affect both flood and drought preparedness.[27]

4 ELECTRICITY TECHNOLOGY CHOICES FOR ADAPTATION AND DISASTER RISK REDUCTION

Technology embraces a range of areas, including information and communication technologies, roads and infrastructure, food and production technologies, and energy systems. Technology choices can either alleviate disaster risk or significantly increase risks and adaptation challenges. For example, storms can render modern energy systems and centralized communication systems, which are dependent on physical structures, highly vulnerable to damage. It has been suggested that relatively centralized high-technology systems, such as the traditional electricity grids, are 'brittle,' and, while they offer efficiencies under normal conditions, in the event of emergencies they are subject to cascading effects.[28]

Another focus on the positive role of technologies in climate extremes is that they may assist with information collection and diffusion, including by monitoring possible stresses and vulnerabilities and communicating with populations and response agencies in the event of emergencies, although in some developing regions access to such technologies may be limited. Numerical climate models have also been developed in recent

27 Ibid. at 271.
28 Ibid. at 493.

decades to provide multi-month forecasts, which can be used to prepare for floods and droughts.[29]

When considering adaptation responses for electricity adaptation, large dams may both mitigate drought and generate electricity. However, there may be unacceptable social and ecological displacement costs related to constructing dams and, unless built to accommodate future climate change, they may, in fact, present new risks to society.[30]

4.1 The Promise of Smart Grids

The Smart Grid is an integration of advanced, two-way communications systems and sensors with the transmission and distribution network, which enables utilities to optimize grid performance in real time.[31] Through demand responses it provides incentives to consumers for reducing energy consumption and it provides for a better integration of renewable energy sources into the grid. For the purposes of resilience the Smart Grid gives utilities an enhanced ability to identify the location of a failure and quickly re-route electricity to locations where demand is most critical. This could occur during times of climate change-induced crises, or peak demand, and prevents outages through proactive diagnosis of the grid and its individual elements. Importantly, it enhances the ability of the grid to continue to provide power following a catastrophic event and to support vital emergency responses as well as military, economic, and social activities during a crisis.[32] Although much of the literature to date assesses the interface between the Smart Grid and climate change mitigation, there is barely any mention of the adaptation benefits emanating from Smart Grid technology. If the Smart Grid improves efficiency and demand response and encourages distributed energy sources, its mitigation benefits are clear. Yet the fragility of electricity networks to climate change impacts suggests that the Smart Grid might also assist utilities to respond to blackouts, and other climate change-induced crises, more effectively than is currently possible.

---

29 Ibid.

30 Ibid.

31 Rosemary Lyster, 'Smart Grids: Opportunities for Climate Change Mitigation and Adaptation' (2010) 36(1) Monash University Law Review 173–191.

32 Kenneth R Nahigian, *The Smart Alternative: Securing and Strengthening Our Nation's Vulnerable Electric Grid* (The Reform Institute 2008) 2, 8.

Hendricks[33] claims that a truly national clean-energy Smart Grid comprises two distinct components: an interstate 'sustainable transmission grid', which will transport clean utility-scale renewable energy to the market across long distances; and a digital 'smart distribution grid' to deliver this electricity efficiently to local consumers.[34]

4.1.1 The sustainable transmission grid

The electricity industry traditionally includes generators,[35] market customers (electricity retailers and end-use customers),[36] network service providers, who own, operate or control either a transmission (TNSPs)[37] or distribution (DNSPs) system,[38] market network service providers (MNSPs),[39] and special participants, who may be appointed to perform various functions such as taking responsibility for operations during power system emergencies. Within this sector, large power plants are connected to the transmission system, which supplies local distribution networks and large industrial users. Electricity flows downstream, and primarily in one direction, from high-voltage transmission networks to low-voltage distribution feeders.

As mentioned above, the traditional grid might be regarded as a 'brittle' type of infrastructure in the face of climate extremes and disasters. The Smart Grid, however, allows the interconnection of distributed generation at almost any point in the transmission grid. Distributed

---

[33] Bracken Hendricks, *Wired for Progress 2.0: Building a National Clean-Energy Smart Grid* (Center for American Progress 2009).

[34] Ibid. at 10.

[35] Generation is the process used to create electricity.

[36] Market customers comprise both electricity retailers and end-use customers. Retailers purchase wholesale electricity through the spot market, or from local generators who sell their entire output to them. The electricity is then sold to customers, increasingly within a contestable retail market. End-use customers purchase electricity directly from the spot market or retailers.

[37] Transmission is the process of transporting electricity at high voltages from where it is generated, often over long distances, to groups of electricity consumers.

[38] TNSPs control the high voltage transmission assets that carry electricity between generators and distributors, while DNSPs operate the low voltage substations and wires that transport electricity from these substations to customers. Distributors hold a franchise over the regions in which their poles and wires are installed but must also be given access to customers outside their regions by using rival distribution networks.

[39] MNSPs are entrepreneurial interconnectors, with a minimum capacity of 30 MW, that offer their capacity to transport power into the market through a bidding process similar to that used by generators.

generation includes small-scale generation (including wind, landfill gas, biomass, and hydro), cogeneration or combined heat and power plants, small stand-alone diesel generators, and domestic or small commercial photovoltaic solar generation.[40] There are many benefits associated with distributed generation. These include:

- reducing demand from the grid by installing a distributed generation plant close to the point of consumption. This can reduce distribution and transmission losses, reduce constraints on power lines and defer the need to upgrade the grid;
- improving the security of supply during peak demand;
- improving power supply to remote communities;
- facilitating the uptake of large-scale new renewable energy sources. Currently high-voltage transmission grids are not situated where many renewable energy sources are being developed;
- locating power supply closer to end users.[41]

4.1.2 The digital smart distribution grid

Smart distribution grids involve: integrating new technology into local electricity distribution networks, such as smart meters in individual homes and businesses to improve energy efficiency; grid monitoring and control devices for improving the efficiency of electricity distribution within local networks by utility companies; better tools for information sharing with consumers; and pricing and control systems to integrate distributed energy resources, such as solar panels, energy storage devices, and electric vehicles. The electric vehicles would be charged off peak and then feed power back into the grid when they are parked either at work or at home during the day.[42]

4.2 Proposals for a Smart Grid in Australia, the European Union and the United States

The development of Smart Grids has become a priority action in Australia, the European Union and the United States as part of each

---

[40] See Nirmal-Kumar C Nair and Lixi Zhang, 'SmartGrid: Future Networks for New Zealand Power Systems Incorporating Distributed Generation' (2009) 37 Energy Policy 3418, 3419.

[41] Ibid.

[42] See Hendricks, above note 33 at 12; see also *SMART 2020: Enabling the Low Carbon Economy in the Information Age* (The Climate Group 2008).

jurisdiction's commitment to reducing greenhouse gas (GHG) emissions, improving energy efficiency and encouraging the uptake of renewable energy sources. Each jurisdiction has only recently (in 2009) released its policy platform for the establishment of Smart Grids.[43] The policy documents provide encouraging signs that the Smart Grid is likely to become a reality within the electricity sectors in these jurisdictions, where it will play a meaningful role in delivering mitigation and adaptation to climate change.

4.2.1 Australia

The principal policy document for Australia's transition to a Smart Grid is *Smart Grid, Smart City: A new direction for a new energy era.*[44] Consistent with the discussion above, the document envisages both *grid-side* applications to reduce line loss and improve fault detection and restoration, and *customer-side* applications to assist consumers to manage their energy consumption. It is estimated that the adoption of Smart Grid technologies could deliver at least AUD 5 billion of gross annual benefit to Australia.[45] The customer-side applications would be enabled using smart metering infrastructure and would include: Information (on energy use or GHG emissions provided by website or in-home display); Controls

---

[43] China has also adopted and begun to construct Smart Grids. China's 12th Five-Year Plan sets the goal of accelerating progress on Smart Grid development supported by the Decision of the State Council on Accelerating the Fostering and Development of Strategic Emerging Industries. Local councils have also begun to adopt Smart Grid policies such as Jiangsu Provinces' *Smart grid industry in Jiangsu Province Development Plan Outline.* State Grid Corporation of China (SGCC) is the principal promoter of Smart Grids in China and has begun to develop a standardization policy for Smart Grids. In the past year, SGCC has launched 228 pilot projects across a range of areas, including: smart substations; electric vehicle charging infrastructure; and a technical support system for Smart Grid Deployment. The most comprehensive pilot project is the Sino-Singapore Tianjin Eco-City; see Yanshan Yu, Jin Yang and Bin Chen, 'The Smart Grids in China – a Review' (2012) 5 Energies 1321–1338. See also Fu Liwen, Zhao Huiry, Guo Sen 'An Analysis on the Low-carbon Benefits of Smart Grid of China' (2012) 24 Physics Procedia 328–336; Yuan Jiahia, Hu Zhaoguang, 'Low carbon electricity development in China – An IRSP perspective based on Super Smart Grid' (2011) 15 Renewable and Sustainable Energy Review 2707–2713.

[44] Australian Government, *Smart Grid, Smart City: A New Direction for a New Energy Era* (2008), http://www.ret.gov.au/energy/Documents/smart-grid/smartgrid-newdirection.pdf, accessed 7 November 2012.

[45] Ibid. at 7.

(using in-home displays, automated controls for appliances and programmable thermostats with communications); and Tariffs (which fluctuate according to time of use, critical peak pricing and real-time pricing).[46]

The grid-side applications would include: integrated Volt-VAR control;[47] fault detection, isolation and restoration; substation and feeder monitoring and diagnostics; and wide-area management. Other key elements would include distributed energy storage, distributed generation (such as solar panels on residential rooftops), and electric vehicle support.[48] The Smart Grid communications network will comprise three key elements: homes are networks to connect smart meters and intelligent devices and appliances with residences; Wide Area Networks which allow for two-way communications between a residence and the distributor substation; and Backhaul Networks, which link the distributor substation to the utility.[49] The policy document presciently notes: 'the increased data that such a communications network would produce requires careful consideration of the physical and cyber-security measures required to effectively manage the smart grid'.[50] Synergies between Smart Grid deployment and the National Broadband Network (NBN) programme have also been noted given that the NBN will create a robust, high-bandwidth communication network to most homes and commercial premises.[51]

Australia's first commercial scale Smart Grid trial is known as *Smart Grid, Smart City*. It is led by the utility Ausgrid and is an AUD 100 million initiative across five sites in Newcastle, Sydney and the Upper Hunter Valley. At least 30,000 households will participate in the three-year trial, which will gather information about the benefits and costs of different Smart Grid technologies in an Australian setting. The technologies being tested include: a web portal that allows users to actively

---

46 Ibid. at 17.

47 Currently, utilities use on-line power flow estimates which rely on stale data. If utilities were to use an integrated volt-VAR control system they could determine the optimal settings for all distribution voltage regulating and control devices, which would assist with their obligation to maintain a set voltage to their customers. By collecting feeder voltage measurements from selected advanced geospacial intelligence (AGI) nodes at strategic feeder locations (for example, at feeder extremities), the nodes can report these measurements on a periodic basis. The integrated volt-VAR system can use this information to maintain the required level of service to customers.

48 Australian Government, above note 44.

49 Ibid. at 19.

50 Ibid.

51 Ibid. at 37.

monitor their energy use online, calculate their energy costs and GHG emissions; a variety of household energy monitoring systems, including a wireless control system that allows appliances to be operated remotely from a computer or smart phone; improved reliability and efficiency of the electricity network through the adoption of better monitoring and measuring devices; and distributed storage and generation devices, including fuel cells and battery storage, and electric cars.[52]

4.2.2 European Union

Under the EU's Energy and Climate Policy, the EU has adopted the goals of reducing GHG emissions by 20 per cent by 2020 and increasing the proportion of renewable energy in the EU's energy mix by 20 per cent by 2020. In the longer term, the EU has committed to reducing its GHG emissions by 80 per cent by 2050 compared with 1990 levels. The EU's Strategic Energy Technology Plan (SET-Plan) is the EU's response to the need to accelerate the development of low carbon technologies to ensure their widespread market take-up. Seven 'Technology Roadmaps 2010–2020' have been developed for the implementation of the SET-Plan in the areas of wind energy, solar energy, bioenergy, Carbon Capture and Storage, the electricity grid, sustainable nuclear energy, and Smart Cities.[53]

The EU has identified three interrelated challenges for electricity networks including: creating a real internal market; integrating a massive increase of intermittent energy sources; and managing complex interaction between electricity suppliers and consumers. The total public and private investment needed in Europe is estimated to be EUR 2 billion.[54]

The objective of the European Industrial Initiative on the electricity grid is to enable 35 per cent of the EU's electricity to be supplied by dispersed and concentrated renewable sources by 2020 and for electricity production to be completely decarbonized by 2050. The Initiative will also develop a market-based pan-European network by integrating national networks and engaging customers as active participants in energy efficiency, while also anticipating new developments such as the electrification of transport. Twenty large-scale Smart Grid demonstration projects, involving 1.5 million customers, will be initially supported, whereby smart meters will be rolled out and the automation and control

---

52 See http://www.smartgridsmartcity.com.au, accessed 7 November 2012.

53 European Commission, Communication, Investing in the Development of Low-Carbon Technologies, COM(2009) 519 final.

54 Ibid. at 5.

of whole network will be trialled.[55] Other significant features include: electricity storage; dynamic pricing mechanisms and appropriate information technology tools.[56] Local renewable energy sources, especially photovoltaic (energy generated from solar radiation) and wind applications, will also be fostered.[57]

Meanwhile, under the Smart Cities Initiative, the development and deployment of ten Smart Grids in cities will be established while the large deployment of low carbon transport systems and alternative fuel vehicles will be developed and tested. The Smart Cities Initiative, which also includes provisions for zero energy buildings, is likely to cost EUR 10 to 12 billion over the next ten years.[58]

Three recent European Commission Mandates are facilitating various elements of the Smart Grid. They include standardization mandates for smart meters,[59] electric vehicles,[60] and Smart Grids, which must all be coordinated with each other. The objective of the *Smart Grid Mandate*[61] is to develop or update a set of consistent standards within a common European framework, to integrate a variety of digital computing and communication technologies and electrical architectures, and associated processes and services.[62]

---

55 European Commission, Commission Staff Working Document, Accompanying document to the Communication 'Investing in the Development of Low-Carbon Technologies'*:* A Technology Roadmap SEC(2009) 1295 at 10.

56 Ibid. at 39.

57 Ibid. at 49.

58 Ibid. at 12.

59 European Commission, Standardisation Mandate to CEN, CENELEC and ETSI in the Field of Measuring Instruments for the Development of an Open Architecture for Utility Meters Involving Communication Protocols Enabling Interoperability (12 March 2009) M/441 EN, http://ec.europa.eu/energy/gas_electricity/smartgrids/doc/2009_03_12_mandate_m441_en.pdf, accessed 8 November 2012.

60 European Commission, Standardisation Mandate to CEN, CENELEC and ETSI Concerning the Charging of Electric Vehicles (4 June 2010) M/468 EN, http://ec.europa.eu/energy/gas_electricity/smartgrids/doc/2010_06_04_mandate_m468_en.pdf, accessed 8 November 2012.

61 European Commission, Standardization Mandate to European Standardisation Organisations (ESOs) to support European Smart Grid deployment (1 March 2011) M/490 EN, http://ec.europa.eu/energy/gas_electricity/smartgrids/doc/2011_03_01_mandate_m490_en.pdf, accessed 8 November 2012.

62 Ibid. at 2.

4.2.3 United States

Under the Federal Power Act (FPA), the transmission of electricity in interstate commerce by public utilities, and the reliable operation of the bulk-power systems across most of the United States, falls under the jurisdiction of the Federal Energy Regulatory Commission (FERC).[63] The Commission also has a new responsibility under Title XIII of the Energy Independence and Security Act of 2007 (EISA)[64] to issue a rulemaking to adopt standards and protocols to ensure Smart Grid functionality and interoperability[65] in interstate transmission of electricity and in regional and wholesale electricity markets.[66] Furthermore, the American Recovery and Reinvestment Act of 2009 (Recovery Act)[67] provided the US Department of Energy with USD 4.5 billion to implement Title XIII of the EISA, which requires the modernization of the electric power grid.

In March 2009, FERC released its *Smart Grid Policy*[68] for comment. The Policy points out that Section 1301 of the EISA states that it is the policy of the United States to support the modernization of the nation's electricity transmission and distribution system to maintain a reliable and secure electricity infrastructure that can meet future demand growth, and to achieve each of several goals and characteristics, which together characterize a Smart Grid.[69] These goals and characteristics are defined by the Act as:

---

[63] 16 U.S.C. 824.

[64] Pub. L. No. 110–140, 121 Stat. 1492 (2007) (EISA).

[65] According to the GridWise Architecture Council, the term 'interoperability' refers to the ability to: exchange meaningful, actionable information between two or more systems across organizational boundaries; assure a shared meaning of the exchanged information; achieve an agreed expectation for the response to the information exchange; and maintain the requisite quality of service in information exchange (i.e., reliability, accuracy, security); see GridWise Architecture Council, Interoperability Path Forward Whitepaper (2005) 1–2, http://www.gridwiseac.org/pdfs/interoperability_path_whitepaper_v1_0.pdf, accessed 8 November 2012.

[66] Federal Energy Regulatory Commission, Smart Grid Policy (2009) 6–7, http://www.ferc.gov/whats-new/comm-meet/2009/031909/E-22.pdf, accessed 8 November 2012.

[67] H.R. 679; see Title IV – Energy & Water Development.

[68] Federal Energy Regulatory Commission, above note 66.

[69] EISA s. 1301, to be codified at 15 U.S.C. 17381, cited by Federal Energy Regulatory Commission, above note 66 at 6.

- increased use of digital information and controls technology to improve reliability, security, and efficiency of the electric grid;
- dynamic optimization of grid operations and resources, with full cyber-security;
- deployment and integration of distributed resources and generation, including renewable resources;
- development and incorporation of demand response, demand-side resources, and energy efficiency resources;
- deployment of 'smart' technologies (real-time, automated, interactive technologies that optimize the physical operation of appliances and consumer devices) for metering, communicating concerning grid operations and status, and automating distribution;
- integration of 'smart' appliances and consumer devices;
- deployment and integration of advanced electricity storage and peak-saving technologies, including plug-in electric and hybrid electric vehicles, and thermal storage air conditioning;
- provision to consumers of timely information and control options;
- development of standards for communication and interoperability of appliances and equipment connected to the electricity grid, including the infrastructure serving the grid; and
- identification and lowering of unreasonable or unnecessary barriers to the adoption of Smart Grid technologies, practices, and services.[70]

FERC's Smart Grid Policy discusses each of these aspects and invites comments on the Policy. FERC also advises that Section 1305(a) of the EISA directs the National Institute of Standards and Technology (the Institute) 'to coordinate the development of a framework that includes protocols and model standards for information management to achieve interoperability of smart grid devices and systems'.[71] The National Institute for Standards and Technology (NIST) has developed a Smart Grid Conceptual Reference Model, which will be further developed and maintained by a Smart Grid Architecture Board.[72]

FERC indicated in its Smart Grid Policy that once NIST's work has led to sufficient consensus on interoperability standards, it is directed by legislation to adopt, through the rulemaking process, the standards and

---

[70] Federal Energy Regulatory Commission, above note 66 at 6.

[71] EISA s. 1305(a), to be codified at 15 U.S.C. 17385(a), cited by Federal Energy Regulatory Commission, above note 66 at 6.

[72] National Institute of Standards and Technology, NIST Framework and Roadmap for Smart Grid Interoperability Standards Release 1.0 (Draft) (2009).

protocols necessary to insure Smart Grid functionality and interoperability in interstate transmission of electricity, and regional and wholesale electricity markets. However, in July 2011, FERC determined that there was insufficient public consensus on the adoption of federal standards. Consequently, NIST has developed the *NIST Framework and Roadmap for Smart Grid Interoperability Standards, Release 2.0*.[73] The Framework comprises voluntary standards to be adopted by all Smart Grid stakeholders including utilities and suppliers, testing laboratories, and certification organizations, and State government regulators. One of the important agencies that will continue to develop these standards is the Smart Grid Interoperability Panel, which is a public–private partnership dedicated to continuing coordination, acceleration, and harmonization of standards for the Smart Grid.[74]

The two largest Smart Grid programmes being managed by the US Department of Energy (DOE) are the Smart Grid Investment Grant (SGIG) programme and the Smart Grid Demonstration Program (SGDP). The aim of the SGIG is to accelerate the modernization of the nation's electricity transmission and distribution systems and promote investments in Smart Grid technologies, tools, and techniques that increase flexibility, functionality, interoperability, cyber-security, situational awareness, and operational efficiency. There are currently 99 SGIG projects, with a total budget of about USD 8 billion, which have received federal financial assistance for up to 50 per cent of eligible costs.[75] The SGDP aims to demonstrate how a suite of existing and emerging Smart Grid concepts can be innovatively applied and integrated to prove technical, operational, and business-model feasibility. The aim is to demonstrate new and more cost-effective Smart Grid technologies, tools, techniques, and *system* configurations that significantly improve on those currently in use. SGDP projects are cooperative agreements with the DOE, whereas the SGIG projects are grants. There are two types of Smart Grid projects selected for the SGDP – regional Smart Grid demonstrations and energy storage technologies. The regional Smart Grid project seeks to verify Smart Grid viability, quantify Smart Grid costs and benefits, and validate new Smart

---

[73] National Institute of Standards and Technology, NIST Framework and Roadmap for Smart Grid Interoperability Standards, Release 2.0 (2012), http://www.nist.gov/smartgrid/upload/NIST_Framework_Release_2-0_corr.pdf, accessed 8 November 2012.

[74] Ibid. at 9–10.

[75] US Department of Energy, SmartGrid.gov, http://www.smartgrid.gov/recovery_act/overview/smart_grid_investment_grant_program, accessed 8 November 2012.

Grid business models at scales that can be readily replicated across the country. Energy storage technologies include batteries, distributed applications, and the grid integration of renewable resources, such as wind and solar power.[76]

5 FLOOD RESILIENCE MEASURES FOR ELECTRICITY INFRASTRUCTURE IN THE FACE OF CLIMATE CHANGE

This chapter has identified the many climate change threats to electricity infrastructure and has suggested that a Smart Grid might prove to be an appropriate technological adaptation response to climate extremes and disasters. However, it is also appropriate to consider the adaptation responses that need to be adopted to try to protect existing and future critical infrastructure from climate extremes and disasters in the future. It is not feasible to identify the many legal responses required to facilitate adaptation responses to each type of threat to electricity infrastructure. However, it is instructive to learn from one of the most recent disasters – the 2010 Queensland floods – to elucidate the vulnerability of electricity infrastructure to climate extremes. As indicated above, the Queensland Floods Commission of Inquiry's Final Report can be relied upon to provide some generic adaptation responses with respect to only one of the many types of climate extremes which might impact on electricity infrastructure. The Final Report highlights the need for electrical infrastructure to be more resilient, flexible, and efficient to address the needs of vulnerable communities as a result of flooding linked to climate change.

The Commission identifies two main reasons why the location of electricity infrastructure is vitally important in areas susceptible to flood.[77] Firstly, existing infrastructure that cannot be viably moved to a more flood resilient site so must be protected, and secondly, as there is a statutory requirement to provide electricity supply, development in flood susceptible areas may be unavoidable.[78]

Before turning to the recommendations in the Final Report in detail (in section 6 below), there are a number of preliminary, theoretical questions

---

[76] US Department of Energy, SmartGrid.gov, http://www.smartgrid.gov/recovery_act/overview/smart_grid_demonstration_program, accessed 8 November 2012.

[77] Queensland Floods Commission of Inquiry, above note 13 at 245.

[78] Ibid.

that arise for consideration in the context of climate adaptation law. Each deserves detailed treatment but is raised here to flag a number of important considerations for further research and discussion:

- *Coordinating multi-jurisdictional responses*. In most countries there are different levels of government, including federal, State, and local, using different methods to mitigate and control damage from disasters. These include strategic land use planning, town planning, building codes, and a range of engineering standards, warning systems, disaster responses, and community education.[79] One of the questions that arise is whether it is desirable for a federal government, where it has constitutional authority, to attempt to mandate consistency across all jurisdictions with regard to flood adaptation responses. Where it lacks authority, should this be managed through a process of cooperative federalism such as exists in the Australian Council of Australian governments, comprising the federal, State and Territory governments? Is it even feasible within the borders of a State jurisdiction to require all local governments to adopt model flood adaptation provisions developed by the State government? Or are the geophysical and socio-economic components of each area so unique that model provisions would serve no purpose?
- *Questioning the efficiency of the law-making process*. Another question when confronting adaptation to climate extremes and disasters is the law-making process. In the face of climate disasters should legislators remain bound by the procedures of Green Papers, White Papers, Bills, regulatory impact statements, draft regulations, and review by Parliamentary Committees? What is the role of public participation and notice-and-comment procedures? Does participation lose its place in climate change governance and knowledge formation when there is 'little time to make detailed projections or wait until better information is available'?[80] Or does it become even more crucial?[81] If flexible and responsive regulation is required to meet the demands of climate disaster risk management is a democratic deficit justifiable?

---

[79] Diane U Keogh et al., 'Resilience, Vulnerability and Adaptive Capacity of an Inland Rural Town Prone to Flooding: A Climate Change Adaptation Case Study of Charleville, Queensland, Australia' (2011) 59 Natural Hazards 699, 700.

[80] Nick Pidgeon and Catherine Butler, 'Risk Analysis and Climate Change' (2009) 18(5) Environmental Politics 670.

[81] Pidgeon and Butler suggest that an analytic-deliberative process might be preferable; ibid. at 681.

- *Do climate extremes and disasters require a new risk regulation and legislative response?* It has been suggested that we have entered a new era in which human-generated catastrophic risks become normal, global in scale, and seemingly unpredictable or at least highly uncertain. In this view, traditional means for governing risks, ranging from insurance and conventional expertise through to the nation state, cannot deal with these new problems. Entirely new political and legal forms are required to govern this environment.[82] Does climate change elucidate some of the very characteristics that render conventional risk analysis tools problematic[83] and demand a regulatory agenda 'that is far more ambitious and radical than anything risk regulation has needed to contemplate up to now?'[84] Risk regulation typically substitutes acceptable risk for unacceptable risk by delineating to discrete manageable segments an otherwise 'amorphous totality of risks'.[85] Risk governance responsibility is then individualized, while actual regulation is preceded by information gathering, classification, and risk assessment so justifying the regulatory activity and delineating decision-making authority to responsible agencies and individuals.[86] In the context of climate change, the stabilizing yet cumbersome and time-consuming law-making process must become agile, flexible, and responsive. Yet with climate risks, the regulator is faced with indications of possible harm, even catastrophic harm, with insufficient data to conduct a 'proper' risk assessment.[87] How can legislation facilitate decision-making in this context?
- *Should climate adaptation and disaster laws adopt a Catastrophic Harm Precautionary Principle?* Integral to a new risk management framework for dealing with climate extremes and disasters are concerns about making decisions in the context of 'uncertainty' rather than 'risk'.[88] Environmental law, especially planning and assessment legislation, is a body of law which emerges as being

---

[82] Veerle Heyvaert, 'Governing Climate Change: Towards a New Paradigm for Risk Regulation' (2011) 74(6) Modern Law Review 817 at 835.

[83] Pidgeon and Butler, above note 80 at 671.

[84] Heyvaert, above note 82 at 835.

[85] Ibid. at 820.

[86] Ibid.

[87] Ibid.

[88] See, for example, Cass R Sunstein, 'The Catastrophic Harm Precautionary Principle' in Daniel A Farber and Michael G Faure (eds), *Disaster Law* (Edward Elgar 2010) 141.

particularly helpful in trying to enhance the resilience of electricity infrastructure. Yet in this area, risk discourse has proliferated as a means of framing and managing potential threats, particularly in the area of environmental impact assessment.[89] Faced with 'uncertainty' about the 'risks', the precautionary principle,[90] found in all Australian environmental statutes, holds that 'if there are threats of serious or irreversible environmental damage, lack of full scientific certainty should not be used as a reason for postponing measures to prevent environmental degradation'.[91] Does catastrophic risk[92] require a different approach? Should a Catastrophic Harm Precautionary Principle be adopted which holds that '[i]n deciding whether to eliminate the worst-case scenarios under circumstances of uncertainty, regulators shall consider the losses imposed by eliminating that scenario, and the magnitude of the difference between the worst-case scenario and alternative scenario?'[93]

6 LEARNING FROM THE 2010 QUEENSLAND FLOODS

6.1 Floodplain Management

The Queensland Floods Commission of Inquiry made a number of recommendations regarding floodplain management, which translate easily into generic recommendations in other jurisdictions, with the obvious proviso that a similar legislative framework exists and there is

---

[89] Pidgeon and Butler, above note 80 at 671.

[90] See Rosemary Lyster and Eric Coonan, 'The Precautionary Principle: A Thrill Ride on the Roller Coaster of Energy and Climate Law' (2009) 18(1) Review of European Community and International Environmental Law 38; Rosemary Lyster, 'The Relevance of the Precautionary Principle: Friends of Hinchinbrook v Minister for the Environment' (1997) 14 Environmental and Planning Law Journal 390–401; Lee Godden and Jacqueline Peel, *Environmental Law: Scientific, Policy and Regulatory Dimensions* (OUP 2010); Jacqueline Peel, *The Precautionary Principle in Practice: Environmental Decision-making and Scientific Uncertainty* (The Federation Press 2005).

[91] Ibid.

[92] See W Kip Viscusi and Richard Zeckhauser, 'Addressing Catastrophic Risks: Disparate Anatomies Require Tailored Therapies' in Cary Coglianese (ed.), *The Regulatory Breakdown. The Crisis of Confidence in U.S. Regulation* (University of Pennsylvania Press 2012) 21.

[93] Sunstein, above note 88 at 165.

capacity to adopt the recommendations. The Commission recommended that a recent flood study of the most flood-affected catchments and all urban areas in the State should be initiated and made available. The flood study should be comprehensive and collate or create data on: rainfall, stream flow, tide levels, inundation levels and extents, the operation of dams, and river channel and floodplain characteristics.

Suitable hydrologic models should be produced and validated. A suitable hydraulic model should also be produced which is able to determine flood heights, extents of inundation, velocities, rate of rise, and duration of floods.[94] Local councils should keep these studies up to date. The body conducting a study should consult with surrounding councils and others to check whether they are doing any work that might assist with the development of the study. Elected council representatives should be informed of the flood study so they can consider its consequences for land use planning and emergency management.[95] The data used in flood studies should be maintained in a repository so that all levels of government and dam operators can access it. Councils with requisite resources should develop floodplain management plans, and, where development is expected to occur, they should develop flood behaviour maps which indicate at least three 'zones of risk'. Where they lack resources, councils should develop a flood map which shows at least three examples of the likely extent of flooding. Similar recommendations are made for non-urban areas where limited development is likely to occur. Where no assessment of the likelihood of flooding has been done, councils should inquire into whether a development site could be subject to flooding.

The State government should ensure that all councils have sufficient guidelines for understanding best practice in the performance of flood studies and the production of flood maps. All flood mapping should be displayed on council and government websites and property specific information should be available to the general public, including by searching an online database. Prospective purchasers should be alerted to the risk of flooding by standard contract conditions drawn up by the Real Estate Institute of Queensland and the Law Society of Queensland.[96]

---

[94] Queensland Floods Commission of Inquiry, above note 13 at 12.
[95] Ibid. at 13.
[96] Ibid. at 14.

6.2 Local Planning Instruments

The Commission recommended that the State government should draft 'model flood planning controls' that councils can adapt for local conditions and it should require these controls to be reflected in new planning schemes. The government should include in the controls a model planning scheme policy and a requirement that councils have a flood overlay map in their planning schemes.[97] These should be included by local councils even if the Queensland government does not require them. A flood overlay map should identify the areas that are: known to be affected by flood and on which councils can impose different planning controls; and areas for which there is no flood information available. Assessment criteria relating to floods should be consolidated by councils in their flood overlay codes. Councils should be allowed to amend planning schemes to update flood mapping information by way of the minor amendment process as long as adequate public consultation has occurred.[98]

6.3 Development Assessment in Practice

Local councils, as development consent authorities, should maintain flood maps and overland flow maps for use in development assessment, which should also be made available to the applicants.[99] Whether or not this is included in a model planning scheme policy, councils should require stormwater and flooding information to be included in development applications (DAs). Information about how flood risks will be assessed should be provided to development applicants either at pre-lodgement meetings, or at the time of receiving the DA.[100]

6.4 Floods and Electricity Infrastructure

The Queensland Floods Commission of Inquiry made very specific recommendations regarding planning for climate change adaptation and risk reduction for the electricity sector.[101] These recommendations are generally related to new development of infrastructure.

---

97 Ibid. at 15.
98 Ibid. at 16.
99 Ibid. at 19.
100 Ibid. at 20.
101 Ibid. at 246.

Firstly, the State government should draft assessment criteria to be included in the 'model flood planning controls', which require substations to be built to remain operational during and immediately after a flood of a particular magnitude. That magnitude should be determined by an appropriate risk assessment.[102] If substation development is captured during the planning stage this will allow for better locational planning, which will provide better placement of such infrastructure. In some cases, building infrastructure to withstand the probable maximum flood would be prohibitively expensive.[103] Government, in consultation with the community, should determine the magnitude of the flood that the infrastructure should be able to withstand, and a risk assessment should be conducted to determine that level.[104] This risk assessment should be done when 'model flood planning controls' are being tailored to take account of local circumstances. Once the magnitude of the flood level has been chosen, steps should be taken to make the infrastructure resilient to it.[105]

Secondly, if 'model flood planning controls' are not developed by the State government, local councils should nevertheless require in their local planning schemes that substations are built to remain operational during and immediately after a flood of a particular magnitude.[106]

Thirdly, electricity distributors should consider installing connection points where generators could be installed so as to provide electricity supply to non-flooded areas that have had their supply cut during floods. This would help to ensure that where supply has been cut, places further down the line that may be unaffected by flood can quickly regain supply of electricity.[107]

Fourthly, the State government should consider whether there should be a legislative requirement that customer dedicated assets be built at or above the applicable defined flood level and, if so, which piece of legislation within the panoply of environmental and electricity infrastructure legislation should contain such a requirement.[108] A guideline produced by the Queensland Reconstruction Authority (QRA) proposes that in new high-rise buildings electrical equipment should be raised and

---

[102] Ibid.
[103] Ibid.
[104] Ibid.
[105] Ibid.
[106] Ibid.
[107] Ibid.
[108] Ibid. at 248.

located out of any flood's reach, which may exclude basement locations in some areas. This will improve resilience against flooding.[109]

Finally, the State government should consider implementing mandatory requirements to ensure the sealing of all conduits for the purpose of providing electrical supply below the applicable defined flood level to prevent floodwaters from entering them.[110] This is because electrical connections are supplied through this cabling to commercial and industrial premises, with the cables running from the footpath through conduits to the substation inside customer's premises.[111] During the 2010 floods, water entered buildings via conduits,[112] which form part of a shared network of infrastructure cabling. Precautionary disconnections were made to ensure the safety of human life and to reduce damage to electrical infrastructure down the line.[113]

Currently Energex's (one of Queensland's major utilities) manual for commercial and industrial substations, requires the conduit to be sealed to prevent the ingress of dirt but not water or water under pressure. However, after new products are trialled, it is intended to update the manual by mid-2012.[114] In addition, Energex is working with the owners of basements to have existing conduits sealed, as it believes building owners are responsible for the location, design, installation, and maintenance of electrical conduits.[115] A draft standard relating to the waterproofing of conduits and cabling for buildings in flood hazard areas is anticipated to be included in the 2013 edition of the Building Code of Australia.[116] Nonetheless, there must be clear guidance and understanding of who has the responsibility to ensure conduits to basements are flood proofed. At present, building owners are responsible for maintenance and design; however, the electricity distributor has the expertise to safely and effectively provide sealed conduits, so the Commission found that the electricity distributor should also have responsibility.[117]

---

109 Ibid. at 247.
110 Ibid. at 249.
111 Ibid.
112 Ibid. at 248.
113 Ibid. at 241.
114 Ibid. at 249.
115 Ibid.
116 Ibid.
117 Ibid.

6.5 Utilities Adopt Flood Resilience Measures Ahead of Regulation

Following the Queensland floods, the two major electricity utilities, Energex and Ergon Energy, have proposed voluntary flood resilience measures. These measures are not currently mandated in legislation. Ergon Energy, for example, has revised flood level standards for setting up new zone substations and bulk supply.[118] New zone substations are to be built at or above the 0.5 per cent annual exceedance probability (AEP)[119] flood level and any infrastructure that would be below this level must now incorporate flood resilience measures to ensure supply is maintained during and immediately after a flood.[120] Ergon Energy has also proposed that if a flood height is not mapped for an area where a new substation is proposed then it will undertake a hydrological assessment performed by an external consultant.[121] Ergon Energy also reported that although local authorities do not advocate above-ground overhead assets in new urban developments, it believes that these are more flood resilient than the preferred underground or on-ground assets.[122]

7 CONCLUSION

The IPCC's latest scientific report (SREX Report) draws the international community's attention to impacts of climate extremes on electricity infrastructure. It also analyses the economic costs of adaptation measures and global and regional climate disasters; implications for development, recovery, and reconstruction; and technologies for extreme events. It seems that now is the time for governments to have a cooperative and transparent debate on the implications of climate change and the energy sector.[123] Governments across local, State and federal levels have an important role to play in adaptation to climate change. These roles include government as an adaptor, a catalyst and facilitator, an intervener,

---

118 Ibid. at 244.

119 Bureau of Meteorology, Australian Government, *Australian Water Information Dictionary*, http://reg.bom.gov.au/water/awid/id-703.shtml, accessed 8 November 2012.

120 Queensland Floods Commission of Inquiry, above note 13 at 244.

121 Ibid.

122 Ibid. at 244.

123 Ian Bailey, 'Australian Climate Politics 2010' (2011) 110 Arena Magazine 32, 35.

or as a rule-setter,[124] as governments have the ability to facilitate collective action, setting rules and guiding behavioural patterns that shape social interactions.[125] There is a need for a shared responsibility between all stakeholders to build resilient communities. This includes moving forward by planning development that includes legislative changes, and linking pieces of legislation for cohesive control and regulation across all levels of government. Specifically, governments must ensure essential services, such as electricity infrastructure, are provided and protected; adequate legislation and policy instruments are coordinated; land use planning instruments and building codes support disaster resilience; disaster risks are negotiated and determined; and communities are educated for awareness.[126] Likewise business and communities have a responsibility to work with government to build disaster resilience, understand risk, be prepared and be responsive. Building disaster resilience is a shared responsibility.

Continuity of electrical supply is essential in managing the severity of effects during and after flood events. This can be facilitated by new technologies such as the Smart Grid concept. New technology built into existing systems and used to replace aged infrastructure has the potential to improve the safety, reliability, and efficiency of the current inefficient supply of electricity. Moreover, utilization of new technologies, including alternative sources of electricity generation, has the potential to mitigate the effects of climate change in reducing GHG emissions while at the same time addressing adaptation needs. There are many dynamic, often interrelated factors that influence exposure and vulnerability of human populations and natural ecosystems to extreme events. Moreover, individual communities are unique and often require site specific consideration. Disaster risk management coupled with quality infrastructure reduces exposure and vulnerability and improves resilience. It is imperative that decision makers at all levels understand these factors and associated disaster risks to inform appropriate disaster risk management and integrate it into climate change adaptation strategies. Furthermore,

---

[124] Brent C Jacobs and Peat Leith, 'Adaptive Capacity for Climate Change: Principles for Public Sector Managers' (2010) 23 (July–September) Public Administration Today 49 at 51.

[125] Ibid. at 55.

[126] Queensland Reconstruction Authority, *Rebuilding a Stronger, More Resilient Queensland: The Capacity to Prepare for, Withstand, Respond to and Recover from Disasters* (Queensland Government 2012) 32, http://qldreconstruction.org.au/u/lib/cms/Resilience-doc.pdf, accessed 8 November 2012.

there is a necessity for legislation to support and facilitate flexible, resilient, and efficient provision of electricity infrastructure to protect and empower vulnerable communities.

Index